PLANT COLD HARDINESS
and FREEZING STRESS

MECHANISMS *and* CROP IMPLICATIONS

Academic Press Rapid Manuscript Reproduction

*Proceedings of an International Plant Cold Hardiness
Seminar Held in St. Paul, Minnesota,
November 2-4, 1977*

*Sponsored by
The United States National Science Foundation
The Japan Society for the Promotion of Science
The College of Agriculture, University of Minnesota*

PLANT COLD HARDINESS
and FREEZING STRESS

MECHANISMS
and CROP IMPLICATIONS

Edited by

P. H. LI
Laboratory of Plant Hardiness
Department of Horticultural Science and Landscape Architecture
University of Minnesota
St. Paul, Minnesota

A. SAKAI
Laboratory of Frost Injury in Plants
The Institute of Low Temperature Science
Hokkaido University
Sapporo, Japan

ACADEMIC PRESS   New York   San Francisco   London   1978
A Subsidiary of Harcourt Brace Jovanovich, Publishers

ACADEMIC PRESS, INC.
111 Fifth Avenue, New York, New York 10003

United Kingdom Edition published by
ACADEMIC PRESS, INC. (LONDON) LTD.
24/28 Oval Road, London NW1 7DX

Library of Congress Cataloging in Publication Data

Main entry under title:

Plant cold hardiness and freezing stress.

    1.   Plants—Frost resistance—Congresses.
2. Plants, Effect of cold on—Congresses.
3. Crops and climate—Congresses.   I. Li,
Paul H., Date   II. Sakai, Akira, Date
III. United States. National Science Foundation.
IV. Nippon Gakujutsu Shikōkai. V. Minnesota.
University. College of Agriculture.
OK756.P53   582'.01'9165   78-7038
ISBN 0-12-447650-3

Contents

Contributors   *ix*

Preface   *xi*

PART I   FREEZING STRESS: AN OVERVIEW

An Overview of Freezing Injury and Survival, and Its Interrelationships to
Other Stresses   3
   J. Levitt

Freezing Processes and Injury in Plant Cells   17
   E. Asahina

Analyses of Freezing Stresses and Plant Response   37
   C. R. Olien

Frost Hardening and Freezing Stress in Tuber-Bearing *Solanum* Species   49
   P. H. Li and J. P. Palta

PART II   MEMBRANES

Plasma Membrane Alterations Following Cold Acclimation and Freezing   75
   P. L. Steponkus and S. C. Wiest

Cell Membrane Properties in Relation to Freezing Injury   93
   J. P. Palta and P. H. Li

Phospholipid Degradation and Its Control during Freezing of Plant Cells   117
   S. Yoshida

PART III   ACCLIMATION

Mechanism of Cold Acclimation in Herbaceous Plants   139
   A. Kacperska-Palacz

Protein Synthesis Mechanisms Relative to Cold Hardiness   153
   G. N. Brown

The Role of Water in Cold Hardiness of Winter Cereals   165
   P. Chen and L. V. Gusta

v

Studies on Frost Hardiness in *Chlorella ellipsoidea*: Effects of Antimetabolites, Surfactants, Hormones, and Sugars on the Hardening Process in the Light and Dark   175
    S. Hatano

Cold Acclimation of Callus Cultures of Jerusalem Artichoke   197
    Y. Sugawara and A. Sakai

PART IV   SUPERCOOLING

The Occurrence of Deep Undercooling in the Genera *Pyrus, Prunus,* and *Rosa*: A Preliminary Report   213
    C. Rajashekar and M. J. Burke

Low Temperature Exotherms in Xylems of Evergreen and Deciduous Broadleaved Trees in Japan with References to Freezing Resistance and Distribution Range   227
    S. Kaku and M. Iwaya

Resistance to Low Temperature Injury in Hydrated Lettuce Seed by Supercooling   241
    C. Stushnoff and O. Junttila

The Role of Bacterial Ice Nuclei in Frost Injury to Sensitive Plants   249
    S. E. Lindow, D. C. Arny, C. D. Upper, and W. R. Barchet

PART V   SURVIVAL AND BREEDING

Adapting Cold Hardiness Concepts to Deciduous Fruit Culture   267
    E. L. Proebsting

Effect of Solar Radiation on Frost Damage to Young *Cryptomerias*   281
    T. Horiuchi and A. Sakai

Freeze Survival of Citrus Trees in Florida   297
    G. Yelenosky

Breeding and Selecting Temperate Fruit Crops for Cold Hardiness   313
    H. A. Quamme

Breeding Frost-Resistant Potatoes for the Tropical Highlands   333
    R. N. Estrada

PART VI   CRYOPRESERVATION AND CRYOPROTECTION

Survival of Plant Germplasm in Liquid Nitrogen   345
    A. Sakai and Y. Sugawara

Ultracold Preservation of Seed Germplasm   361
    P. C. Stanwood and L. N. Bass

Effects of Combinations of Cryoprotectants on the Freezing Survival of Sugarcane Cultured Cells   373
    B. J. Finkle and J. M. Ulrich

PART VII SPECIAL TOPICS

Plant Cold Hardiness Seminar Summary and General Remarks 391
   C. J. Weiser

Summary of the Papers Presented at the First European International
Symposium on "Winter Hardiness in Woody Perennials," September 1977 395
   T. Holubowicz

Index 407

Contributors

Numbers in parentheses indicate the pages on which the authors' contributions begin.

D. C. ARNY (249), Department of Plant Pathology, University of Wisconsin, Madison, Wisconsin

E. ASAHINA (17), The Institute of Low Temperature Science, Hokkaido University, Sapporo, Japan

W. R. BARCHET (249), Department of Meteorology, University of Wisconsin, Madison, Wisconsin

L. N. BASS (361), National Seed Storage Laboratory, USDA-ARS, Fort Collins, Colorado

G. N. BROWN (153), Department of Forestry, Iowa State University, Ames, Iowa

M. J. BURKE (213), Department of Horticulture, Colorado State University, Fort Collins, Colorado

P. CHEN (165), Crop Development Centre, University of Saskatchewan, Saskatoon, Saskatchewan, Canada

R. N. ESTRADA (333), The International Potato Center, Lima, Peru

B. J. FINKLE (373), Western Regional Research Center, USDA-ARS, Berkeley, California

L. V. GUSTA (165), Crop Development Centre, University of Saskatchewan, Saskatoon, Saskatchewan, Canada

S. HATANO (175), Department of Food Science and Technology, Kyushu University, Fukuoka, Japan

T. HOLUBOWICZ (395), Academy of Agriculture, Poznan, Poland

T. HORIUCHI (281), Ibaraki Prefectural Forest Experimental Station, Naka, Ibaraki, Japan

M. IWAYA (227), College of General Education, Kyushu University, Ropponmatsu, Fukuoka, Japan

O. JUNTTILA (241), Institute of Biology and Geology, University of Tromso, Tromso, Norway

A. KACPERSKA-PALACZ (139), Institute of Botany, University of Warsaw, Warsaw, Poland

S. KAKU (227), College of General Education, Kyushu University, Ropponmatsu, Fukuoka, Japan

J. LEVITT (1), Department of Plant Biology, Carnegie Institution of Washington, Stanford, California

P. H. LI (49, 93), Department of Horticultural Science and Landscape Architecture, University of Minnesota, St. Paul, Minnesota

S. E. LINDOW (249), Department of Plant Pathology, University of Wisconsin, Madison, Wisconsin

C. R. OLIEN (37), Crop and Soil Science Department, Michigan State University, East Lansing, Michigan

J. P. PALTA (49, 93), Department of Horticultural Science and Landscape Architecture, University of Minnesota, St. Paul, Minnesota

E. L. PROEBSTING (267), Irrigated Agriculture Research and Extension Center, Washington State University, Prosser, Washington

H. A. QUAMME (313), Research Station, Agriculture Canada, Harrow, Ontario, Canada

C. RAJASHEKAR (213), Department of Horticulture, Colorado State University, Fort Collins, Colorado

A. SAKAI (197, 281, 345), The Institute of Low Temperature Science, Hokkaido University, Sapporo, Japan

P. C. STANWOOD (361), National Seed Storage Laboratory, USDA-ARS, Fort Collins, Colorado

P. L. STEPONKUS (75), Department of Agronomy, Cornell University, Ithaca, New York

C. STUSHNOFF (241), Department of Horticultural Science and Landscape Architecture, University of Minnesota, St. Paul, Minnesota

Y. SUGAWARA (197, 345), The Institute of Low Temperature Science, Hokkaido University, Sapporo, Japan

J. M. ULRICH (373), Western Regional Research Center, USDA-ARS, Berkeley, California

C. D. UPPER (249), USDA and Department of Plant Pathology, University of Wisconsin, Madison, Wisconsin

C. J. WEISER (391), Department of Horticulture, Oregon State University, Corvallis, Oregon

S. C. WIEST (75), Department of Agronomy, Cornell University, Ithaca, New York

G. YELENOSKY (297), South Regional U.S. Horticulture Research Laboratory, USDA-ARS, Orlando, Florida

S. YOSHIDA (117), The Institute of Low Temperature Science, Hokkaido University, Sapporo, Japan

Preface

This volume is based on the proceedings of an international seminar on Plant Cold Hardiness, which was held at the University of Minnesota, Saint Paul Campus, November 2–4, 1977. It contains a collection of valuable articles on recent advances in plant cold hardiness research. The contributors are eminent international scientists, renowned for their significant contributions to our current knowledge of plant stress physiology.

The United States National Science Foundation and the Japan Society for the Promotion of Science under the auspices of the United States–Japan Cooperative Science Program, as well as the College of Agriculture, University of Minnesota, jointly sponsored the seminar. The intention was to update plant hardiness knowledge, to exchange viewpoints on research priorities, to foster collaborative research, and to apply research findings to the increasing of food production. The focus of the seminar was on the fundamental biological processes of freezing survival in plants. The seminar was therefore composed of six sessions, consisting of General Review, Membranes, Mechanisms of Cold Acclimation, Supercooling and Freezing Processes, Plant Survival to and Breeding for Freezing Stress, and Cryopreservation and Cryoprotection. The subject matter encompassed agronomic, forest, and horticultural food crops including beans, cereals, citrus, corn, deciduous tree fruits, lettuce, rape, Jerusalem artichokes, potatoes, and also Japanese and North American forest species. Sixty-nine participants representing the nine nations of Canada, Colombia, Iran, Japan, Mexico, New Zealand, Norway, Poland, and the United States of America attended the seminar; among them were 25 invited speakers.

The importance of plant cold hardiness research has become increasingly apparent in agriculture and food production. An untimely frost or a severe winter in a major production area can influence the world food reserve, and affect economic and human health conditions. In many cases the difference between a good crop and crop failure is determined by only a 2–3°C increase in frost resistance. Such subtle differences are highly amenable to research solutions. While significant research progress has been made, it is apparent that appropriate emphasis and support of plant cold hardiness research are needed for two simple reasons. First, a lack of understanding of the basic mechanisms by which plants are injured by freezing, and by which some species can resist freezing stress, limits our ability to develop crops and production systems that will attenuate losses and expand production. The success of physiological and/or genetic approaches to increasing cold hardiness will be highly dependent upon continued basic research to establish the nature of freezing injury and resistance in plants. Second, the pressure of an increasing population requires crops to be grown in areas that are marginal because low temperatures limit production.

It is our sincere hope that the seminar will make a substantial contribution to the efforts involved in the improvement of human living quality in the future.

We would like to express our gratitude to Dr. William F. Hueg, Jr., Dr. James F. Tammen, and Dr. Jim L. Ozbun, all from the University of Minnesota, for their interest and support. We are also deeply indebted to Dr. Jacob Levitt, Carnegie Institution of Washington, to Dr. Conrad J. Weiser, Oregon State University, and to our colleagues from the staff of the Department of Horticultural Science and Landscape Architecture, University of Minnesota, for their help in organizing the seminar, particularly Dr. Jiwan P. Palta, Dr. Lawrence R. Parsons, Dr. John V. Carter, Dr. Cecil Stushnoff, Dr. Harold M. Pellett, and the graduate students. A special word of thanks goes to Ms. Angie Klidzejs for her impeccable secretarial work.

We would like to acknowledge the help of Dr. Jiwan P. Palta for his editorial work, and of Ms. Angie Klidzejs for her preparation of a camera-ready manuscript for delivery to the publisher. Finally, we wish to thank the cooperative efforts of the contributors, and the contributions of the staff of Academic Press in bringing the volume to rapid publication.

Symposium participants (left to right). First row: A. Sakai, M. Burke, J. Levitt, M. Westwood, J. Carter, L. Parsons, A. Hatch, C. Doughty, P. Li. First row extension: S. Hatano. Second row: T. Holubowicz, C. Weiser, N. Estrada, P. Steponkus, M. Mostafavi, R. Layne, E. Hewett, S. Seeley, P. Lombard, P. Chen. Third row: B. Finkle, E. Mielke, F. Dennis, J. Tammen, G. Yelenosky, P. Stanwood, J. Lyons, E. Proebsting, J. Anderson. Fourth row: C. Upper, T. Niki, S. Kaku, C. Rajashekar, D. Ketchie, J. Seibel, K. Rapp, E. Stadelmann, S. Yoshida. Fifth row: Y. Sugawara, R. Olien, H. Quamme, O. Smith, E. Asahina, J. Palta, A. Kacperska-Palacz, F. Widmoyer, T. Raese.

Part I
Freezing Stress: An Overview

AN OVERVIEW OF FREEZING INJURY AND SURVIVAL,
AND ITS INTERRELATIONSHIPS TO OTHER STRESSES

J. Levitt

Department of Plant Biology
Carnegie Institution of Washington
Stanford, California

I. INTRODUCTION

This brief review will deal only with the freezing of
plants in nature or in the field, or with results of laboratory experiments intended to explain freezing in nature or in
the field. Artificial cryoprotection, cryopreservation, etc.
will not be discussed. The review of each topic will include
a statement of the most important questions that still need to
be attacked. Hopefully, some of them will be answered by the
succeeding speakers. Only one topic--intracellular freezing
avoidance--will be treated in detail, because of very recent,
pertinent investigations.

II. FREEZING INJURY

It has long been known that under laboratory conditions,
plants may be induced to freeze either intracellularly or
extracellularly. Yet, to this day, it is still not known
with certainty whether or not intracellular freezing occurs
in nature. Circumstantial evidence indicates that it probably
does occur in the case of sunscald of moderately hardy plants
and mild freezing of tender plants. Nevertheless, the first
question that needs to be answered categorically is: 1) Does
intracellular freezing of plants occur in nature or in the
field?

3

Regardless of the answer to this question, it is of fundamental importance to compare intracellular with extracellular freezing (Figure 1). Most important is the fact that intracellular freezing is essentially always instantly fatal, at least under laboratory conditions intended to duplicate those in nature. Extracellular freezing, on the other hand, is tolerated in at least a considerable quantity by all overwintering plants in temperate climates.

The instant and essentially invariable nature of the injury due to intracellular freezing, points to a physical process--probably a piercing of the cell membranes by the growing ice crystals inside the protoplasm. Nevertheless, this explanation has not been proved, nor is it clear whether one or several of the cell's membranes is likely to be damaged. The second question then is: 2) Exactly how does intracellular freezing kill the cell?

Injury due to extracellular freezing is even more difficult to understand than intracellular freezing injury. Unlike the latter, an extracellularly frozen cell has no way of detecting the freezing, since the ice crystals occur in the intercellular spaces and therefore without contacting the living protoplasm. The cell is, however, dehydrated in the process, in exactly the same manner as by evaporative dehydration or desiccation. At least some components of the injury must, therefore, be identical in freeze and evaporative dehydration. In both cases the injury must occur at a certain degree of dehydration specific to the plant. The third question, then, is: 3) How does the dehydration induced by extracellular freezing injure the cell?

Freezing

Intracellular	Extracellular
1. Occurs relatively rarely if at all in nature.	1. Occurs in all plants in temperate climates during winter freezes.
2. Results from rapid freezing, usually following marked supercooling.	2. Results from slow freezing, following slight supercooling.
3. Is essentially always fatal.	3. May or may not injure, depending on the freezing temperature.
4. Injury is probably physical (membrane piercing?).	4. Nature of the injury not fully understood.

FIGURE 1. The two kinds of freezing of plants.

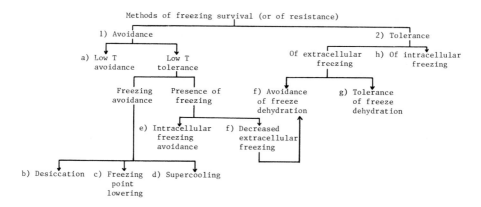

FIGURE 2.   The eight possible kinds of freezing survival or resistance.

Other kinds of freezing injury, such as ice pressure, have been proposed. Due to the brevity of this review, they will not be considered here.

III.   FREEZING SURVIVAL

Freezing survival requires the prevention or repair of both intracellular and extracellular freezing injury. This is due to the plant's resistance and must, therefore, include both avoidance and tolerance (Figure 2).

1. Avoidance

A. Low Temperature Avoidance. Due to the poikilothermic nature of plants, this method is rare and of limited value to the plant, occurring only when the freezing temperature persists for short periods. A recent example is *Espeletia schultzii*, a large rosette species in the parma zone of the Venezuelan Andes. It has strongly nyctinastic leaves which close around the apical bud at night, protecting it from freezing temperatures and preventing the rapid warming of young leaves just after sunrise (26). If the nyctinastic leaf movement is prevented, leaf wilting and death occur. Artificial low temperature avoidance, by covering plants with a liquid foam that rapidly solidifies, and reliquefies later during rewarming, was introduced by Siminovitch and has been used on plants such as strawberries (1).

   B. Freezing Avoidance by Desiccation is a method available only to dormant structures (seeds, buds) among higher
plants, the cells filling with dry matter at the expense of
water.

   C. Freezing Avoidance of Normally Hydrated Cells can also
be accomplished colligatively--by the accumulation of solutes.
Among overwintering higher plants, the concentration probably
rarely permits the complete avoidance of freezing below
about -4° C.

   D. Freezing Avoidance by Supercooling was not considered
to be important as a method of survival in nature, until a
few years ago. Now, due to the extensive investigations at
the University of Minnesota and elsewhere (see later papers
this symposium), it is known to be the limiting factor in the
survival of specific tissues in many woody plants. The question is: 4) How do these specialized tissues succeed in
remaining supercooled for hours, days, or months, in the same
plant in which the vast majority of cells and tissues freeze
extracellularly?

   *E. Intracellular Freezing Avoidance in Plants that do
Freeze*, can be a method of survival only if accompanied by
tolerance of extracellular freezing. The method of freezing
avoidance, and the questions raised will be discussed in
detail below.

   F. Decreased Extracellular Freezing is merely a quantitative variation of 3. It is identical with the avoidance of
freeze dehydration and will therefore be considered below.

2. *Tolerance*

   All of the above avoidance methods of survival are ex-
plainable without any understanding of the mechanism of
freezing injury, because, in the case of any stress, the
nature of avoidance by definition, is independent of the nature of the injury. In opposition to avoidance, tolerance can
be understood only if the nature of the freezing injury is
known. Among the above six methods of avoidance, only avoidance methods 1 and 3 can permit survival of the vegetative
plant unaccompanied by any freezing tolerance. Since these
two kinds of avoidance can occur only during mild (and brief
in the case of 1) freezes, they are of limited importance.
All the other kinds of avoidance must be accompanied by
freezing tolerance of some tissues, in order for the plant to

survive. The most important question for freezing survival
therefore remains: What is the mechanism of extracellular
freezing tolerance? The question, however, can be posed more
specifically, because extracellular freezing is, by definition, a freeze-dehydration (see above). The plant may survive
the freeze-dehydration in two ways.

G. *Avoidance of Freeze-Dehydration* is a decrease in the
amount of ice formed at any one freezing temperature. It is
due to the colligative effect of the accumulated solutes.
Some differences in freezing tolerance (e.g. among some
grains) are apparently due to differences in the quantities
of solutes accumulated, and therefore, in avoidance of
freeze-dehydration.

H. *Tolerance of Freeze Dehydration* is the property
responsible for superior hardiness among other tolerant species and varieties that fail to show a correlation between
freezing tolerance and solute accumulation. This is demonstrated by the greater amount of ice that must form in the
tissues of the hardier plants before they show freezing injury
(6, 13). The above question, stated more specifically,
becomes: 5) What is the mechanism of tolerance of freeze
dehydration?

IV. METHODS OF ACHIEVING RESISTANCE

Of the eight methods of survival (or resistance) listed
in Figure 2, the first three are of limited importance in the
vegetative overwintering plant (see above) and tolerance of
intracellular freezing is non-existent in nature as far as
we know. This leaves four important kinds of freezing re-
sitance requiring explanations.
Freezing avoidance by supercooling has been discussed
above.
Avoidance of freeze dehydration is readily explained,
since it has long been known that the amount of solute (primarily sugar) accumulated is very often characteristic
of the hardening process (see above). The question is: 6)
What is the specific enzyme behavior leading to this sugar
accumulation during hardening? One possible answer to this
question is suggested by results of Dear (8). He showed that
cooling cabbage seedlings to their hardening temperature produces a hydrolysis of starch within minutes. This indicates
an immediate effect of the low temperature on relative enzyme
activities, rather than a synthesis of enzyme protein. Such

an immediate effect could be produced if the slope of the
Arrhenius plot is steeper (and the energy of activation
greater) for starch synthesizing than starch hydrolysis enzymes.   The result would be a net hydrolysis of starch.

Tolerance of freeze-dehydration is perhaps the most complicated and least understood of the four kinds of resistance
and is probably the most important since it always accompanies
the other three kinds of resistance.   It is believed to depend
on membrane properties, and more specifically properties of
membrane proteins.   The most important question is, therefore,
the fifth above: What is the mechanism of tolerance of freeze
dehydration?   I shall leave this question for the succeeding
speakers to answer.

Avoidance of intracellular freezing was first explained
some years ago.   It requires a sufficiently large specific
surface of the cell combined with a high enough cell (plasma
membrane) permeability to water, to permit a rapid enough
efflux of the cell water to the extracellular ice loci.
There are two recent sources of information that respectively
oppose and support this concept.

Stout *et al.* (28) have concluded that "under natural
freezing conditions, membrane resistance does not limit water
efflux."   They base this conclusion on measurements of ice
formation in the bark of *Hedera helix*, as a result of inoculation after supercooling to -10° C.   From these measurements
and calculations of the water potential difference, they estimate that the membrane represented only 0.5% of the barrier
resistance to ice formation.   They further point out that
other tissue structures provide even less resistance than the
membrane.   The only source of the remaining 99.5% of the
resistance that they can suggest is that the heat removal is
slow enough to retard the inward progress of the ice formation.   This explanation is difficult to accept, since there is
no reason why the heat removal should allow ice formation to
progress exactly to the cell surface but no further, unless
there is a barrier at that surface.   Furthermore, it is incapable of explaining the increase, on hardening, of the ability of cells to prevent intracellular freezing, observed by
Siminovitch and Scarth (24).   The evidence must, therefore,
be reexamined.

A barrier to extracellular freezing can be measured only
if extracellular freezing occurs, and the barrier slows down
the process.   If freezing is mainly intracellular, the ice
forms *in situ*, and the water molecules have not moved through
the plasma membrane, which therefore cannot be acting as a
barrier to ice formation.   Unfortunately, Stout *et al.* (28)
fail to report whether the freezing was intracellular or
extracellular, or even whether the cells were alive or dead on

thawing. These questions must, therefore be explored on the basis of the results that they do present. We will assume that their calculations are correct.

Freezing at -10° C under a driving force of 31 bars resulted in a $T_{1/2}$ (the time required for freezing half the plant's water) of 147 sec. Since the membrane represented only 0.5% of the barrier resistance, if we removed all other barriers yet permitted freezing to occur extracellularly, the movement through the membrane would occur in 1/200 the measured time and the $T_{1/2}$ would be 1/200 of 147 = 0.74 sec. Assuming a Q_{10} of 1.5, a similar diffusion process at room temperature (by plasmolysis) would be $(1.5)^3$ times as rapid and would require less than 0.2 sec. This is, or course, 2-3 orders of magnitude too rapid for plasmolysis of any cell to half its normal volume, under a gradient of 31 bars. Therefore, the ice formation measured could not have been extracellular, but must have been intracellular.

In agreement with this conclusion, the rate of ice formation was only 1.2 to 1.3 times as rapid in dead as in living acclimated and non-acclimated tissue respectively. Since the membranes were freely permeable in the dead tissue, they must also have been freely permeably in the "living" tissue during all except the first few seconds of the freeze. In other words, the living tissue must have frozen in essentially the same manner as the dead tissue--intracellularly. This clearly explains why the cell membrane represented only 0.5% of the barrier resistance. This simply means that only 0.5% of the cell's water froze extracellularly before freezing was initiated intracellularly and all the rest of the ice formed intracellularly, killing the cells.

It is obvious from this brief analysis that the results of Stout *et al.* (28) do not support their conclusion that "under natural freezing conditions, membrane resistance does not limit water efflux." On the contrary, their results show that when the bark tissue of *Hedera helix* is supercooled to -10° C then inoculated with ice, freezing is so rapid that the bulk of the cell water cannot diffuse through the membrane to the extracellular ice loci and nearly all the ice forms intracellularly. This is not surprising since the rate of cooling (31° C/min) was 600 x as rapid as the rate used normally to insure extracellular freezing in standard freezing tests.

The only plausible conclusion, therefore, is that "under natural conditions membrane resistance *does* limit water efflux," and in order for extracellular freezing to occur, the cell membrane must be sufficiently permeable to prevent intracellular freezing. The question that remains to be answered is: 7) How does the cell achieve and maintain a sufficiently high permeability to permit avoidance of intracellular freezing? In the present state of our knowledge of

membrane structure, more than one plausible answer is available. One method of increasing cell permeability to water is to increase the unsaturation of the membrane lipids (25).
In agreement with this relation, it has long been known that the unsaturation of fatty acids increases during the exposure of some plants to hardening low temperatures (14). Many recent investigations have supported this relation (9-12, 27). This response to hardening temperatures, however, is a general one, and occurs to the same degree in the less hardy as in the hardier varieties of a species (16, 20).

A possible clue to these apparently contradictory results is the lowering of the phase transition temperature associated with the increase in lipid unsaturation, and the fact that chilling resistance depends on this lowering of the phase transition temperature (15). The phase transition of a lipid membrane from the liquid crystalline to the solid (gel) state has been shown to lower the permeability of a lipid membrane to about 1/3 its original value (3). This would markedly increase the danger of intracellular freezing, since the cell water would move more slowly through the plasma membrane to the extracellular ice loci. An increase in unsaturation of the membrane lipids would lower the phase transition temperature, protecting the plant against this danger.

How much of a lowering of the phase transition temperature would be necessary to achieve this protection? Table I summarizes an analysis of the problem.

A plant with a freezing point of -1° C is cooled at a rage of 1° C/hr, a rate as rapid as is commonly found in nature but gradual enough to permit near equilibrium conditions throughout the freezing process. We will assume that the phase transition of the membrane lipids to the solid (gel) state decreases the water permeability of the cell to 1/4 its rate in the absence of phase transition (slightly more than the value for liposomes) (3). In order for the cells to to avoid intracellular freezing, 50% of their water must diffuse out of the cells to the extracellular ice loci within the first hours (Table I). After that, half as much (25%) must leave within the next two hours and half as much again (12.5%) within the next four hours. This means that in order for extracellular freezing to continue, the exosmosis during the cooling from -4 to -8° C needs to occur at only 1/16 the rate of the exosmosis between -1 and -2° C.

The conditions necessary for intracellular freezing avoidance become clearer if we assume that the membrane in the fluid state is just permeable enough to permit the relative exosmosis rate of 16 during the cooling from -1 to -2° C. A drop in temperature from -2 to -8° C would have a negligible effect *per se* on the permeability of the cell (not more than

TABLE I. *Relative Rates of Exosmosis of Water from Cells into Intercellular Spaces of Freezing Tissues. Original Freezing Point of Cells -1° C. Rates of Cooling are Assumed to be 1° C/hr and to Permit Near Equilibrium of the Cells with the Different Freezing Temperatures.*

Tissue temp.	Time below freezing point (hrs)	% of original cell water unfrozen	% of original water exosmosed to extracellular ice crystals	Relative rates of exosmosis
-1	0	100		
-2	1	50	50 in 1 hr	16
-4	3	25	25 in 2 hr	4
-8	7	12.5	12.5 in 4 hr	1

a decrease of about 30% if the Q_{10} is 2 or less). If, however, the drop in temperature induces a phase transition of the membrane lipids, and therefore a four-fold decrease in permeability, the most rapid possible rate of exosmosis would now be reduced by a relative value of four. This permeability value is therefore, 4X too low to support the rate of exosmosis of 16 during the freezing of the first 50% of the cell's water, but would be fully adequate to support the exosmosis of the subsequent 25% or any subsequent amount. Therefore, a phase transition of the membrane lipids at or above -1° C would result in intracellular freezing and death of the cell; but if the phase transition temperature were lowered to -2° C or lower, no intracellular freezing would occur at any temperature.

Therefore, an unhardened plant such as the above, with a phase transition temperature of its membrane lipids above -1° C, needs to accumulate only enough unsaturated fatty acids to lower this temperature of -2° C, in order to have fully adequate intracellular freezing avoidance for the rates of cooling found in nature. In support of this conclusion, direct measurements of the phase transition temperature have demonstrated that it is below the freezing point in hardy herbaceous and woody plants (as compared to chilling sensitive plants whose phase transition temperature is well *above* 0° C), but it showed no relation to the degree of hardiness among the hardy plants tested (25).

There is a practical aspect to this concept. It is usually assumed that all the gradations in hardiness of a series of

varieties are due to a greater and greater development of the
same one or more factors. On the basis of the above hypothesis there are at least two distinct kinds of factors which
must develop successively. The plant must first lower the
phase transition temperature of its membrane lipids to slightly below its freezing point (about -2 to -4° C), in order to
prevent intracellular freezing under natural rates of cooling.
It must then develop the factors that lead to a greater
avoidance and/or tolerance of extracellular freezing. This
is important information for the breeder; for if a plant already has the ability to increase the unsaturation of its
membrane lipids during the hardening period, sufficiently to
lower the phase transition temperature to about -2 to -4° C,
there is no need to breed into it a greater degree of unsaturation. In fact, this may do more harm than good. Instead,
the breeder must attempt to increase those factors that improve the plant's avoidance and tolerance of extracellular
freezing, such as cell sap concentration and reduction capacity. In the case of wheat, he has apparently done this un-
knowingly, for both the varieties with low freezing resistance and high freezing resistance have the same capacity to
increase the unsaturation of their lipids during hardening
(see above), but they differ markedly in cell sap concentrations and in ability to maintain SH groups in the reduced
state (14). Conversely, if a species possesses little or no
intracellular freezing avoidance, there is no point in trying
to breed into it an increased cell solute concentration, or
any other factor that increases tolerance of extracellular
freezing. It is first necessary to transfer to it an ability
to increase the degree of unsaturation of its membrane
lipids. This will confer sufficient avoidance of intracellular freezing to permit utilization of factors that increase
avoidance and/or tolerance of more severe extracellular
freezing.

V. INTERRELATIONS BETWEEN STRESSES

A. *Cryostress and Water Stress*

The living cell suffers the same physical strain whether
it is subjected to evaporative dehydration (desiccation) or to
an equal freeze dehydration, since the water diffuses out of
the cell in both cases. Therefore, freezing injury and desiccation injury should both occur at the same degree of dehydration. Experimental evidence, in the past, has repeatedly
supported this conclusion (14). This has now been corroborat-
ed for animal cells. The limpet showed 50% survival of

freezing at -10 to -12° C for 24 h, resulting in freezing of
60 to 80% of the body water. In agreement with this result,
the LD_{50} for water loss by desiccation was 76.5% (21).
Similarly, some of the limpets failed to survive immersion in
450% seawater for 6 h, which produced the same concentration
as by 70 to 80% loss of body water.

It has long been known that moderate desiccation can
harden the plant, increasing its tolerance both of drought and
freezing. In the case of red osier dogwood (*Cornus stolonifera*), the lowering of the freeze-killing temperature is pronounced: from -3° C in the control to -11° C in the plants
subjected to a water stress for 7 days (4-6). Even this
evidence does not adequately emphasize the importance of the
water stress in the freeze-hardening of plants. In the case
of cabbage, a water deficit is a strict requirement. When
kept at full turgor, the seedlings were unable to harden at
all, even under optimum conditions of temperature and light
(7). This absolute requirement is apparently very small, and
in order to establish it, zero water potential had to be
maintained by immersion in continuously aerated water. The
plant is apparently aware of this dependence of hardening on
at least a partial desiccation, and actually gets rid of its
excess water during the fall hardening. In the case of red
osier dogwood, it does this by opening its stomata wider during early acclimation at low temperature and under short days,
and by increasing its transpiration rate (18). Root resistance increased simultaneously. Stem water content decreased
43% during this acclimation.

B. *Other Stresses*

Although the interrelations between the cryostress and
the water stress are the best established, there is considerable evidence of other interrelations. The above results
with the limpet indicate that a salt stress may injure at the
same degree of osmotic concentration as that produced by the
cryostress. Similarly, in the case of lilac (*Syringa
vulgaris*), twigs receiving NaCl lost hardiness in proportion
to the increase in content of Cl^- and Na^+ (30).

One kind of heat tolerance is correlated with freezing
tolerance, since both reach a maximum, in at least some
plants, during mid-winter (14).

Similarly, among 25 strains of wood-inhabiting fungi,
heat and desiccation tolerance were highly correlated (31).
The same correlation was found among 78 strains of wheat (19).

Santarius (23) suggests that the sensitivity of ATPase
may underlie the interrelations between the three stresses:
the cryo-, thermo-, and water stress. In support of this

concept is the frequently observed sign of injury in all three—an efflux of ions. Until recently, this efflux was interpreted as simply a measure of dead cells. Recent results, however, indicate that it is the first sign of injury in stressed cells, and that it may be due to an inactivation of ion pumps, and therefore possibly an ATPase inactivation (17).

It must be realized that not all kinds of stress injury and resistance are correlated. This is partly because some resistances are due to avoidance, others to tolerance. Some kinds of resistance are, in fact, mutually exclusive. For instance, adaptation of photosynthesis to one stress may lead to a poor performance under another (2). Similarly, no correlation was found between the temperature optimum and the pH optimum among 21 species of thermotolerant fungi (22).

In summary, an understanding of cold hardiness is not possible until the following questions are answered:

(1) Does intracellular freezing occur in nature?

(2) How does intracellular freezing kill the cell?

(3) How does extracellular freeze-dehydration injure the cell?

(4) How do some living cells succeed in remaining supercooled for long periods, in the presence of extracellular ice?

(5) What is the specific enzyme behavior leading to solute (mainly sugar) accumulation during hardening?

(6) How does the cell achieve and maintain a sufficiently high permeability to permit avoidance of intracellular freezing?

(7) What is the basis of the interrelations between resistances to different kinds of stress?

Other questions will certainly arise as these are answered.

REFERENCES

1. Bartholid, J. F., *Proc. Fla. State Hort. Soc. 85*, 299 (1972).

2. Berry, J. A., *Science 188*, 644 (1975).

3. Blok, M. C., van Deenen, L. L. M., and de Gier, J., *Biochim. Biophys. acta 433*, 1 (1976).

4. Chen, P., Li, P. H., and Weiser, C. J., *HortScience 10*, 372 (1975).

5. Chen, P. M., and Li, P. H., *Plant Physiol. 59*, 240 (1977).

6. Chen, P. M., Li, P. H., and Burke, M. J., *Plant Physiol.* *59*, 236 (1977).
7. Cox, W., and Levitt, J., *Plant Physiol. 57*, 553 (1976).
8. Dear, J., *Cryobiology 10*, 78 (1973).
9. Grenier, G., and Willemot, C., *Cryobiology 11*, 324 (1974).
10. Grenier, G., and Willemot, C., *Can. J. Bot. 53*, 1473 (1975).
11. Grenier, G., Maxliak, P., Tremolieres, A., and Willemot, C., *Physiol. Veg. 11*, 253 (1973).
12. Grenier, G., Hope, H., Willemot, C., and Therrien, H.-P., *Plant Physiol. 55*, 906 (1975).
13. Gusta, L. V., Burke, M. J., and Kapoor, A. C., *Plant Physiol. 56*, 707 (1975).
14. Levitt, J., "Responses of Plants to Environmental Stresses." Academic Press, New York (1972).
15. Lyons, J. M., *Ann. Rev. Plant Physiol. 24*, 445 (1973).
16. Miller, R. W., de la Roche, I., and Pomeroy, M. K., *Plant Physiol. 53*, 426 (1974).
17. Palta, J. P., Levitt, J., and Stadelmann, E. J., *Plant Physiol. 60*, 393 (1977).
18. Parsons, L. R., *Soc. for Cryobiol. 14th Ann. Meeting,* 23 (1977).
19. Polimbetova, F. A., Mamonov, L. K., Kim, G. G., Vorob'eva, E. A., Skripka, S. A., Zykin, V. A., and Kulichenko, N. A., *S-KH Biol. 9*, 235 (1974).
20. Roche, I. A. de la, Pomeroy, M. K., and Andrews, C. J., *Cryobiology 12*, 506 (1975).
21. Roland, W., and Ring, R. A., *Cryobiology 14*, 228 (1977).
22. Rosenberg, S. L., *Can. J. Microbiol. 21*, 1535 (1975).
23. Santarius, K. A., *Planta 113*, 105 (1973).
24. Siminovitch, D., and Scarth, G. W., *Can. J. Res. C16*, 467 (1938).
25. Siminovitch, D., *Soc. for Cryobiol. 14th Ann. Meeting,* 28 (1975).
26. Smith, A. P., *Biotropica 6*, 263 (1974).
27. Stein, W. D., "The Movement of Molecules Across Cell Membranes." Academic Press, New York (1967).
28. Stoller, E. W., and Weber, E. J., *Plant Physiol. 55*, 859 (1975).
29. Stout, D. G., Steponkus, P. L., and Cotts, R. M., *Plant Physiol. 60*, 374 (1977).
30. Sucoff, E., and Hong, S. G., *Can. J. Bot. 54*, 2816 (1976).
31. Zimmermann, G., and Butin, H., *Flora (Jena) 162*, 393 (1973).

FREEZING PROCESSES AND INJURY
IN PLANT CELLS

E. Asahina

The Institute of Low Temperature Science
Hokkaido University
Sapporo

I. INTRODUCTION

Direct observation of freezing plant cells under a microscope was repeatedly made by many investigators since the
nineteenth century (see IV-A). Based on the observations a
number of works on frost hardiness have been done (see IV-B).
Even at present, however, detailed description on the freezing
process of plant cells appears very important for the workers
on frost hardiness, since most of the original papers on this
subject have been out of print. Although some of the observed
results are still available in the excellent comprehensive
reviews written by Levitt (8-10), the descriptions are usually
insufficient to show detailed freezing processes in various
plant cells. The freezing of a small piece of plant tissue
under a microscope, of course, not always represents the actual process of the freezing in an intact whole plant under
natural climatic conditions. However, knowledge obtained from
the observations under artificial freezing conditions may
certainly be applicable to the freezing natural plants provided that a careful consideration is made about freezing
conditions, especially on cooling rates. The purpose of the
present paper is to describe in detail some of the typical
freezing processes of plant cells in the hope that this will
contribute to the elucidation of various problems involved
in the mechanism of freezing injury.

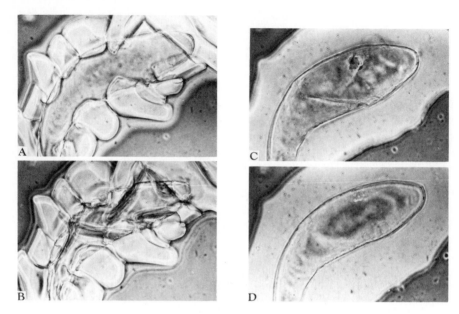

FIGURE 1. Slow freezing of a cultured cell of tobacco.
X290. Cooling rate: 0.25° C/min. A: Freezing of surrounding medium at -2° C. B: Extracellular freezing to -6° C.
C: Immediately after thawing from extracellular freezing
for 20 min. D: Recovery of the same cell as in C, 20 min
after thawing at 5° C.

II. EXTRACELLULAR FREEZING

     The freezing process of plant cells is strikingly influ-
enced by many factors both external and internal, such as
the grade of supercooling, cooling rate, the hardiness of the
cell itself and the amount of freezable liquid moistening the
surface of the material. The prevention of excessive supercooling of the material is necessary to clearly observe the
process of cell freezing, since an introduction of freezing
in a well supercooled plant tissue invariably results in an
intense darkening in cell appearance because of the formation of very fine ice crystals both within and without the
cell. In the following observation, therefore, the materials
which were covered with silicone oil were, as a rule, inoculated with ice at a slightly supercooled state.
     Isolated cultured cells are the best materials to demon-
strate the freezing process of plant cells. When inoculated
with ice onto the cell, the liquid moistening the cell freezes
first. As cooling proceeds, ice crystals outside the cell

grow larger and larger withdrawing water from the cell. As
a result of such an extracellular growth of ice crystals cells
undergo dehydration and contraction. However, if the cooling
is sufficiently slow, the final temperature is not too low,
and the total length of freezing period is short, then the
frozen cells, even unhardy ones, can survive freezing (Figure
1).

As an example of a very tender plant, internodial cells
of fresh water algae *Nitella* were subjected to freezing (Figure 2). An extracellular ice formation occurred at a locus
on the cell surface where an inoculation with ice was made at
-0.6° C. This causes little injury in most of the frozen
cells at least for several minutes. After thawing from such
a slight extracellular freezing, the characteristic streaming
of protoplasm can be observed in the cell. If the *Nitella*
cells frozen in this way are cooled further to a temperature
lower than -1° C, a distinctly rapid growing of ice column
is always observed on the cell surface. This always resulted
in a remarkable disintegration or clotting of protoplasmic
layer, at least in the area above which a rapid growth of
extracellular ice took place (1).

By the use of the staminal hair of *Tradescantia*, a special
type of extracellular freezing can be observed. Under slightly supercooled conditions ice forms between the protoplast and
the cell wall, withdrawing water from the retreating protoplast (Figure 3). This phenomenon may be referred to as
"frost plasmolysis" in a true meaning. The same word has
often been used in literature (8) to indicate a contracted
figure of destroyed protoplast which ensued after an extracellular freeze-thawing. For this type of dead cells, the
term "frost pseudoplasmolysis" might well be employed. When
"frost plasmolysis" takes place, the protoplast decreases in
volume as a result of ice growing, sometimes being reduced by
one-half of its original amount. Ice seems to grow easily
along the inner side of the cell wall (Figure 3B), since some
semisphere ice masses attached to the cell wall are frequently
formed in the cell even at the distal end of the "frost
plasmolysing" cell. Such formation of an ice mass, when it
occurs in the middle part of the cylindrical hair cell, often
results in the complete devision of the protoplast (Figure
3C). After thawing the divided cell parts expend to unite
with each other and recover the original appearance of protoplast with active streaming of protoplasm over the whole
cytoplasmic layer (1). "Frost plasmolysis" was known to
occur since early ages (23) and was occasionally observed not
only in *Tradescantia* cells but also in various plant cells,
particularly in epidermis under suitable slow cooling conditions (1).

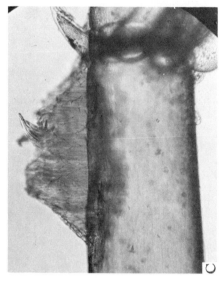

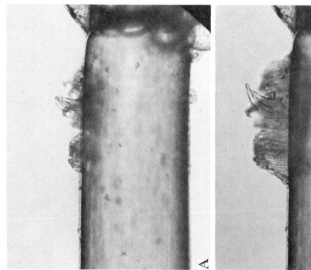

FIGURE 2. Extracellular freezing of an in-ternodial cell of Ntella. X100. (Asahina, 1956). A: Initiation of extracellular freez-ing at -0.60 C, one minute after ice seeding on to the cell surface. B: 3 min later, at -1° C. C: 5 min later, at -1° C.

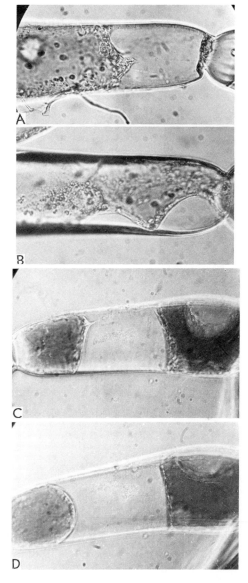

FIGURE 3.   Frost plasmolysis in the staminal hair cells of Tradescantia at about -2° C. (Asahina, 1956).   A: Ice formation at one end of the cell which is adjacent to an already frozen cell.   Note the protoplasmic strands within the ice mass.   X300.   B: Formation of semisphere ice masses along the cell wall.   X370.   C: Division of protoplast results from freeze plasmolysis.   X270.   D: The same as in C, immediately after melting of ice.   X270.

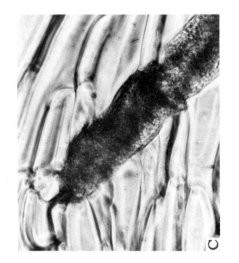

C

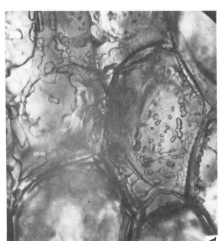

A

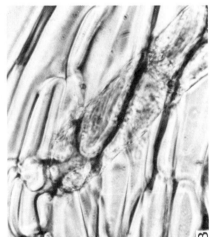

B

FIGURE 4. A: Extracellular freezing of the parenchymatous tissue of watermelon fruit at -3º C, 2 min after an ice seeding. X100. (Asahina, 1956). B: Extracellular freezing of cultured cells of tobacco, at -6.5º C. X450. C: Same as B, cooled to -9º C at a rate of 1º C/min. Immediately after the occurrence of intracellular freezing. X450.

Dehydration of a cell caused by an extracellular freezing has long been regarded, at least by some authors, as a withdrawal of water due to the osmotic gradient between cell fluid and outside medium concentrated by ice formation. In the "frost plasmolysis" as described here, however, no detectable amount of any solution was apparently observed between the protoplast and cell wall. If the surface of the protoplast is in direct contact with ice, it is reasonable to suppose that water can pass through the surface layer of protoplast without any hypertonic outside medium to arrive at the surface of an extracellular ice. In other words, dehydration in extracellularly frozen plant cells may be a result of the simple movement of cell water due to the difference in chemical potential of water molecule (25) between supercooled water within the cell and the ice outside.

When a slice of plant tissue is subjected to freezing by ice inoculation the liquid moistening the cut surface freezes first and then dendritic ice branches rapidly spread over the whole cell surface in the slice. These ice branches on the cell surface frequently are divided into many small lumps (Figure 4A). In the case of unhardy plants, the rate of growth of such ice branches or lumps soon becomes very slow. With further cooling the cells become very apt to freeze internally (Figure 4B and 4C). This suggests that during a previous extracellular freezing a poor supply of water from unhardy cells to outside ice may accelerate the occurrence of intracellular freezing, since there is still enough freezable water remaining within the cooling cells at a given low temperature.

A fatal injury in extracellularly freezing cells is not caused only by the following intracellular freezing but also by extracellular freezing itself. During very slow extracellular freezing, if freezing cells are cooled beyond a certain limit of tolerable low temperature, unhardy or moderately hardy cells are easily injured irreversibly without any ice formation within the cells. Death occurs in these cells under dehydrated and contracted conditions. In the cells killed in this way the denatured nucleus appears very clear and dark immediately after thawing (Figure 5). Sometimes "frost pseudoplasmolysis," which means a contracted figure of denatured protoplast, is observed after thawing. In less injured cells by extracellular freezing, there are found hardly any appreciable changes in cell appearance at least for a short period of freezing. However, the cells gradually exhibit various pathological changes such as slow darkening, fine or coarse granulation, partial or entirely frothy appearance of the protoplasm, and the occurrence of apparently defined outline of the nucleus. Besides, it is often the case of these cells that they appear intact soon after thawing

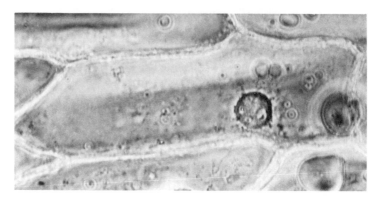

FIGURE 5.   An epidermal cell from moderately hardy onion,
killed by an extracellular freezing.   A denatured nucleus is
clearly observed in the thawed cell.   X500.

without complete loss of their semipermeability. Such cells
are capable of plasmolysing at least immediately after thawing, but a following deplasmolysis usually leads the cells to
a fatal coagulation of the protoplast. If such plasmolysing
cells are left to themselves without any treatment, they sooner or later lose some of their contents which diffuse out into
the surrounding medium. Sometimes a rupture of the protoplast is clearly observed during deplasmolysis of these cells.
    Extracellular ice on hardy cells, on the other hand,
appears to form continuously withdrawing water from the cell
interior resulting in a remarkable growth of ice crystals
on the cell surface (Figure 6). The cells frozen in this
way consequently undergo a severe dehydration and contraction,
and as a rule, none of them freeze internally as long as the
cooling is slow. In a freezing plant tissue, intercellular
spaces were observed to be the best locus to form the extracellular ice crystals (Figure 6A). As ice crystals occupy
the intercellular spaces, a squeezing out of the air from the
intercellular space results. This follows a translucent
appearance of the tissue during freezing and more remarkably
after thawing when the intercellular spaces are filled with
melted water. If the cells in the tissue have not been
seriously injured by the previous freezing, they can soon
reabsorb the intercellular water and regain the original
appearance.
    In the case of very hardy cells of woody plants, extra-
cellularly freezing cells are capable of shrinking remarkably.
If they have coloured vacuoles, the center area of these
cells, therefore, becomes quite light, as the vacuole is
squeezed into the periphery of the cells (1, 27) (Figure 7).

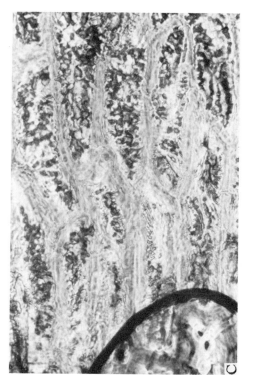

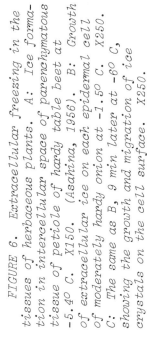

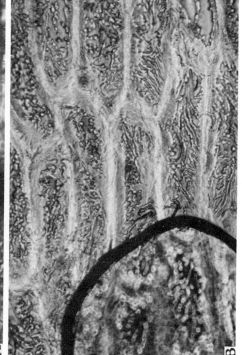

FIGURE 6. Extracellular freezing in the tissues of herbaceous plants. A: Ice formation in intercellular space of parenchymatous tissue of petiole of hardy table beet at -5.4° C. X150. (Asahina, 1956). B: Growth of extracellular ice on each epidermal cell of moderately hardy onion at -1.5° C. X250. C: The same as B, 9 min later at -6° C, showing the growth and migration of ice crystals on the cell surface. X250.

FIGURE 7.   Extracellular freezing in cortical tissue of hardy mulberry tree at -7⁰ C.   The vacuoles, which were artificially stained, are squeezed into the periphery of the contracted cells.   X420.   (Asahina, 1956).

Upon warming, the contracted cells absorb water and expand as surrounding ice melts to recover the normal appearance and activity.   No "psuedoplasmolysis" is observed in most of the hardy cells after freeze-thawing.

III.   INTRACELLULAR FREEZING

For the observation of intracellular freezing the staminal hair of *Tradescantia* is perhaps one of the best materials for the following reasons:

(1) Cell surface can be easily kept free from the moisture which disturbs a clear observation of the process of intracellular freezing.

(2) By freezing one of the hair cells, any cell of the hair can be inoculated with ice only at a very small area of the cell end connected with the neighbouring cell.

(3) Consequently, the order of the propagation of cell freezing can always be predicted.

The filament and the hair cell can be frozen at temperatures below -1.2⁰ C.   When inoculated with ice onto the cut end of the filament at a moderately supercooled state, the cells of the filament freeze in rapid succession, then the

freezing propagates through the proximal hair cell towards
the distal end of the staminal hair. The cell once frozen
inoculates the cell next to it until all the cells in a hair
freeze in succession, with some short time interval between
the freezing of each cell (Figure 8). The process of cell-
to-cell freezing in *Tradescantia* cells has been strikingly
recorded by the use of cinematography (18). In the distal
end of the newly frozen hair cell, a growing of very clear
ice mass always takes place, resulting in a slight projection of the frozen cell end into the neighbouring unfrozen
cell. With slightly faster cooling, a jet of ice crystals
spouts out from this ice mass into the unfrozen cell without
any appreciable destruction of the cell wall. The ice crystals newly formed in the neighbouring cell first grow in
the cytoplasmic layer with dendritic branches towards the
distal end of the cell resulting in a formation of ice layer
outside the tonoplast (Figure 8D). From this ice layer a
rapid growth of newly formed ice fronts into the tonoplast
is observed. It is of note that ice grows in sap vacuole
distinctly faster than in cytoplasm. The fact that cytoplasm is observed to freeze previous to the freezing of the
vacuole, of course under a suitable temperature condition, was
already reported with hardy tree tissues by Siminovitch and
Scarth (27).

The manner of intracellular freezing of parenchymatous
cells is distinctly separable into two types, i.e. flash and
non-flash types. The former is a sudden freezing characterized by an instantaneous darkening of the whole cell (3, 13),
whereas the latter is a slow freezing with clearly visible
ice growing in the cell (1). The rates of ice growth within
these cells clearly indicate the degree of supercooling of
the cell contents at the time of the introduction of cell
freezing. A high rate of cooling is, of course, favorable
to the cell freezing of the flash type. However, the difference in the manner of cell freezing is not brought about
only by the difference in cooling rates of the materials,
but also by the character of the cell itself. Cells from
non-hardy plants, such as the fruit of tomato, melon and
pod of bean usually freeze in flash type manner, even if
they are cooled slowly. While parenchymatous cells from various vegetables such as cabbage, turnip, sugar beet and spinach, in which plant frost hardening is applicable, exhibit
intracellular freezing of non-flash type usually in summer
season even at a moderate cooling rate, say 4° C per minute.
When cells freeze in non-flash manner, the cell freezing
usually propagates continuously yet with a short time
interval between their respective congelations. In such a
case, intracellular ice does not grow from a point but from
a broad area or from one side of the cell surface adjacent

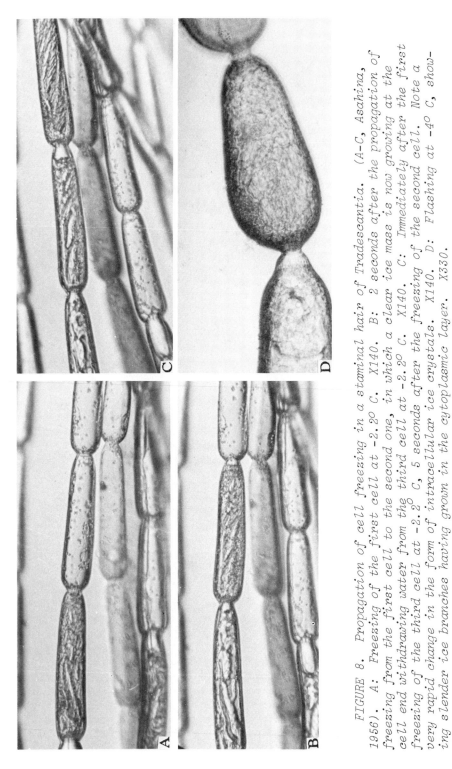

FIGURE 8. Propagation of cell freezing in a staminal hair of Tradescantia. (A–C, Asahina, 1956). A: Freezing of the first cell at -2.2° C. X140. B: 2 seconds after the propagation of freezing from the first cell to the second one, in which a clear ice mass is now growing at the cell end and withdrawing water from the third cell at -2.2° C. X140. C: Immediately after the first freezing of the third cell at -2.2° C, 5 seconds after the freezing of the second cell. Note a very rapid change in the form of intracellular ice crystals. X140. D: Flashing at -40° C, show-ing slender ice branches having grown in the cytoplasmic layer. X330.

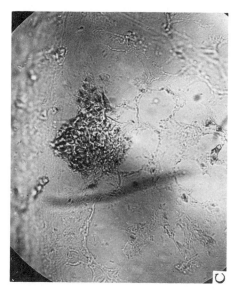

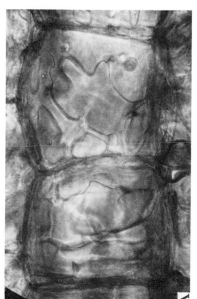

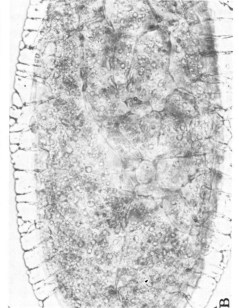

FIGURE 9. Intracellular freezing in parenchymatous cells. (Asahina, 1956). A: Pith cells of
the petiole of dehardened table beet frozen at -6° C.
X315. B: Isolated cell from tomato fruit, 18 min
after flashing at -10° C. A number of spherical
drops of condensed cell sap embedded in ice and a
net work of destroyed cytoplasm can be seen within
the frozen cell. X280. C: Denatured nucleus with
a network of denatured cytoplasm in a thawed cell of
tomato fruit soon after flashing at -15° C. X440.

to already frozen cell. From this area several smooth ice fronts grow in parallel with each other towards the opposite side of the cell in a period from approximately one-half to a few seconds. The cell contents are entirely concentrated between the masses of these ice crystals. Cells frozen in this way appear quite clear and light during freezing (Figure 9A) contrary to the remarkable darkness in "flashed" cells (Figure 10).

Thin strips of epidermis from various herbaceous plants also provide a suitable material for the observation of cell freezing. In fact, striking observations of dynamic process of cell freezing were made with onion epidermis by Chambers and Hale, and Luyet and Gibbs many years ago (3, 13). In a moderate supercooled state, for instance, at several degrees centigrade below the freezing point of the squeezed sap of the material cell freezing takes the manner of typical flashing, a darkening begins at one end of the cell and proceeds in a very rapid wave-like fashion to the other end, usually in a fraction of a second. Having started from the inoculated point in the tissue, such flashing in the cells occurs one by one and propagates all over the strip, gradually in rapid succession. At the moment of flashing the cell appears intensely opaque and dark; however, it soon becomes appreciably lighter (Figures 9 and 10). If the temperature at which cells are kept frozen is very low, the high opacity of the cell continues for a long time.

At the instant of intracellular freezing the structure of the protoplast is entirely destroyed. Being coarsened by dendritic ice crystals, the cytoplasmic layer presents a coarsely granular appearance and then becomes lighter as the destroyed cytoplasm is rapidly concentrated around the developing ice mass, sometimes resulting in a formation of the network. The nucleus deforms as a dense granular mass. The vacuole transforms into many globules of concentrated cell sap and gas bubbles, both embedded in ice mass (Figures 9B and 10). The lower the temperature at which freezing occurs, the more the branching of the ice front in the cell. This results in the formation of a larger number of smaller sap droplets packed in the ice mass. This is the reason why a rapid intracellular freezing, i.e. flashing, causes an intense darkening in cell appearance. Immediately after thawing of a "flashed" cell, destroyed nucleus represents a remarkable spongy structure (Figure 9C) which is distinctly different in appearance from those denatured by extracellular freezing (Figure 5).

It was well known that very hardy cells are very apt to freeze extracellularly (8). However, if they are cooled rapidly or inoculated with ice at a considerable supercooled state, they too undergo flashing. In such a case,

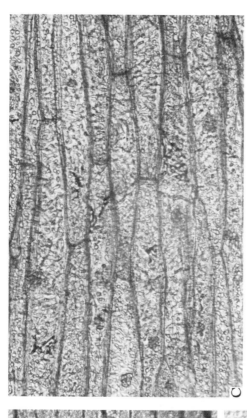

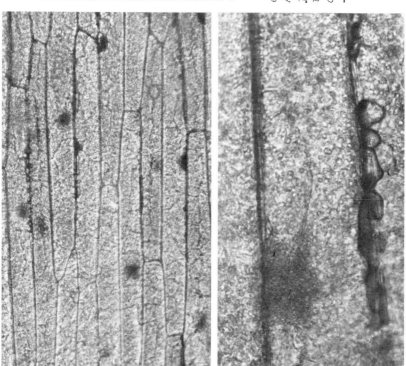

FIGURE 10. Intracellular freezing in the epidermis of Welsh onion. (Asahina, 1956). A: Freezing of unhardy tissue, 4 min after flashing at -11° C. X150. B: Details of the same sample as in A. X660. C: Freezing of hardy tissue, 7 min after flashing at -13.5° C. X150.

31

intracellular formation of fine ice crystals is sometimes
confined to the cytoplasmic layer; around the unfrozen tonoplast in a foamy network of coagulated protoplasm is observed
in frozen cells (Figure 11B). After thawing from such a
type of cell freezing, the tonoplast occasionally keep semipermeability to some extent at least for a short period of
time (1, 27).

IV. FREEZING INJURY

    As pointed out by Levitt (10), there are four moments of
injury in frozen-thawed plant cells: 1) during freezing,
2) after freezing equilibrium has been reached, 3) during
thawing, and 4) after thawing. The following consideration
only deals with the first.
    From the observations described before, at least two
moments of freezing injury may be noted during freezing as
far as the artificial freezing under a microscope is
concerned.

A. *Intracellular Freezing Injury*

    Cells are, as a rule, killed at the time of intracellular
freezing perhaps because of the mechanical destruction of
membrane systems resulting from the growth of ice crystals
in the protoplasmic structure. It now becomes clear that
cells from various organisms can survive intracellular freezing provided that both freezing and thawing are rapid enough
to form very fine intracellular ice crystals which are innocuous to the cells, and then melt these fine ice crystals
before they grow to the dangerous size (2, 15, 21, 24).
However, such an innocuous ice formation in the cell may only
occur by means of an extremely rapid cooling above 1000° C/
min. It may be safely said, therefore, that no plant cells
in nature can survive intracellular freezing.
    It has long been believed that intracellular freezing
rarely takes place under natural temperature conditions. On
the other hand, the extracellular freezing in tender unhardy
plant tissues is very frequently followed by an intracellular
freezing even at a relatively slow rate of artificial cooling,
as observed under a cryomicroscope. This may suggest the
assumable occurrence of intracellular freezing in nature when
the plant tissue is exposed to an unusually rapid cooling
(8). However, this assumption has never been confirmed ex-
perimentally. Weiser (29) reported that evergreen foliage
which is not injured at -87° C during slow freezing in the

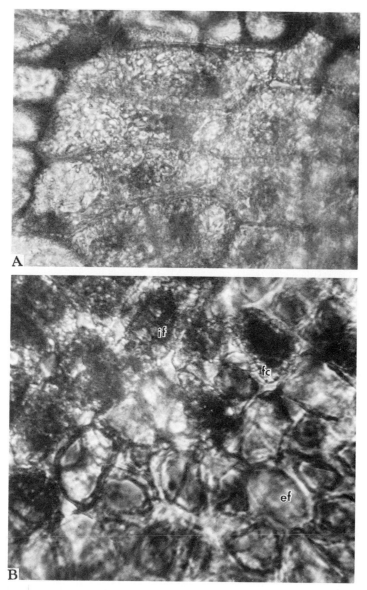

FIGURE 11.   Freezing of cortical parenchyma of woody
plants.   (Asahina, 1956).   A: Intracellular freezing in the
tissues of red-berried elder, 8 min after flashing at -8.6° C.
X490.   B: Extra (ef) and intracellular (if) freezing in hardy cells of apple at -8.2° C, 6 min after an ice seeding.
Frozen cytoplasm (fc) appears foamy around the unfrozen tonoplast in intracellularly frozen cells.   X665.

winter is killed at -10° C when thawed tissues are frozen to
-10° C at a cooling rate of 8 to 10° C/min. He suggested
that sunscald which is a special winter injury occurring on
south-facing branches of many tree species, is probably
caused by rapid temperature change which gives rise to intracellular freezing. However, Sakai and Horiuchi's work on the
temperature change in trunks of various trees in Hokkaido
suggested that natural occurrence of a rapid cooling at a
higher velocity than 0.5° C/min in tree branches is very difficult (20). An alternative interpretation of sunscald, at
least in some parts, may be a dehydration injury resulting
from a combination of freezing of soil and exposure of south
facing branches to sunshine (19). On the other hand, a possible occurrence of intracellular freezing in natural plant
tissue has recently been reported with flowers of apple (17)
and azalea (6). The whole plant freezes extracellularly
without fatal injury, while some parts of flowers or flower
buds can remain supercooled. When the plant is cooled beyond
a critical low temperature, fatal intracellular freezing may
finally result only in flower tissues, even if the cooling is
very slow.

B. *Extracellular Freezing Injury*

Cells are fatally injured by extracellular freezing alone
when they are cooled beyond a tolerable limit of low temperature and depending on the length of freezing time. Concerning
the mechanism of this type of freezing injury, a number of
theories have been expounded using materials from a variety
of organisms (5, 7, 8, 16). At present, however, few of them
are generally applicable to various types of plant cells.
The present writers' observation on the behavior of plant
cells during freezing (1) appears to support the view that one
of the common causes of injury in extracellular freezing cells
may certainly be freeze-induced dehydration (12, 22, 26).
Levitt pointed out two basic factors in freeze-dehydration
injury (8): 1) the mechanical stress, which increases with
the degree of cell contraction and 2) the resistance of the
protoplasm to this stress which is inversely related to
protoplasmic dehydration, but which decreases with the decreasing temperature. In relation to Levitt's view mentioned
in the last phrase, it seems interesting to note that the
cortical cytoplasm has been known to become fluid (14) and
the internal cytoplasm to become viscous (4) with the decreasing temperature, at least in sea urchin eggs. Takahashi
very recently demonstrated that sea urchin eggs in hypertonic
solutions exhibited a remarkably higher resistance at 0° C
than at 18° C to the cell contraction beyond a critical cell

size which causes a fatal injury (28). Among three kinds of sea urchin eggs, the higher the resistance to the dehydration-induced cell contraction, the higher the tolerance to extra-cellular freezing at lower temperatures. At any rate, the fundamental mechanism of freezing injury may probably remain to be solved until a fuller knowledge is obtained about the protoplasmic structure, especially the structure of the sur-face membrane.

REFERENCES

1. Asahina, E., *Contrib. Inst. Low Temp. Sci. 10*, 83 (1956).
2. Asahina, E., Shimada, K., and Hisada, Y., *Exptl. Cell Res. 59*, 349 (1970).
3. Chambers, R., and Hale, H. P., *Proc. Roy. Soc. B 110*, 337 (1932).
4. Costello, D. P., *J. Cell Comp. Physiol. 4*, 421 (1934).
5. Farrant, J., and Morris, G. J., *Cryobiol. 10*, 134 (1973).
6. George, M. F., Burke, M. J., and Weiser, C. J., *Plant Physiol. 34*, 29 (1974).
7. Heber, U., and Santarius, K. A., *In* "Temperature and Life" (H. Precht, J. Christophersen, H. Hensel and W. Larchers, eds.), p. 232. Springer-Verlag, Berlin (1973).
8. Levitt, J., "Responses of Plants to Environmental Stresses." Academic Press, New York (1972).
9. Levitt, J., *In* "Cryobiology" (H. T. Meryman, ed.), p. 495. Academic Press, New York (1966).
10. Levitt, J., *Protoplasmatologia VII 6*, Springer-Verlag, Wien. (1958).
11. Levitt, J., "Frost Killing and Hardiness of Plants." Burgess, Minneapolis (1941).
12. Levitt, J., and Siminovitch, D., *Can. J. Res. C18*, 550 (1940).
13. Luyet, B. J., and Gibbs, M. C., *Biodynamica 25*, 1 (1937).
14. Marsland, D., *J. Cell. Comp. Physiol. 36*, 205 (1950).
15. Mazur, P., *Cryobiol. 14*, 251 (1977).
16. Meryman, H. T., Williams, R. J., and Douglas, M. St. J., *Cryobiol. 14*, 287 (1977).
17. Modlibowska, I., *Cryobiol. 5*, 175 (1968).
18. Modlibowska, I., and Rogers, W. S., *J. Exp. Bot. 6*, 384 (1955).
19. Sakai, A., *Ecology 51*, 657 (1970).
20. Sakai, A., and Horiuchi, T., *J. Jap. Forest. Soc. 54*, 379 (1972).
21. Sakai, A., Otsuka, K., and Yoshida, S., *Cryobiol. 4*, 165 (1968).

22. Scarth, G. W., *Plant Physiol. 16*, 171 (1941).
23. Schander, R., and Schaffnit, E., *Landwirt. Jahrb. 52*, 1 (1919).
24. Shimada, K., and Asahina, E., *Cryobiol. 12*, 209 (1975).
25. Shumskii, P. A., "Principles of Structural Glaciology" (Trans. from the Russian by David Kraus). Dover Publications, New York (1964).
26. Siminovitch, D., and Levitt, J., *Can. J. Res. C19*, 9 (1941).
27. Siminovitch, D., and Scarth, G. W., *Can. J. Res. C16*, 467 (1938).
28. Takahashi, T., *Contrib. Inst. Low Temp. Sci. 19*, in press (1977).
29. Weiser, C. J., *Science 169*, 1269 (1970).

ANALYSES OF FREEZING STRESSES
AND PLANT RESPONSE[1]

C. R. Olien

Crop and Soil Science Department
Michigan State University
East Lansing, Michigan

I. INTRODUCTION

Winter hardiness is a complex trait with a low heritability. Freeze hardiness, as evaluated from plant response to specific controlled stress tests, is determined by a simpler set of distinguishable traits that have a higher heritability. The following outline relates the cryoprotective traits of hardy plants, that have been described by stress physiologists, to the stage of development where they exert their effect.

Initial plant growth: Hardy plants generally have simpler morphology (e.g., cereal plants with few tillers) and protected meristems (not associated with tissues easily injured by freezing).

Hardening: Hardy plants adapt their metabolism to low temperature (e.g., changes in membrane lipids). Substances that protect tissues from various forms of freezing stress are synthesized.

Recovery: Hardy plants accumulate metabolic reserves for development of new tissues from uninjured meristems.

[1]*Cooperative investigations of the Agricultural Research Service, U.S. Department of Agriculture and Michigan Agricultural Experiment Station.*

Disease resistance is important, especially of plants predisposed by freeze injury.

Hardy plants need protection from toxicity of degenerating freeze-injured tissues as a factor for recovery.

STRESS ANALYSIS AND EFFECTIVE CRYOPROTECTANTS: See the diagram on the following page.

Stress energies develop as the free energy of water transition is dissipated: ($\Delta G_\mathbf{l}$) Free energy for crystal growth into the protoplast; ($\Delta G_\mathbf{ll}$) Free energy for crystal growth along the cell wall; (U_{adh}) Potential energy of ad-hesion that draws ice and hydrophilic plant systems into matrices; (π) Osmotic activity; (Ψ) Water potential. Water redistribution occurs as ice forms until the ΔG is zero.

ΔG *is balanced* by rise in temperature of supercooled tissue (ΔT), irreversible work (ΔA), shifts in activation limits of transitions ($\Delta \xi_{lim}$), shifts in concentration of solute in liquid at the ice interface (ΔC), or shifts in the density of water molecules in a gas phase at the interface (Δn).

Effective temperature is that which results in injury of hardened 'Hudson' barley leaf tissue (*Hordeum vulgare* L.) frozen under various test conditions in which the different stress energies predominate.   Each form of stress causes a unique pattern of injury.

Cryoprotectants that affect each form of stress are listed.

II.   CRYODYNAMICS

Gibbs free energy (G) or chemical potential (dG/dn), that is the driving energy of transitions, is especially useful as a measure of one form of stress.   The free energy of freezing is determined by the displacement from equilibrium and can be estimated from the degree of supercooling (ΔT) or from the vapor pressure ratio of ice to liquid (P/P_o).

$$\Delta G = \Delta T - T\Delta S \simeq 5.2 \ \Delta T$$

$$= RT \ in \ P/P_o \simeq 5.2 \ \Delta T$$

$$\log P/P_o \simeq 1.11 \ \Delta T/T$$

It can be expressed in terms of the activity of liquid (M) and ice (F) or simply as the activity of water for phase transition (A) (12).   The net free energy equals the free

Stress Analysis and Effective Cryoprotectants

	Freezing			
	Nonequilibrium		Equilibrium	
	Nucleation	Propagation	Adhesion	Desiccation
Stress energy	$\Delta G_\top$	$\Delta G_{\|}$	U_{adh}	$\pi \qquad \Psi$
ΔG balanced by	ΔT	ΔA	$\Delta \xi_{lim}$	$\Delta C \qquad \Delta n$
Effective T	$-50\ C$	$-50\ C$	$-100^\circ\ C$	$-200\ C$
Cryoprotectants	1. Membrane stabilizers e.g. proteins 2. Nucleators e.g. bacteria 3. Freezing inhibitors or supercooling (5) 4. Freezing point depressors e.g. glycols	1. Water content 2. Equilibrium transition patterns 3. Kinetics inhibitors e.g. cell wall mucilages	1. Fluid barriers to adhesions e.g. cell wall glycols	Intracellular composition

energy of freezing minus the free energy of melting, so ΔG = $RT\ln$ F/M = $RT\ln A$ (12, 16). For small displacement from equilibrium, the difference between the activities of freezing and melting, can be substituted for $\ln A$ ($\ln A$ = $\ln\{1+\Delta A\}$ $\simeq$ ΔA). ΔA has been used as an index of the activity of water with respect to freezing (A_{wf}) (12).

A. Free Energy for Crystal Growth into Protoplast (ΔG_1)

The primary distinction between tender and hardened cells involves the ease with which ice can grow from the outer free space into the protoplasts (9, 28). The free energy for crystal growth across the plasmalemma (ΔG_1) is determined by the degree of supercooling below the freezing point of the protoplasm. The lipid component of the plasmalemma may, in part, account for the slight resistance of tender plants. Hardening involves synthesis of cryoprotectants, especially hydrophilic proteins (22, 24). The protoplasmic membrane is not a simple porous structure. Pore size, proposed to account for the barrier effect, involves so few molecular diameters that there are no bulk effects. Molecular interactions of the interface, negligible in macroscopic systems with large pores, become an essential parameter (27).

Liquid water is characterized by a structure or molecular ordering that is complex and dynamic with a half-life in the range of 10^{-11} seconds (4). This molecular ordering is affected by association with other substances at interfaces. Understanding liquid water structure and how it is affected at interfaces is essential for characterization of cryoprotectants (7, 8). This is a complex and developing science with diverging viewpoints being expressed in the literature. The most exciting aspect has been the development of a means for critical evaluation of new concepts of water structure and its interactions with plant substances.

This evaluation of liquid structure theory has been based on partition of the Helmholtz free energy, which is a function of the number of molecules in various patterns of association (3, 10, 21, 26). The distribution of water among various forms depends on the energy parameters used to relate structure theory to thermodynamic properties, and these calculated values can be compared with empirical data.

B. *Free Energy for Crystal Growth along Cell Wall ($\Delta G_{\parallel}$)*

Histological aspects of hardening help prevent growth of ice into critical tissues (11, 25). Meristematic cells, essential for plant survival, are protected by freezing inhibitors (12). Certain mucilaginous polymers along the cell wall form a film on ice crystals as they grow. The ice crystals then grow more slowly and imperfectly. Large crystals develop only in spaces where deformation does not cause serious injury. Stress is further diminished in the critical tissues of a hardy plant as the water diffuses and distills from vital regions where freezing is inhibited to less critical regions where ice crystals can grow freely. Freezing kinetics distinguish factors that affect the equilibrium pattern of phase transition that occurs as a function of temperature (dM/dT), from factors that affect the displacement from equilibrium that occurs as a function of the rate of temperature change (12). The displacement from equilibrium, or free energy for growth of ice in the outer free space ($\Delta G_{\parallel}$), also is a function of the rate of heat transfer per rate of freezing. The free energy equals the excess heat transferred per mole frozen as the rate is increased from an infinitely slow process.

The rate of crystal growth ($-dM/dT$) equals the product of three factors. The first describes the equilibrium transition pattern (dM/dT), the second, inhibition of freezing kinetics, and the third, the energy driving the transition (A_{wf}).

$$dM/dT = dM/dT \; dT/dt$$

$$dM/dT = ae^{bt} \; (1 - d\Delta T_{eq}/dT)$$

ae^{bt} describes the transition pattern for pure water on a hydrophilic solid such as cellulose. The rest of the equation describes the modifying effect of solutes, such as glycerol, on the equilibrium transition pattern.

$$dT/dt = A_{wf}/(1 + Is/s)$$

for freezing rates that do not significantly change the pattern of crystal growth.

A_{wf} - the net activity of water for phase transition-- in plants, a measure of displacement from equilibrium in the outer free space, not across the plasmalemma. Is/s represents the ratio of ice interface blocked by adhesion with inhibitor (Is) to that exposed (s). Freezing and melting normally occur only at an ice interface, the interface acting as a catalyst. Substances that block the interface inhibit freezing in the sense described by the

Langmuir absorption isotherm or by Michaelis-Menten theory
of enzyme activity. Here, the rate of reaction (v) is considered proportional to the ratio of activated complex, interface where reaction is occurring (Ms), to inactive
uninhibited interface (s). So,

$$v = (V\{Ms/s\})/(1 + Is/s)$$

when Ms/s is small, as in growth of normal crystals; the
maximum velocity (V) corresponds with dM/dT, and Ms/s corresponds with A_{wf}. The principal difference between this
study of freezing kinetics and other studies of catalytic
reactions involved control. A change in temperature was used
to displace the partially frozen interfacial system from equilibrium rather than isothermally manipulating the activity of
reactants.

Certain arabinoxylans have been found to inhibit freezing
kinetics. In plants that produce this type of cryoprotectant,
ice crystals grow more slowly and smaller crystals form in
critical tissues (13, 14, 23).

C. *Potential Energy of Adhesion that Draws Ice and Hydrophilic Plant Systems into Matrices* (U_{adh})

Two direct effects of freezing have been observed in
biological systems: displacements caused by crystal growth
and restrictions caused by crystal adhesion. Displacements
require free energy, excess energy of lattice bonding over
disruptive effects of kinetic energy and opposition of tissue
structure. Restrictions arise from competitive structuring
of interfacial liquid water that then becomes an adhesive
between ice and hydrophilic substances (16).

Partition of the chemical potential of water was necessary to account for transitions observed in freezing. For
example, although the chemical potential of ice-liquid water
at 0° C is zero, recrystallization occurs. Although there is
no net change in the proportion of ice to liquid, with time
ice crystals in the liquid change in size and shape. So
there must be a free energy of melting and of freezing and
they must be in balance.

The individual free energies can be evaluated by partitioning the chemical potential (u) on a basis of the frequency distribution of exchangeable kinetic energy (ξ) over
activation limits (16, 18). Then the transition components
are integrated.

$$U = (dG_>/dn) + (dG_o/dn) + (dG_</dn)$$

$$dG_>/dn = \xi - \xi_h = \xi_>, \; dG_</dn = \xi - \xi\ell = \xi_<$$

$$\Delta G_>/n_t = \int (dG_>/dn) \; (dn/n_t) = \int^\infty (\xi - \xi_h) \; F \; (\xi) \; d\xi$$

$$= (\Delta G_>/\Delta n) \int din \; n = \overline{\xi}_> \; in \; n/n_{eq}$$

$$\Delta G_</n_s = \int_o^{\xi\ell} (\xi - \xi_\ell) \; F \; (\xi) \; d\xi$$

The free energy of melting ($\Delta G_>$) is represented by the region above the activation limit of melting (ξ_h), the free energy of freezing ($\Delta G_<$) by the region below the activation limit of freezing (ξ_1), and the latent heat (ΔH) is the difference between the two activation limits. The free energy of freezing involves exchangeable kinetic energy below the activation limit because kinetic energy is only an index. The energy of freezing is the potential energy of vectorized bonding that draws water molecules into an ice lattice structure. This is opposed by temperature, a measure of the exchangeable kinetic energy. Freezing and melting involve resonance between potential and kinetic energy, kinetic energy being expended by a water molecule in escape from the lattice while acquiring potential energy and, conversely, for a molecule from the liquid approaching a lattice site (16). Density functions (n_t for water in the lattice, n_s for water in liquid at the interface) must be evaluated and incorporated because concentration gradients frequently occur across an interface.

$$\Delta G_{trs} = (\Delta G_>/n_t) \; n_t + (\Delta g_</n_s) \; n_s$$

The activation limits, especially for associations of water with hydrophilic substances, are functions of temperature and the amount of liquid water, as are viscosity, freezing point, latent heat, entropy, and chemical potential.

Calculation of the individual free energies of transitions by partition of chemical potential involves coordination of phase transition data, thermal transition data, and vapor pressure-temperature vector analysis (10, 15, 16). The free energies of freezing and melting that cause recrystallization in ice-liquid water at $0°$ C were estimated to be 50 cal/mole (16).

The free energy of freezing becomes greater than the free energy of melting as the temperature decreases. The free energies of freezing and melting return to balance as equilibrium freezing progresses principally by a decrease in the activation energy of melting (17, 18). Competitive

equilibrium freezing is quite different from a system in which
the ice is separated from the polymer by a gas phase (frost
desiccation). In frost desiccation, the free energies are
balanced by a shift in vapor pressure, the density function,
rather than a shift in activation energies. The shift in
activation energy for the ice-liquid-polymer interface results
in development of an adhesion energy between ice and the hydrophilic substance as they compete for the intervening
liquid. Competitive structuring of the interstitial liquid
causes is to bind the hydrophilic substance to the ice. The
energies of activation and adhesion can be calculated (17,
18). Because the energy of adhesion is determined mainly by
the shift in the activation energy of melting ($\Delta \xi h_{ice}$), the
energy of adhesion per mole of liquid water approximately
equals the reduction in latent heat of equilibrium freezing
($\Delta^2 H$) (2, 17, 18). The energy of adhesion per area of interface (U_{adh}) also depends on the bonding density with the poly-
mer (n_{tpm}). $U_{adh} = \Delta \xi h_{ice} (n_{tpm}/n_{tice}) \simeq \Delta^2 H_{eg} (\Delta H_{pm}/\Delta H_{ice})$.
Surface tension contributes to stability of an adhesion
between ice and a hydrophilic substance as does intertwining
of ice crystals in plant micro structures such as fibrils of
the cell wall.

The physiological effect of an adhesion in a plant depends on the function of the substance with which ice inter-
acts. Adhesions between ice and the protoplasmic membrane
becomes a serious form of stress. Conversely, cryoprotectants
such as soluble proteins or arabinoxylans in the cell wall,
can adhere to an ice lattice, interfere with crystal growth,
and act as inhibitors of freezing kinetics (12, 18). Control
of crystal growth provides a means of avoiding stress by preventing the extension of crystal growth into critical regions.

Below -8° C, the density functions of competitive equilibrium freezing are about equal, and the net latent heat of
water transition from a hydrophilic substance to ice, as from
one ice crystal to another, is about zero. The adhesion
energy estimated from the reduction in the partial latent heat
of ice formation, although not a linear function, approximately equals 20 cal/mole in the temperature range between
-8° C to -20° C (17, 20).

D. Water Potential (Ψ)

Freezing is a means of inducing simple desiccation when
ice is separated from the living cells by a vapor phase.
The water potential (Ψ) in the protoplasm, which is the sum
of the osmotic (Ψ_π), matric (Ψ_m), and hydrostatic (Ψ_h) components (1), equilibrates with the water potential in the

outer free space, which in turn equilibrates with ice by
vapor diffusion. When a plant tissue is cooled, the internal
vapor pressure decreases approximately along the line for
liquid water. Solutes and colloids interact with water to
displace the vapor pressure slightly. Freezing reduces the
vapor pressure to that of ice, and with continued cooling,
the internal vapor pressure follows the ice line. This line
is the resultant of a temperature vector and a vapor pressure
vector which can be independently manipulated to distinguish
associative effects of ice from desiccation through the vapor
phase. The equilibrium water potential is proportional to
the log of the vapor pressure ratio of ice to supercooled
liquid. The ratio decreases as the temperature is lowered,
and the cells contract as water diffuses to the ice lattice.
Irreversible changes and injuries of various types have been
proposed to occur as the cell is reduced to a critical volume.
Death from frost desiccation occurs near -20° C for young
mesophyll cells in hardened plants of 'Hudson' barley
(*Hordeum vulgare* L.). Other forms of freezing injury that
involve direct interactions of ice closely associated with
plant substances kill earlier in the freezing process at a
higher temperature, Figure 1 (15, 19).

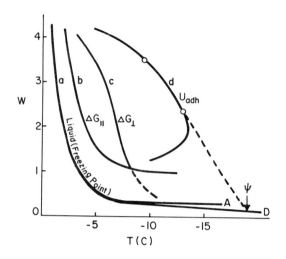

FIGURE 1.   *Water transition patterns and responses of Hudson barley to freezing and desiccation. W-water (gm/gm dry matter) vs T-temperature.*

(a) Intercellular liquid water content at equilibrium with ice as a function of temperature. A - Ice closely associated with tissue, crystals throughout space between protoplasts, D - Ice separated from tissue by a vapor phase.
(b) Supercooling required to provide free energy for crystal growth throughout the space between protoplasts ($\Delta G_{\parallel}$).
(c) Supercooling required to provide free energy for crystal growth across the plasmalemma ($\Delta G_{\perp}$). $\Delta G_{\perp}$ is determined by displacement of temperature from the freezing point for isothermal freezing of a thin tissue supercooled in an efficient heat sink. Freezing point of the protoplasm is evaluated from the temperature plateau caused by adiabatic freezing of a supercooled tissue.
(d) LD-50 caused by equilibrium freezing as a function of total water content. Freezing is initiated with minimum supercooling (line b) followed by a slow (1° C/h) decrease in temperature. Adhesion energy (U_{adh}) is estimated from calorimetric data ($\Delta^2 H_{eq}$ {$\Delta H_{pm}/\Delta H_{ice}$}). Adhesion energy also depends on the extent of the ice-cell interface, so survival decreases with increase of water from the vapor desiccation limit (Ψ). Survival deviates from the adhesion energy line at both high and low water contents. At high water contents (W>3 gm water/gm dry matter), initial crystallization energy is high and injurious (although not lethal) because of the large amount of water that freezes with slight supercooling. At low water contents (W<2 gm water/gm dry matter), initial

freezing energy is high because of the minimum supercooling (line b) required to initiate crystallization in the space between protoplasts. Circles in line d represent standard tests for comparison of cultivars.

REFERENCES

1. Boyer, J. S., *Ann. Rev. of Plant Physiol. 20*, 351 (1969).
2. Dorsey, N. E.(ed.), "Properties of Ordinary Watersubstance." Reinhold Publishing Co., New York (1940).
3. Eyring, H., Ree, T., and Hirai, N., *Proc. Nat. Acad. Sci., U.S., 44*, 683 (1958).
4. Fletcher, N. H., "The Chemical Physics of Ice." Cambridge University Press (1970).
5. George, M. F., Burke, M. J., and Weiser, C. J., *Plant Physiol. 54*, 29 (1974).
6. Gullord, M., Olien, C. R., and Everson, E. H., *Crop Science 15*, 153 (1975).
7. Jellinek, H. H. G. (ed.), "Water Structure at the Water-Polymer Interface." Plenum Publishing Co., New York (1974).
8. Kavanau, J. O., "Water and Solute-Water Interactions." Holden-Day, Inc. (1964).
9. Mayland, H. F., and Cary, J. W., *Adv. in Agron. 22*, 203 (1970).
10. Nemethy, G., and Scheraga, H. A., *J. Chem. Physics 36*, 3382 (1962).
11. Olien, C. R., *Crop Sci. 4*, 91 (1964).
12. Olien, C. R., *Cryobiology 2*, 47 (1965).
13. Olien, C. R., *Crop Sci. 7*, 156 (1967).
14. Olien, C. R., Marchetti, B. L., and Chomyn, E. V., *Mich. Agric. Exp. Stn. Quarterly Bull. 50*, 440 (1968).
15. Olien, C. R., *Cryobiology 8*, 224 (1971).
16. Olien, C. R., *J. Theoretical Biol. 39*, 201 (1973).
17. Olien, C. R., *Plant Physiol. 53*, 764 (1974).
18. Olien, C. R., *Mich. State Univ. Agric. Exp. Stn. Res. Rpt. 247*, 1 (1974).
19. Olien, C. R., *U.S. Dept. of Agric. Tech. Bull. No. 1558* (1977).
20. Olien, C. R., and Smith, M. N., *Plant Physiol. 60*, 499 (1977).
21. Rahman, A., and Stillinger, F. H., *J. Chem. Physics 55*, 3336 (1971).
22. Santarius, K. A., and Heber, U., *Cryobiology 7*, 71 (1970).

23. Shearman, L. L., Olien, C. R., Marchetti, B. L., and Everson, E. H., *Crop Sci. 13*, 514 (1973).
24. Siminovitch, D., and Briggs, D. R., *Plant Physiol. 28*, 177 (1953).
25. Single, W. V., *Aust. J. of Agric. Res. 15*, 869 (1964).
26. Vand, V., and Senior, W. A., *J. Chem. Physics 43*, 1878 (1965).
27. Viaud, P. R., *Cryobiology 9*, 233 (1972).
28. Weiser, C. J., *Science 169*, 1269 (1970).

FROST HARDENING AND FREEZING STRESS
IN TUBER-BEARING SOLANUM SPECIES

P. H. Li
J. P. Palta

Laboratory of Plant Hardiness
Department of Horticultural Science
and Landscape Architecture
University of Minnesota
St. Paul, Minnesota

I. INTRODUCTION

The potato crop currently ranks fourth after wheat, rice
and corn on a world production basis, but ranks first in yield
per unit of land among the world's major food crops (9). In
many parts of the world, the potato serves as a staple food.
It is an excellent dietary source of protein containing
all of the essential amino acids. It is one of the few crops
having the potential of providing large populations a nutri-
tionally balanced diet. However, the potential contribution
of potato to the qualitative as well as the quantitative
enrichment of the human diet has not been fully recognized.
The potato originated in the Andean region of South
America. It was introduced to Europe during the 16th century.
The potato is considered to be a cool season crop and produces
higher yields at relatively cool temperatures. Most potatoes
are grown in the Northern Temperate Zone and in the high
elevations of the Andean tropics where frost is often a major
factor limiting potato production. Depending on the intensity
of the frost, potato plants can be killed or the foliage
severely damaged, resulting in crop failure or reduced yield.
An example of frost kill is illustrated in Figure 1.
In many cases the difference between a good potato crop
and crop failure is determined by 2 or 3° C of frost resistance. Such subtle differences are highly amenable to research
solutions such as breeding better adapted varieties. The

49

FIGURE 1.   Potato crop frost-kill in the Andean Range of
Peru.   (Courtesy of Dr. R. N. Estrada, the International
Potato Center, Lima, Peru).

successful approaches to increasing hardiness will be highly
dependent upon physiological and genetic research to establish
the nature of frost injury and resistance in potatoes.   In
addition, population pressure requires not only the utilization of cultivated lands at maximum efficiency but also requires potato to be grown in areas which are marginal because
of environmental stress.

II.   METHODS OF DETERMINING FROST HARDINESS

Many techniques have been introduced for evaluating the
cold hardiness of plants (8, 24, 27, 39, 47, 54).   However,
none of these methods except plasmolysis (37, 38, 46) actually
measure the living activities after freezing.   The research
objectives and plant species and tissues to be evaluated dictate which is the best method to follow.   The ideal method
should be rapid, simple, repeatable and non-destructive.   Unfortunately, to date there is no test available which meets
all of these criteria.

Injury can be quantitatively estimated by measuring the amounts of electrolytes which leak out from cells following freezing exposure (8). Electrolytes diffuse more freely from the injured cells. The lethal temperature for potato leaves is generally considered to be the temperature when 50% leakage is observed (49). For example, 50% leakage of *Solanum tuberosum* and *S. acaule* occurs at -2.5° and -6.0° C, respectively, which are the killing temperatures for these species (49). The viability after freezing can also be visually distinguished (5). Normally, the freeze-injured potato leaves are darker than non-injured leaves. The freezing temperature at which a leaf has fully lost turgor, completely darkened, and has a water-soaked appearance could be designated as the frost killing point. The validity of this technique agrees well with conductivity measurements. However, some plant species, which may have water-soaked appearances initially, will eventually recover from freezing stress (36-38). Following are some of the methods used for exposing potato plants to desired freezing stress for determining their frost hardiness.

A. *Field Tests*

Exposure to frost in the natural environment is perhaps the oldest and ultimate test of potato frost hardiness. Unfortunately, when populations are exposed to the unreproducible natural frosts, results are often difficult to interpret. Nevertheless, a number of hardy potato species have been reported on after field observations (7, 23, 45).

B. *Intact Potted Plants*

Potato plants, which are grown either in flats (30) or pots (43), can be subjected, in controlled freezing chambers at a predetermined rate, to a series of freezing temperatures. This method is perhaps the simplest one to prepare. The length of exposure to a given freezing temperature may be varied among the experiments (5, 30, 43). Upon completion of freezing, plants are slowly thawed and their injuries rated either visually (5) or by determining leakage of cell constituents (49). The drawback in using intact plants is that the material being tested is sacrificed. Thus a large plant population is needed. Large plant numbers may be impossible to obtain among certain segregating populations. After freezing chambers became available, these were used to evaluate intact tuber-bearing *Solanum* hybrids (43).

In many plant species, such as winter cereals, roots can
be killed at a warmer temperature than leaves (42). Precau-
tion should be taken when intact potato plants are tested to
insulate the pots before subjecting the plants to freezing
temperatures. Thus, plant growth can be sure if potato
foliage has survived certain low temperatures.

C. Leaflet Tests

The test with excised leaves is relatively simple and,
more important, is non-destructive to the whole plants.
However, resistant and non-resistant potato clones might not
be distinguishable unless supercooling was prevented (19).
For instance, Solanum tuberosum potato tuber sprouts could
remain supercooled for 18 hours at -5.5^o C and 4 hours at
-7.5^o C (1). When foliage was inoculated at -6.0^o C freezing
occurred intracellularly resulting in cell death. When leaves
were nucleated at -2.0^o C on the upper surface, freezing
occurred extracellularly, and the tissue resisted freezing
(49). With proper inoculation at about -1.5 to -2.0^o C,
good correlations can be obtained between intact plant and
excised leaf tests (49); these are also consistent with field
frost observations (2). Recently, Li (29) reported the frost
hardiness of 60 tuber-bearing Solanum species; the tests were
carried out in a controlled freezing apparatus with excised
leaves collected from plants grown in a warm temperature and
long day regime. None of the commonly Minnesota-grown cultivars of Solanum tuberosum can survive beyond -2.5^o C. The
Alaska Frostless (4) can survive only to 2.7^o C. However,
a large number of clones of non-cultivated potato species,
which are normally poor yielders, possess frost resistance.
They can survive below -4.0^o C. Table I lists some frost
resistant potato species tested. The excised leaf test appears to be a reliable method of assessing potato frost hardiness. Ultimately, observations should be made under natural
frost conditions.

III. FREEZING PROPERTIES

Two kinds of freezing can easily be observed in potato
leaf tissue in the laboratory. They are intracellular and
extracellular freezing, as shown in Figure 2. However,
under natural conditions, when plants cool slowly, all of
the ice forms in the intercellular spaces, resulting in extracellular freezing. During freezing, water moves from the
cells (mainly vacuole) to the intercellular ice, causing the

TABLE I.   *Frost Killing Temperatures of Several Tuber-
Bearing Hardy Solanum Species Evaluated by the
Controlled Freezing Excised Leaf Tests*

Species	Killing temp. °C	Species	Killing temp. °C
S. acaule		*S. chomatophilum*	
HOF 1509[a]	-4.5	PI 266387	-5.3
HOF 1584	-4.5	*S. commersonii*	
EBS 1781	-5.0	OKA 4583	-5.0
EBS 1809	-4.5	PI 243503	-5.3
OKA 3718	-6.0	*S. megistacrolobum*	
OKA 3878	-6.5	OKA 3787	-5.0
HHC 4116	-4.5	*S. multidissectum*	
HHC 4243	-4.5	PI 210044	-4.7
PI 210044	-6.0	*S. sanctae rosae*	
S. bolviense		PI 230464	-5.5
PI 310974	-4.5	*S. toralapanum*	
PI 310975	-4.5	HOF 1851	-5.0
S. canasense		HHC 4556	-5.0
PI 283074	-4.5	*S. vernei*	
		PI 320329	-5.0

[a]*Identification number at the Potato Introduction Station,
Sturgeon Bay, Wisconsin.*

cells to dehydrate. Levitt (26) pointed out that some plants
resist freezing injury by either tolerating or avoiding
cellular dehydration. As freezing proceeds, ice crystals
cause cellular contraction. Dehydration and contraction can
thus lead to mechanical stress on the cells.

In barley plants, Olien (33) identified five basic water
redistribution patterns during freezing. The patterns are
dependent on freezing rate, protoplasm resistance to ice
penetration, cell moisture content, initial ice growth pattern, and other factors contributing to cell hardiness. He
was able to demonstrate that freezing occurred either as an
equilibrium or non-equilibrium process depending upon the
above conditions. The non-equilibrium process was associated
with tender or semi-tender plants. In 1972, Sukumaran and
Weiser (50) reported that *S. tuberosum* had non-equilibrium
freezing and *S. acaule* a semi-equilibrium freezing process.
Olien (32) also suggested that the association of water
molecules with plant hydrophilic compounds is essential for

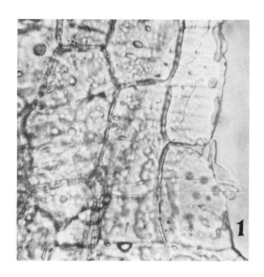

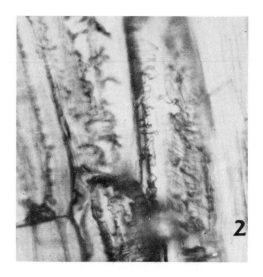

FIGURE 2.   Laboratory observations of intracellular (1) and extracellular (2) freezing in potato leaf tissues. Nucleations were made at -6° C (1) and -2° C (2).

survival of plants subjected to frost. Carbohydrates, proteins and nucleic acids tend to associate with water molecules. If the temperature doesn't drop continuously, an equilibrium can be established between ice and plant compounds for the remaining water. With a constant temperature decrease, the equilibrium will shift in favor of ice growth. Thus, the more hydrophilic compounds a cell has, the greater the cell's capacity to survive freezing stress.

Solanum tuberosum possesses very little frost tolerance (-2.5° C) in contrast to a number of non-cultivated tuber-bearing species which are considered to be frost tolerant (-4.0° C or lower). The difference in frost tolerance between the two groups, however, is only a few degrees. In some plant species a small variation in tissue moisture content could result in such a difference in a plant's capacity to survive frost. In barley tissue with a high moisture content, water essentially has a single freezing point; the resulting ice crystals can cause disruption of the cells (33). When tissue freezes, the cell water supercools if there are no sites of ice nucleation. A high cell sap concentration may induce a few degrees more of supercooling; thus, the tissue could survive greater freezing stress by the avoidance mechanism.

Hudson and Idle (18) noted that a reduction of protoplast turgor had a marked effect on the *S. acaule* freezing process. The amount of unfrozen water present at sub-freezing temperatures can be measured using a Nuclear Magnetic Resonance (NMR) spectrometer. It is a useful tool for increasing our understanding of water freezing properties of tissue. With a pulsed NMR, Chen *et al.* (5) studied the freezing behavior of six genotypes including *S. tuberosum*, Alaska Frostless (7), *S. multidissectum, S. commersonii, S. chomatophilum* and *S. acaule* (frost killing temperatures ranging from -2.0° C to -6.0° C). They found a highly significant correlation between frost hardiness of leaves and percentage of unfrozen tissue water at frost killing temperatures. The amount of unfrozen tissue water at the killing temperatures averaged 44, 43, 32, 32, 22 and 20% of total liquid water in partially frozen leaf tissues of *S. tuberosum*, Alaska Frostless, *S. multidissectum, S. commersonii, S. chomatophilum* and *S. acaule*, respectively (5). No correlations were observed between frost hardiness and cell sap concentrations or tissue water content. These data provide evidence that the major difference between hardy and tender genotypes is the ability of hardy potato plants to tolerate more frozen water than tender plants.

IV.   ENVIRONMENTAL FACTORS

In general, herbaceous plants cannot withstand temperatures below -20° C, and the hardiness mechanism may be quite different from that in woody perennials. An adapted plant species has internal mechanisms that permit it to survive and reproduce in its ecological niche. These mechanisms are extremely complex and involve adaptive responses to temperature, photoperiod, light intensity and quality, water availability, nutrients, competition from neighbors, and other factors. Frost hardening is a term describing the transition of a plant species from a tender to a hardy state. Shorter days, decreasing temperatures, reduced water supply and other factors usually trigger the hardening. Some species harden extensively in response to environmental factors while others harden only a few degrees and some do not harden at all. In most woody plants which harden, water contents decreases with increasing hardiness. However, this was not found to be true in various *Solanum* species. Under a frost hardening conditions (4), the water content of *S. acaule* leaves remained unchanged (Table II).

It has been well documented that long days and warm temperatures encourage plants to grow without hardening (20). Short days and warm temperatures cause growth cessation and enhance hardiness in deciduous woody perennials (20, 34) but not in certain evergreens (13) or winter cereals (40). Some species can be hardened by low temperatures with a sufficient amount of light, regardless of the photoperiod (22). Prior to 1976, there was some dispute as to whether the *Solanum* species will harden to frost. Mastenbrock (30) and Richardson and Estrada (43) reported that a 2 to 3 week period of cool temperatures can differentiate hardy from non-hardy potatoes; however, Firbas and Ross (10), and Hayden *et al.* (15)

TABLE II. *Changes of Leaf Moisture Content in S. acaule during Frost Hardening*

Days of Frost Hardening	Moisture Content (%)		Killing Temperature (°C)	
	Control[a]	Hardened	Control	Hardened
0	85.7	85.7	-5.5	-5.5
7	85.2	85.1	-5.5	-6.5
14	84.0	85.1	-5.5	-7.5
21	84.0	84.5	-5.5	-8.5

[a]*Control: warm temperature/long day regime.*

indicated that potatoes possess a stable frost hardiness level and do not harden. In 1976, Chen and Li (4) reported that several non-cultivated tuber-bearing *Solanum* species hardened in response to shortened photoperiods and lowered day/night temperatures, while *S. tuberosum* cultivars and Alaska Frost-less (7) failed to harden. Under frost hardening conditions, these non-cultivated species increased from 3^0 to 6^0 C in frost resistance during three weeks of treatment. The inability of *S. tuberosum* cultivars to harden suggests that these cultivars do not have the physiological bases for developing frost hardiness.

Potato frost hardening appears to be an inducible genetic system triggered primarily by low temperature. Low temperature regime with either short or long days can harden the hardy species; the former induces more hardiness than the latter. Warm temperature combined with short days failed to harden any of the species tested. It appears that low temperature functions as a primary factor, short days being secondary, in the induction of frost hardening. Some plant species, however, require short day preconditioning prior to gaining maximum hardiness induced by low temperature (51).

V. BIOCHEMICAL CHANGES

The direct influence of environmental factors on biochemical changes in potato plants has been studied much less extensively than the influence on morphology and development. It is still not known how environmental conditions induce frost hardiness in potatoes. Certain biochemical studies on such environmental effects may provide 1) clues for a better understanding of the frost hardening process, 2) reliable bases for designing further studies and interpreting results, and 3) the foundation for devising a practical means of attenuating frost injury on a field scale. A comprehensive review of nucleic acid and protein changes in relation to frost hardiness has been reviewed by G. N. Brown in this volume. A short account of the biochemical changes in relation to potato frost hardiness is given here.

A. *Nucleic Acids*

During frost hardening the primary sites of biochemical changes are most likely at the levels of transcription and translation of genetic information controlled by low temperature. Therefore, nucleic acid metabolism under these conditions is of specific interest. Since no specific metabolites are known to bear a causal relationship to environmentally induced frost hardiness in the potato, it seems logical to look for such causal relationships at this level.

In potato plants, temperature had a marked influence on ribosomal RNA (rRNA) metabolism while the photoperiodic response is not as great (35). The quantitative RNA variation under low temperatures is probably due primarily to the RNA synthesis rate rather than to changes in RNase activity (13, 35). Marked quantitative increases in rRNA following exposure to low temperatures have been reported but no significant changes in nucleotide composition of rRNA and other RNA species has been found appreciably (34). In these studies no overall qualitative change in nucleotide composition was observed in potato foliage (34). This, however, does not preclude that the variation of specific aminoacyl-tRNA synthetase activities may not take place under different environments. There is evidence in higher plants of the existence of multiple synthetase for the isoaccepting tRNA species of a single amino acid (21). Differential synthesis of proteins in response to environmental stimuli may depend on the functional concentration of specific aminoacyl-tRNA species which, in turn, are controlled by multiple aminoacyl-tRNA synthetases.

B. *Proteins, Polyribosomes and Free Amino Acids*

Proteins are at the center of frost hardiness studies because of their roles in membrane constituency, hydrophilic properties in binding water, determination of protoplasmic viscosity and elasticity, involvement of enzymatic systems and control of membrane transport systems. Proteins, either structural or enzymatic, are almost certainly involved in some way in frost hardening. Preliminary studies have revealed that soluble proteins increase in potato leaves when the plants are exposed to frost hardening environments.

In a recent study by Vigue and Li (52), leaf polyribosome levels of *Solanum acaule* and *S. tuberosum* plants grown under the cool (15/10° C, day/night) conditions were found to be higher than those of plants grown at warm temperatures (25/ 15° C, day/night). Whether the higher level of polyribosomes in cool grown *S. acaule* has a direct relation to frost hardening is difficult to determine because *S. tuberosum*, which cannot be hardened, also maintains a higher level when plants are grown in cool temperatures. Furthermore, young *S. acaule* leaves had a significantly higher polyribosome level than mature leaves. Both the young and mature leaves of *S. tuberosum* had the same level of polyribosome. The preservation of polyribosomes in low temperatures probably is a mechanism for

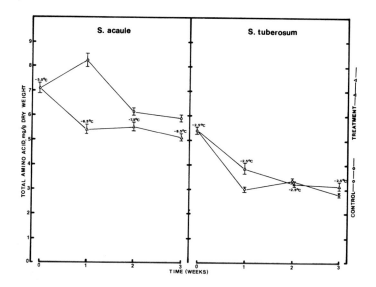

FIGURE 3.   Changes in total free amino acids in S. acaule
and S. tuberosum during a three-week frost hardening.   Degrees
indicated are the killing temperatures after each period of
treatment.   Control: long day/warm temperature.

efficient protein production under cooler conditions.   In the
high Andes, where *Solanum* species evolved, frost can occur
any time of the year; there are many alternating periods of
cool and warm weather during potato growth and development.
It is advantageous for the protein synthesis mechanism (the
polyribosomes) to remain intact during low temperatures.
When the temperatures rise again, plants do not have to
reassembly polyribosomes to continue protein synthesis for
growth and development.
    Figure 3 illustrates the changes in total free amino acids
for both *S. acaule* and *S. tuberosum* during a three-week frost
hardening period.   The frost hardening conditions are described elsewhere (4).   Both species showed general trends of
decreasing of free amino acid contents, but only *S. acaule* increased its hardiness from -5.5° C to -8.5° C.

C.   *Sugars*

    A decrease in starch and a corresponding increase in sugar
have long been thought to be associated with frost hardiness.
This has led many researchers to suggest that sugars have a
causal role in frost hardiness.   Steponkus (48) suggested that

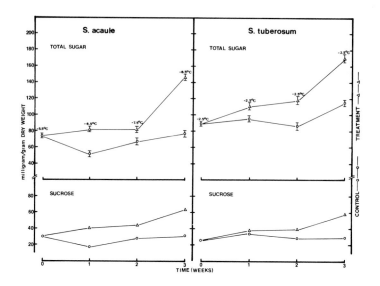

FIGURE 4. Changes in total sugar and sucrose in S. acaule and S. tuberosum during a three-week frost hardening period. Degrees indicated are the killing temperatures after each period of treatment. Control: long day/warm temperature.

an alteration in protein structure occurs during frost hardening which increases the affinity for sugars. The bound sugars are then able to protect against protein denaturation during freeze-induced dehydration. The commonly detectable sugars in potato leaves are sucrose, glucose and fructose (28). During frost hardening (4), our recent studies, however, have shown that sugar increases even in *S. tuberosum* although its hardiness did not increase. Therefore, the role of sugar in frost hardiness of *Solanum* species is questionable.

D. *Enzyme Activity*

Frost hardiness is believed to be genetically-controlled, whereby certain genes could be turned on during a period of low temperature to bring about frost hardening. Elucidation of this control mechanism is essential in providing a basis for tailoring the plant to survive frosts. Identification of the key enzymes and measuring their activities during a period of low temperatures could provide a better understanding of the nature of frost resistance. Glutamate dehydrogenase (GDH) is one of the key enzymes controlling metabolic reactions between organic acids and amino acids, which in turn affect

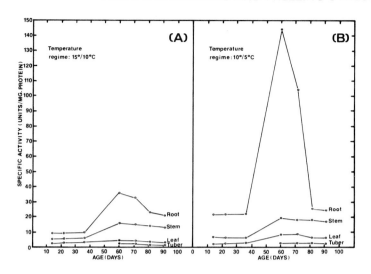

FIGURE 5. Changes and distributions of NADH-dependent GDH specific activity during growth and development in different parts of S. tuberosum grown under 15/10° C (A) and 10/5° C (B) day/night temperature regimes. Assays were carried out with crude enzyme extracts.

carbohydrate and protein metabolism. In a previous communication, Vigue and Li (53) reported that temperature regimes had significant influence on amino acid metabolism in potato plants. This has led to an examination and comparison of GDH activity during growth of *S. tuberosum* under 15/10° C (day/night) and 10/5° C temperature regimes. Figure 5 shows the changes of GDH specific activity (units of enzyme activity per mg protein basis) during growth and development, and activity distributions among organs in both regimes. GDH specific activity, measured from root crude enzyme extracts, indicated that a cool temperature regime resulted in a higher level of activity than did a warm one. For example, at the 60th day after emergence, the activity was four times greater in cool-grown roots than in warm-grown ones. Total protein content in cool-grown roots was twice as high. Therefore, based on total protein content in the roots, the total GDH activity was two times greater in 10/5° C than that in 15/10° C plants.

It is interesting to note that roots contain the highest activity among organs including leaf, stem, root and tuber (Figure 5). The least activity was found in the tuber. For instance, in a 15/10° C regime, GDH activity in roots was ten and four times greater than in leaves and stems respectively,

with a dramatic increase at the 60th day after emergence.
Freezing of plant cells often results in a loss of oxidative
activities (16), suggesting that oxidative enzymes such as
succinate dehydrogenase, glutarate dehydrogenase complex,
etc., may be damaged due to freezing. The oxidation systems
in *Solanum* species located on the mitochondrial membranes
susceptible to freezing deserve investigation.

VI.   ANATOMICAL CHARACTERISTICS

    Morphologically, *Solanum tuberosum* differs a great deal
from hardy species such as *S. acaule* and *S. commersonii*.
One could suspect that these species may have different anatomical characteristics which could be correlated with the
differences in frost resistance. Studies of this kind between
hardy and tender types of potatoes could provide answers
to the basic question on the inability of *S. tuberosum* to
survive frost and to frost harden. Leaf anatomy of hardy
and tender species was compared both with light as well as
electron microscopes.

A. *Light Microscopic Observations*

    An important aspect of frost hardiness research is the
search for plant anatomical characteristics related to hardiness. The aim of such research is to find "markers" that
can be used as criteria for selection purposes in order
to breed for hardy varieties. Some early investigators
found that smaller cell size (3, 14, 25), thicker cell walls
(11) and lower stomatal density (17) are associated with hardiness. In recent years research in this area has been ne-
glected because of inconsistency in results (26).
    Recently Palta and Li (36) examined under a light micro-
scope leaf cross sections from nine hardy species (killing
temperature $\leq$ -4.0° C), twelve non-hardy wild species (killing temperature -2.5 to -3.5° C) and seven cultivars of
Solanum tuberosum (killing temperature -2.0 to -2.5° C). They
found that all of the hardy species had two layers of palisade
parenchyma cells and that non-hardy species, including various
tuberosum cultivars except for two, had only one layer.
Typical leaf cross sections of hardy and non-hardy species are
shown in Figure 6. Figure 6-1 shows the *S. acaule* cross section (killing temperature -4.5° C) and Figure 6-2 shows the
S. tuberosum (killing temperature -2.5° C). Palta and Li
(36) also measured palisade thickness, cell size and the number of stomata on the upper surfaces of the leaves. It was

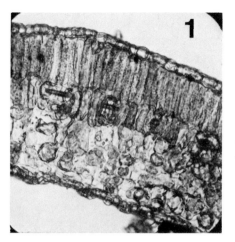 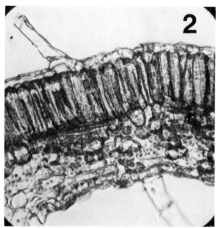

FIGURE 6.   Leaf cross-sections of Solanum acaule (1) and
S. tuberosum (2).   Note the double layers of palisade paren-
chyma cells in S. acaule.

found that hardy species in general had smaller cell size, a
thicker palisade and two to three times greater stomatal in-
dexes than non-hardy species.

Further studies along these lines will provide information
about the usefulness of these anatomical characteristics in
an effective breeding program.

B.   Electron Microscopic Observations

A comparative study of the ultrastructural differences
between a hardy and a tender species grown under normal conditions and frost hardening conditions was reported by Chen
et al. (6).   The typical leaf parenchyma cell contains a large
vacuole and a thin layer of protoplasm with chloroplasts and
other organelles, along the cell wall (Figure 9, control).
The first difference noted between hardy (*S. acaule*) and tender (*S. tuberosum*) species was the leaf cell wall thickness.
In general, the cell wall thickness in *S. acaule* was about
twice that of *S. tuberosum* (Figures 7-1 and 8-3).   The thicker
cell wall may serve as a protective mechanism in avoiding
membrane rupture (55) and protoplasmic distortion (26, 32)
caused by freeze-induced dehydration.   Cell walls may contain
other soluble polysaccharides, in addition to xylans, which
affect freezing (12, 31).

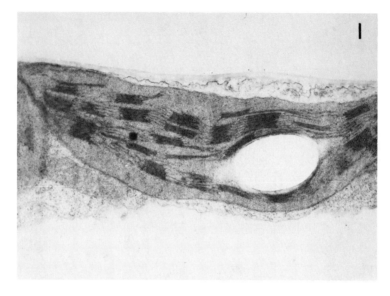

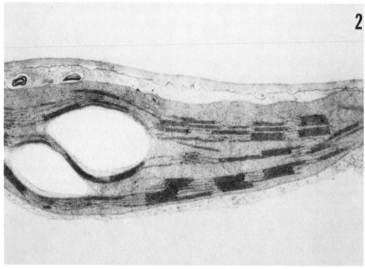

FIGURE 7.   Electron micrograph of the cell wall and the chloroplast in the leaf mesophyll of S. tuberosum before (1) and after (2) frost hardening; X33,250.

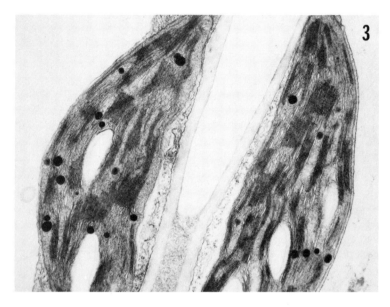

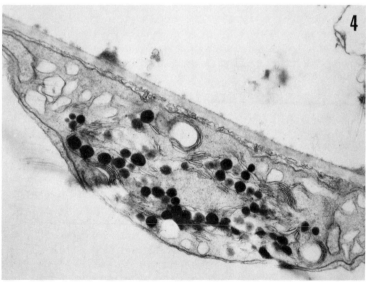

FIGURE 8.   Electron micrograph of the cell wall and the chloroplast in the leaf mesophyll of S. acaule before (3) and after (4) frost hardening; X33,250.

After frost hardening, the grana in *S. acaule* underwent obvious changes (Figure 8-4). They contained fewer thylakoids and the membranes appeared swollen and irregular. There were large patches of stroma in the chloroplasts; some had unusual stroma membranes suggesting the change in chloroplast structure under frost hardening. An increase in lipidic materials (osmiophilic globuli), the disappearance of starch grains in *S. acaule* chloroplasts (Figure 8-4) and the intact starch grains in the chloroplasts of *S. tuberosum* (Figure 7-2) after frost hardening may help to answer the question "Why can't *S. tuberosum* be frost hardened?" In 1971, Pomeroy and Siminovitch (41) reported that the increase in lipidic bodies in the black locust is associated with an increase in frost hardiness. They concluded that the total expression of frost hardiness involves participation of total protoplasmic augmentation, lipid transformation, and starch-sugar transformations.

C. Electron Microscopic Observations of Freeze Injured Cells

In order to understand the frost injury mechanism in plants, one approach that shows promise is the examination of freeze injured cells (37, 38). A typical example of freeze injured cells is shown in Figure 9. These electron micrographs were made of thawed leaves which had been previously frozen to -7° C. Clearly a large vacuole, common to plant cells (Figure 9, control), is absent in freeze injured cells (Figure 9, -7° C). This is most likely due to the breakdown of cell membranes, particularly the tonoplast. However, some cell organelles like mitochondria and chloroplasts remain more or less intact (grana stacks in -7° C, Figure 9). Freeze injured parenchyma cells appear to have what is commonly known as frost plasmolysis (separation of the plasma membrane from the cell wall). This should not be confused with ordinary plasmolysis which occurs when plasma membrane separates from cell wall in a hypertonic medium. During extracellular ice formation, the whole cell collapse occurs as the ice accumulates outside the cell. On thawing, the ice melts and the cell wall, being elastic, comes back almost to its original shape. If the cells are permanently damaged (the cells do not recover from injury and finally die which results in a loss of semipermeability of the membranes) the protoplast does not swell back; rather, it remains contracted, giving the appearance of frost plasmolysis (Figure 9, -7° C).

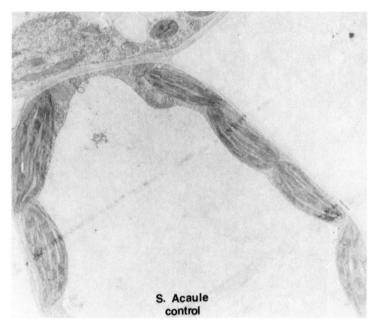

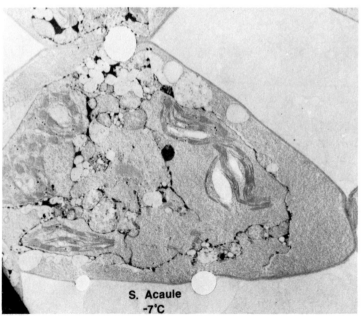

FIGURE 9.   Electron micrograph of a freeze injured cell of S. acaule at -7° C as compared with the control.

VII. SUMMARY

Solanum tuberosum, the most commonly grown potato, possesses very little or no frost tolerance while a number of non-cultivated tuber-bearing species such as *S. sanctae rosae*, *S. vernei*, etc., are considered to be frost tolerant. Excised leaf testing appears to be the most reliable method of assessing potato frost hardiness. Injury can be quantitatively estimated by measuring the electrolyte leakage following freezing exposure. The lethal temperature for potato léaves is generally considered to be the temperature at which 50% leakage occurs. Freezing temperatures at which leaves fully lose turgor, completely darken, and become water soaked, can also be designated as the frost killing points.

The major difference between a tender and hardy species of potatoes is the ability of the hardy species to tolerate more frozen water at frost killing temperatures. For example, *S. acaule* can tolerate up to 80% liquid water frozen before death occurs, while *S. tuberosum* can only tolerate up to 60%. No relationship could be found between the frost hardiness of leaves and cell sap concentrations or tissue water contents in potatoes.

Stepwise lowered day/night temperatures and shortened photoperiods provide the best environment for frost hardening of the hardy *Solanum* species, while *S. tuberosum* fails to respond. Low temperature alone, regardless of photoperiod, can also induce frost hardines. Warm temperatures and short days have failed to increase hardiness. Evidence indicates that low temperature functions as a primary factor in the induction of frost hardiness in potatoes. *S. tuberosum* does not possess the physiological base for developing frost hardiness. During frost hardening, a decrease in starch and total free amino acids, and an increase in sugars, soluble proteins and ribosomal RNA have been observed in potato leaves including *S. tuberosum*.

Electron microscopic observations indicate that the hardy *S. acaule* species undergoes significant structural changes during frost hardening. In all hardy species examined under light microscope, a double layer of palisade parenchyma cells in the leaves was observed, which may serve as one of the markers in breeding programs for screening large populations for frost resistant clones.

VIII.   ACKNOWLEDGMENT

We wish to thank the International Potato Center, Lima, Peru and the Rockefeller Foundation for their support in the study of potato frost resistance.

REFERENCES

1.   Asahina, E., *Low Temp. Sci., Ser. B 11*, 13 (1954).
2.   Blomquist, A. W., and Lauer, F. I., *Amer. Potato J. 39*, 389 (1962).
3.   Chandler, W. H., *Mo. Agric. Expt. Sta. Res. Bull. No. 8*, 171 pp. (1913).
4.   Chen, P. M., and Li, P. H., *Bot. Gaz. 137*, 105 (1976).
5.   Chen, P. M., Burke, M. J., and Li, P. H., *Bot. Gaz. 137*, 313 (1976).
6.   Chen, P. M., Li, P. H., and Cunningham, W. P., *Bot. Gaz. 138*, 276 (1977).
7.   Dearborn, C. H., *Amer. Potato J. 46*, 1 (1969).
8.   Dexter, S. T., Tottingham, W. E., and Graber, L. F., *Plant Physiol. 7*, 63 (1932).
9.   F. A. O. Production Yearbook (1974).
10.   Firbas, H., and Ross, H., *Z. Pflanzenzuchtung 45*, 259 (1961).
11.   Gladwin, F. E., *N. Y. Agric. Expt. Sta. Bull. 433* (1917).
12.   Gould, S. E. B., Rees, D. A., Richardson, N. G., and Steele, I. W., *Nature 208*, 876 (1965).
13.   Gusta, L. V., and Weiser, C. J., *Plant Physiol. 49*, 91 (1972).
14.   Harvey, R. B., *J. Agric. Res. 15*, 83 (1918).
15.   Hayden, R. E., Dionne, L., and Fenson, D. S., *Can. J. Bot. 50*, 1547 (1972).
16.   Heber, H., *Cryobiol. 5*, 188 (1968).
17.   Hirano, E., *Bot. Gaz. 92*, 296 (1931).
18.   Hudson, M. A., and Idle, D. B., *Planta 57*, 718 (1962).
19.   Hudson, M. A., *Euphytica 10*, 169 (1965).
20.   Irving, R. M., and Lanphear, F. O., *Plant Physiol. 42*, 1191 (1967).
21.   Kanabus, J., and Cherry, J. H., *Proc. Nat. Acad. Sci. 68*, 873 (1971).
22.   Kohn, H., and Levitt, J., *Plant Physiol. 40*, 476 (1965).
23.   Kovalenko, G. M., *Amer. Potato J. 9*, 205 (1932).
24.   Larcher, W., and Eggarten, H., *Protoplasma 51*, 595 (1960).
25.   Levitt, J., and Scarth, G. W., *Can. J. Res. C14*, 267 (1936).

26. Levitt, J., "Responses of Plants to Environmental
     Stresses," Academic Press, New York (1972).
27. Li, P. H., Weiser, C. J., and van Huystee, R., *Amer.*
     Soc. Hort. Sci. 86, 723 (1965).
28. Li, P. H., Lauer, F. I., and Weiser, C. J., *Amer. Soc.*
     Hort. Sci. 91, 436 (1967).
29. Li, P. H., *Amer. Potato J. 54*, 452 (1977).
30. Mastenbrock, C., *Euphytica 5*, 289 (1956).
31. Nordin, J. H., and Kirkwood, S., *Ann. Rev. Plant Physiol.*
     16, 393 (1965).
32. Olien, C. R., *Ann. Rev. Plant Physiol. 18*, 387 (1967).
33. Olien, C. R., *Barley Genetics 2*, 356 (1969).
34. Oslund, C. R., and Li, P. H., *Plant and Cell Physiol.*
     13, 201 (1972).
35. Oslund, C. R., Li, P. H., and Weiser, C. J., *Amer. Soc.*
     Hort. Sci. 97, 93 (1972).
36. Palta, J. P., and Li, P. H., *Plant Physiol. 59 (suppl)*,
     196 (1977).
37. Palta, J. P., Levitt, J., and Stadelmann, E. J., *Plant*
     Physiol. 60, 393 (1977).
38. Palta, J. P., Levitt, J., and Stadelmann, E. J., *Plant*
     Physiol. 60, 398 (1977).
39. Parker, J., *Science 118*, 77 (1953).
40. Paulsen, G. M., *Crop Sci. 8*, 29 (1968).
41. Pomeroy, M. K., and Siminovitch, D., *Can. J. Bot. 49*,
     787 (1971).
42. Rammelt, R., *in* "Proceedings of a colloquium on the win-
     ter hardiness of cereals" (S. Rajki, ed.). Agric. Res.
     Inst. of Hungarian Acad. of Sci. (1972).
43. Richardson, D. G., and Estrada, R. N., *Amer. Potato J.*
     48, 339 (1971).
44. Richardson, D. G., and Weiser, C. J., *Hort. Sci. 7*, 19
     (1972).
45. Ross, R. W., and Rowe, P. R., *Amer. Potato J. 42*, 177
     (1968).
46. Siminovitch, D., and Briggs, D. R., *Plant Physiol. 28*,
     15 (1953).
47. Steponkus, P. L., and Lanphear, F. O., *Plant Physiol. 42*,
     1423 (1967).
48. Steponkus, P. L., *Cryobiol. 6*, 285 (1969).
49. Sukumaran, N. P., and Weiser, C. J., *Hort. Sci. 7*, 467
     (1972).
50. Sukumaran, N. P., and Weiser, C. J., *Plant Physiol. 50*,
     564 (1972).
51. van Huystee, R., Weiser, C. J., and Li, P. H., *Bot. Gaz.*
     128, 200 (1967).
52. Vigue, J., and Li, P. H., *Plant and Cell Physiol. 15*,
     1055 (1974).

53. Vigue, J., and Li, P. H., *Amer. Hort. Sci. 101*, 341 (1976).
54. Wilner, J., Plant Res. Inst., Res. Branch, Can. Dept. Agric., Ottawa, Canada. Pp. 1-12 (1962).
55. Young, R., and Mann, M., *J. Amer. Hort. Sci. 99*, 403 (1974).

Part II
Membranes

PLASMA MEMBRANE ALTERATIONS FOLLOWING
COLD ACCLIMATION AND FREEZING[1]

P. L. Steponkus
S. C. Wiest

Department of Agronomy
Cornell University
Ithaca, New York

I. INTRODUCTION

Membrane damage is a universal manifestation of freezing injury to biological systems, and many of the methods commonly used to quantitate freezing damage are based on this fact. As early as 1912, Maximov concluded that freezing damage was the result of the freeze-induced removal of water from the surface of the plasma membrane. However, after 65 years and several thousand papers on the subjects of cold hardiness and freezing injury, it can only be said that freezing results in membrane rupture or loss of semipermeability. While considerable effort has been directed to characterization of the physico-chemical events that occur during freezing and the resultant changes in the cellular environment (33, 34), equal effort has not been directed to understanding what constitutes damage to the plasma membrane. Several factors appear to have contributed to this situation.

First, most efforts have been directed to changes in the cytoplasmic components as both the site of freezing injury and the mechanism of cold acclimation. This incursion into the cytoplasm undoubtedly had its roots in Gorke's observations in 1906 that freezing of plant sap resulted in precipitation of the proteins and Schaffnit's observations in 1910 that proteins in sap from acclimated rye were not readily precipitated by freezing temperatures that precipitated proteins of sap from non-acclimated plants. Furthermore, addi-

[1]*Department of Agronomy Series Paper 1230.*

tion of sugars to the sap of unhardy plants prevented the
frost precipitation of the proteins. The implications of
these observations with respect to both freezing injury and
cold acclimation are rather convincing and justified examination of cytoplasmic changes for many decades. However,
they continued after Heber (14) demonstrated that the "frost-
sensitive proteins" were membranous particles.

A second factor contributing to the situation was the
relative ease with which cytoplasmic changes could be studied
as opposed to the difficulties encountered in studies with
membranes, a factor that still prevails today. However, there
were some notable individuals who pursued the membrane alterations. In 1940, Scarth, Levitt and Siminovitch (49) ad-
dressed changes in plasma membrane structure in relation to
cold acclimation and freezing injury. Drawing upon information presented in several preceeding papers (29, 30, 48,
55), they indicated that both intracellular and extracellular ice formation result in damage specifically to the
plasma membrane, albeit for different reasons. Moreover,
while extracellular ice formation results in dehydration and
plasmolysis and subsequent deplasmolysis on thawing, plasmolysis *per se* was not the injurious event. Rather, rupture
of the plasma membrane occurred upon deplasmolysis. Thus,
nearly 30 years after Maximov's observations, attention was
refocused on the plasma membrane. Unfortunately, after the
observations of Scarth, Levitt and Siminovitch in 1940 (49),
interest in the plasma membrane again gradually diminished.
Another 30 years elapsed when in 1970, Levitt and Dean (28),
acknowledging the cyclical interest in the plasma membrane,
indicated that "...once again many investigators are turning
to the plasma membrane for the answer". However, they concluded that, when taken cumulatively, the evidence indicates
that the loss of semipermeability occurs whenever a cell is
killed by freezing but added that "categorical proof of
direct injury to the plasma membrane during extracellular
freezing is still lacking".

As cellular membranes are the primary site of freezing
injury, it follows that cold acclimation must involve cellular
alterations that allow the membranes to survive lower freezing
temperatures. Such alterations may be in the membranes *per
se*, or in the cellular environment in which they reside, or
a combination of both. Changes in the cellular environment
may either alter the freezing stresses that occur or they may
result in direct protection of the membrane. Both of these
possibilities were identified quite early (3) and have maintained their attractiveness in recent times (21, 27, 33,
34, 35, 36). However, while Scarth and Levitt (48) acknowledged that factors which result in supercooling depression
of the freezing point, or reduction in the amount of freezable

water occur during cold acclimation and reduce the freezing
stresses imposed on the plasma membrane, they are not the only
changes to occur.   Other changes can render the plasma membrane more tolerant to freezing stresses.   Others have suggested that acclimation involves both membrane alterations
and increases in soluble compounds (41, 59).   Recently, Olien
(36) acknowledged that cold hardiness is the result of both
alterations in the freezing stresses and increased tolerance
to the stresses.   Evidence that cold acclimation results in
changes in the tolerance of the plasma membrane has been provided by Scarth, Levitt and Siminovitch (49).   When cells are
sufficiently plasmolyzed, subsequent deplasmolysis causes
rupture and death, but the degree of plasmolysis that produces this effect is much less in non-hardy than in hardy
cells.   In addition, no deplasmolysis injury could be observed in the hardiest cells even following plasmolysis with
solutions of 6 M glucose.   Thus, Siminovitch and Levitt (52)
indicated that the plasma membrane of hardy protoplasts is
more resistant to dehydration and is less easily ruptured by
deplasmolysis or tension.

Scarth (47) and Levitt and Scarth (30) postulated that
cold acclimation could result in a membrane alteration that
altered cellular stresses.   They observed that membrane water
permeability increases with cold hardiness and suggested
that the increased permeability was instrumental in permitting rapid removal of water to extracellular sites of
ice nucleation.   This provided an explanation for the observation that lethal intracellular ice formation occurred in non-
acclimated tissues at slower freezing rates than in acclimated
tissues.   While other aspects of the work of Scarth, Levitt
and Siminovitch on the plasma membrane may have gradually
diminished, this concept has endured and has been cited by
numerous individuals through the years.   However, recent
evidence by Stout, Steponkus and Cotts (62) indicates that
there may not be a direct cause and effect relationship
between the two observations.   Calculations of the water
efflux rate required for extracellular ice formation indicate
that the membrane water permeability, as measured by a nuclear magnetic resonance technique, was not limiting extracellular ice formation in non-acclimated tissue.   Hence, the
decreased indidence of intracellular ice formation in tissue
may not be a result of increases in membrane permeability.

Although Levitt and Dear (28) indicated that there was
renewed interest in the plasma membrane in the late 1960's,
only a few reports were directly concerned with this membrane.   Rather, interest in membranes was stimulated by
studies of alterations in mitochondrial (26, 25) and chloroplast (19) membranes in relation to cold acclimation and
freezing injury.   Numerous papers by Heber and co-workers

(15, 16, 17, 18, 19, 20, 21, 22, 23, 42, 43, 44, 45, 46)
provided a wealth of information on the effects of freezing
on the function of chloroplast thylakoid membranes. Subsequently, Garber and Steponkus (11, 12) and Steponkus *et al.*
(60) extended this work providing information on the reper-
cussion of freezing and cold acclimation on thylakoid structure and function at the molecular level. However, to date,
information at a comparable level for the plasma membrane
remains to be presented.

Efforts by Siminovitch and co-workers (51, 53, 54) were
concerned with the localization of injury and resistance
in cellular membranes and indicated that resistance is an intimate property of the components of the plasma membrane
rather than some property arising from the purely colligative
action of solutes. Evidence for an increase in membrane
structures during cold acclimation was presented and termed
augmentation. Pomeroy and Siminovitch (40) provided electron microscopic evidence that the process of augmentation
was manifested by marked invaginations in the plasma membrane,
with some infoldings appearing to be in the process of
pinching off into vesicles in the peripheral cytoplasm.
The concept of autmentation is specifically concerned with
quantitative increments in the plasma membrane rather than
major qualitative changes (56, 58), although other studies
(61) indicate that functional alterations in the plasma membrane occur following cold acclimation and may be due to an
altered acidic phospholipic composition. Similarly, Wiest and
Steponkus (65) observed differences in the functional properties of membrane fractions containing plasma membrane vesicles
isolated from root tissue and callus cultures which had been
subjected to acclimating conditions. Accurate assessment of
the significance of these reports awaits elucidation of the
specific lesions that result in freezing injury to the plasma
membrane.

A rigorous answer to the question of what constitutes
plasma membrane damage following a freeze-thaw cycle will ultimately require reconstitution experiments with isolated
plasma membranes. However, to our knowledge, no one has yet
been able to isolate plant plasma membranes in both sufficiently pure form and in a state which can be proven to be
identical to their state *in vivo*. These problems prove to be
formidable for several reasons. Firstly, purification of
plasma membranes from any tissue requires the use of specific
markers. Hall and Roberts (13) have discussed the fact that
no such markers have been yet identified as specific to the
plasma membrane, and attempts to label the plasma membrane
with radioactive compounds have failed because of permeation
by the reagents and/or labelling of cell walls rather than the
membrane. While purity of plant plasma membrane preparations

has been estimated by a staining technique (24) which is rather specific for the plasma membrane of oat root tissues *in situ*, it may not be as specific in other plants or *in vitro* (63). Furthermore, it is obvious that purification of the plasma membrane requires dissolution of the membrane so that cytoplasmic components are removed. This dissolution is usually accomplished by homogenization of the tissue and vesicles of the order of 0.5 µ diameter (24), are obtained. Whole cells from which these vesicles were obtained probably had diameters no less than 10 to 20 µ. Such a large change in the radius of curvature may alter the structure of the membrane. The process of lysis in itself may also alter membrane structure. For instance, purified plasma membranes of red blood cells have been available since 1963 (7), but recent evidence (1) suggests that the state of phospholipids in these "resealed ghosts" is quite different from those in intact erythrocytes. Such problems must be solved before meaningful studies of freezing injury with isolated plasma membranes can be performed.

Questions of what constitutes plasma membrane damage can, however, be approached at the present time by studying the plasma membrane *in situ* in either tissue slices or protoplasts. The question of whether to investigate injury to the plasma membrane *in situ* versus in a disrupted but purified form can be argued quite extensively. However, the admonitions of Levitt and Scarth (29, 30) regarding the validity of investigations on the living cell versus extracts and "products of disorganization" remain valid if judged by the lasting value of some of their observations in contrast to the uncertainty of inferences made from studies using tissue extracts. For instance a controversy currently exists in regards to the relevance of lipid changes during cold acclimation (5, 67). Clearly, both approaches are essential to a complete understanding of the nature of freezing injury.

Recent investigations advocate the use of either tissue slices (37, 38, 39) or isolated protoplasts (64, 57, 6, 66) to investigate freezing injury and resistance of the plasma membrane. With protoplasts, cell walls are removed, leaving the plasma membrane exposed directly to the osmoticum so that the external face of the plasma membrane can be probed. Even though the detailed composition and structure of plant plasma membranes is essentially an unchartered region of plant biochemistry, meaningful data can be obtained on the mechanism of freezing injury to protoplasts as well as characterization of sites on the membranes which may be involved in the lytic event.

II.   OSMOTIC SIMULATION OF FREEZE-THAW INJURY

As mentioned previously, Scarth, Levitt and Siminovitch
(31) indicated that injury to the plasma membrane was related
to the plasmolysis and deplasmolysis that occurred during
freezing and thawing and that lysis occurred during deplasmolysis. Our initial concern was to determine how well
osmotic manipulation of protoplasts at room temperature simu-
lates freeze-thaw injury. Simulation of a freeze-thaw cycle
was achieved by exposing protoplasts to solutions of high
osmolalities followed by dilution of the osmoticum. Protoplasts that have been contracted in relatively high osmolalities and subsequently induced to expand upon dilution of the
osmoticum lyse before they regain their original size. This
fact indicates that some membrane alteration occurs when
the protoplasts are contracted and subsequently limits the
protoplast size that can be achieved upon dilution.

We have subsequently shown that the amount of injury to
frozen and thawed spinach protoplasts can be quantitatively
accounted for by injury that occurs when the plasma membrane
is osmotically induced to expand (66). Protoplasts suspended
in binary solutions of $CaCl_2$ and NaCl of varying osmolalities
were frozen to -3.9° C. Thus, during freezing all protoplasts were exposed to a 2.1 osmolal solution, but the amount
of contraction incurred at this concentration depended on
the initial osmolality of the suspending medium. The initial
osmolality also determined the extent to which the protoplasts
expanded during thawing. Following such a freeze-thaw cycle,
the extent of injury in the protoplasts in the various solutions decreased as the initial osmolality of the suspending
was increased, which was as expected since both the extent
of contraction and the extent of expansion were reduced.
Furthermore, the extent of injury in the protoplasts subjected to a freeze-thaw cycle in the various initial osmolalities was equal to the extent of injury predicted from the
simulation experiments.

In addition, the kinetics of freeze-thaw injury were as
predicted by the osmotically-induced contraction during the
simulation experiments. During a freeze-thaw cycle the extent
of injury is maximal and complete after the heat of ice formation has been dissipated from the sample and the temperature
of the sample returned to that of the cooling bath. Thermal
equilibration occurred in 10 to 12 minutes after ice formation
was initiated. The extent of injury was maximal after 10 to
12 minutes and did not increase during the subsequent 60
minutes. At the time of thermal equilibrium, the osmolality
of the extracellular solution is maximal and vapor pressure

equilibrium between the extracellular ice and the extracellular solution and the intracellular solution is established. At this time the protoplasts are maximally contracted. Hence, maximal injury coincides with the time when maximum contraction occurs.

Thus, both the extent and kinetics of injury in protoplasts exposed to a freeze-thaw cycle and in those subjected to osmotic manipulation are similar. These facts strongly suggest that injury to protoplasts during a freeze-thaw cycle is due to the same stresses of contraction and expansion that result from osmotic manipulation in the absence of ice. Therefore, it was desirable to further investigate the two separate events, contraction and expansion, that are associated with osmotically-induced lysis.

III.   PROTOPLAST CONTRACTION

During a slow freeze-thaw cycle intracellular water is converted to extracellular ice and cells are consequently forced to contract. Microscopic observations of this process indicate that collapse of the entire cell, including the cell wall, occurs (55). Later, Siminovitch and Levitt (52) demonstrated that the extent of injury to unhardened cells exposed to high osmolalities and then transferred to hypotonic solutions was dependent upon the manner of contraction--whether the cell walls collapsed with the protoplasm or whether the protoplasm was plasmolyzed and shrunk away from the cell wall. Furthermore, the degree of injury was different whether the cell was convexly plasmolyzed or concavely plasmolyzed. Therefore, additional strains may result from the association of the plasma membrane with the cell wall and a complete investigation of freeze-thaw injury to plant tissue will require investigations concerning the influence of the cell wall. However, the influences of stresses on the plasma membrane and resultant potentially lethal strains must be first established independent of the cell wall effects. Protoplasts provide a system free of potential stresses associated with the cell wall. Furthermore, when a protoplast is exposed to a hypertonic solution it contracts uniformly, that is, it remains spherical. In such a configuration the diameter can be easily determined and the volume and surface area of the protoplast can be accurately calculated. Accurate measurements or calculations of cell volume and surface area are difficult in cells with the cell wall remaining because of non-spherical contraction.

When a protoplast is exposed to a hypertonic solution and contracts in a uniform spherical configuration membrane components of the plasma membrane should become more tightly appressed. One might expect, *a priori*, that this would be a reversible process and that the membrane should be able to return to its original conformation when the osmoticum is diluted. However, as was previously mentioned, when a protoplast is osmotically induced to expand from a contracted state, lysis occurs before it regains its original size. The degree of expansion that can be tolerated before lysis occurs is dependent on the degree of contraction previously incurred by the protoplasts. For instance, when protoplasts are isolated in 0.54 osmolal solutions of salt ($CaCl_2$ + NaCl), the average surface area of population is 3250 μm^2. Upon dilution of the osmoticum, lysis occurs to 50 percent of the population when the average surface area exceeds 4100 μm^2. However, if protoplasts are first contracted to an average surface area of 1300 μm^2 by exposure to 1.61 osmolal salt solutions, lysis occurs to 50 percent of the population upon dilution when the average surface area exceeds 2100 μm^2. Thus, the maximum critical surface area where lysis occurs is established by the degree of contraction previously experienced by the protoplast. Therefore, we suggest that a plasma membrane alteration occurs during contraction which is either irreversible or only slowly reversible and limits the extent to which the plasma membrane can expand.

Evidence that the membrane alteration induced by contraction has a lasting influence (either irreversible or only slowly reversible) on the limit to which the protoplast can expand is obtained from the following observations. When protoplasts are contracted in 1.61 osmolal salt no lysis occurs. If this osmoticum is directly adjusted to 1.05 osmolal, 62% of the protoplasts survive. If the 1.61 osmolal solution is first adjusted to 1.22 osmolal and the protoplasts allowed to equilibrate for 30 minutes before the solution is further diluted to 1.05 osmolal, 63% of the protoplasts survive. Thus, injury is the same whether dilution occurs immediate or sequentially. A rather simplified interpretation of this data is that the protoplasts "remember" having been exposed to a 1.61 osmolal solution and they do not "forget" this event during the 30 minute euqilibration period. This "memory" can be reconciled by some irreversible or only slowly reversible change in the plasma membrane occuring during contraction.

Although the molecular or chemical nature of this membrane alteration has not yet been investigated, Siminovitch and Levitt (52) concluded that the surface membrane of the protoplasts of non-hardy cells stiffened when dehydrated osmotically and, as a result, ruptured more readily when

subjected to tension. We fully concur with this viewpoint--
if stiffening can be envisioned to involve either a deletion
of membrane components or a structural rearrangement, either
of which would lower the expansion potential of the membrane.
This fact that protoplasts survive exposure to rather high
osmolalities for at least several hours implies that the
contraction-induced membrane alteration in itself does not
result in lysis but is potentially lethal because it limits
the expansion potential. Whether it is lethal or not depends
on whether the membrane is induced to expand and to what
extent. Thus, the contraction-induced alteration specifically
affects the resilience--the capability of a strained body to
recover its size and shape after deformation caused especially by compressive stress--of the protoplast.

IV. PROTOPLAST EXPANSION

    Even though Scarth, Levitt and Siminovitch established in
1940 that lysis specifically occurs during deplasmolysis, little information exists in regards to the specific membrane
lesions that result in lysis. Even today attempts to demonstrate what constitutes plasma membrane damage at the molecular level are unfeasible because of the paucity of information regarding both the composition and structure of the plant
plasma membrane. Nevertheless, it is possible to investigate
changes in the spatial contour of the protoplast on lysis
and, in this way, gain some insight into the nature of the
lesions involved in disruption of the plasma membrane.
    In the previous section it was shown that the maximum
critical surface area that can be attained during osmotic
dilution before lysis occurs varies with the extent of contraction previously experienced by the protoplast. For example, protoplasts contracted in 0.54 osmolal salt solutions
had a surface area of 3250 μm^2 while those contracted in 1.61
osmolal salt solutions had a surface area of 1300 μm^2 and
50 percent lysed when the surface area exceeded 2100 μm^2.
This surface area is not only considerably less than that
which they could expand to when diluted from the 0.54 osmolal
solution, it is also considerably less than the surface area
of the contracted state in the 0.54 osmolal solution. Therefore, protoplasts do not possess any single and fixed critical
maximum surface area. However, even though the maximum
critical surface area where lysis occurs varies as a result
of the previous contraction, it is not a constant proportion
of the contracted surface area. When solutions of other
osmolalities were used, lysis of 50 percent of the population
always occurred when the increase in surface area from the

contracted state exceeded approximately 800 μm^2. Thus, there
was an absolute increment in surface area that could be
tolerated which was constant and independent of the degree of
contraction. While there are several speculative ways in
which this can be achieved any such discussion would be very
tenuous at this time because of a lack of information on
plasma membrane structure and composition. Rather than engage
in such a discussion we would prefer to merely establish the
observation at this time.

The fact that lysis occurred when a constant increment in
surface area was exceeded should underline the importance of
considering the repercussions of expansion in cells on the
basis of surface area rather than some other parameter of
size. Siminovitch and Levitt (52) noted that the difference
in injury, depending on whether concave or convex plasmolysis
occurred, was the result of a difference in the surface area
of the plasma membrane even though the extent of dehydration
was the same in both cases. Acknowledging the importance of
cell surface area is relatively important and its determination in spherical cells is relatively easy. However, such is
not the case in cells where the protoplasm is in an irregular
conformation on plasmolysis or in red blood cells which are
concave discoids and can double in volume with little or no
change in surface area (9).

The fact that injury is correlated with the surface area
expansion of the plasma membrane immediately suggests that
injury is related to the disruption of intermolecular forces
occurring in the plant tangent to the membrane surface. It
is well known (2) that the work done on a membrane can be
described as $W=\gamma dA$, where γ is the surface tension of the membrane and dA is the change in surface area. The work done on
the membrane is directly related to the increase in its
kinetic energy (4). These relationships lead to the concept
that the disruption of intermolecular forces in the membrane
causes protoplast lysis. One can conceptually envision the
process as follows. A membrane is composed of many different
components and their association is the result of characterization energies of attraction. As the plasma membrane ex-
pands, the above relationships demand that the kinetic energy
of the membrane increase. When this kinetic energy exceeds
the energy of association at some particular site, the association will no longer be stable and denaturation (lysis
of the cell) will follow.

Such a concept of membrane disruption can be used to
consider the observation that, upon dilution, lysis of the
protoplasts is correlated with the absolute increase in surface area. This implies that disruption of the membrane requires the same work (in a given osmoticum) regardless of the

surface area. Previously it was stated that the fact that
protoplasts lyse at different maximum critical surface areas
depending on the degree of contraction indicates that a membrane alteration which limits the extent of expansion potential is incurred during contraction. Since the expansion
potential is altered due to contraction and since disruption requires the same work regardless of the surface area,
but not the same pressure differential across the membrane,
it is suggested that the work required for disruption is
equal to the magnitude of the weakest intermolecular forces
joining the membrane together and is constant--regardless of
the surface area. Therefore, for lysis to occur at the same
work but at different surface areas implies that the contraction-induced alteration limits the expansion potential at a
site other than the weak link, such as would occur if a deletion of a unit area of membrane occurred. However, such a
deletion would necessitate that the maximum critical surface
area be some constant of the surface area in the contracted
state and this was not observed. Therefore, in addition to,
or instead of, a deletion, the ductility of the remainder of
the membrane must be altered. But even though it was altered
(decreased) it is still not decreased sufficiently to become
the weak link on expansion as accounted for by the constancy
of the work required. Thus the contraction-induced alteration results in a limitation of the expansion potential because either 1) a qualitative reduction in the ductility of
the membrane accompanies a quantitative reduction in membrane substance or 2) only a qualitative reduction in ductility occurs. Furthermore, taken cumulatively, the data
would indicate that the site of the contraction-induced alteration is different from the site where membrane disruption
occurs during expansion.

V. CHARACTERIZATION OF THE SITE(S) OF INJURY

     The environment of the membrane can influence its struc-
ture and therefore its stability. An attempt to characterize
the purported site(s) of injury on the plasma membrane can
therfore be made by investigating the effect of the membrane
environment on the susceptibility of the membrane to osmotically-induced lysis and hence freeze-thaw damage of protoplasts. To this end we have examined the effect of monovalent ions, pH and ionic versus non-ionic osmotica on the
susceptibility of protoplasts to osmotic lysis.

A. *Monovalent Ions*

    Both osmotic lysis and freeze-thaw injury to spinach pro-
toplasts are affected by the type of monovalent ions present
in the osmoticum. When suspended in equiosmolal solutions and
subjected to either osmotic-induced expansion or a freeze-thaw
cycle, more protoplasts lyse when K^+ rather than Na^+ is the
monovalent cation in the osmoticum. A more comprehensive
study of the influence of monovalent cations demonstrated
that protoplast sensitivity to a freeze-thaw cycle followed
the series:

$$Li^+ = Na^+ > K^+ = Rb^+ = Cs^+ \text{ and } Cl^- > Br^- > I^-,$$

in order of decreasing protoplast survival. Such a specificity of ions can be accounted for by the theory of Eisenman
(8), wherein the dimensions of the binding site and the
anhydrous crystallographic radius of the ions dictate the
energy of interaction of the ions with the binding site.
    For Eisenman's theory to be applicable to this situation,
the plasma membrane must possess a specific binding site
which varies in its affinity for these ions. Evidence that
such specific binding sites exist can be obtained from experiments where the lysis of protoplasts in osmoticum containing only a single monovalent cation (Na^+ *vs* K^+) is
compared to the amount of lysis that results when the protoplasts are suspended in osmotica containing varying propor-
tions of Na^+ and K^+. Previously it was stated that lysis
in K^+ is greater than the amount of lysis in Na^+. When the
proportion of Na^+ is less than 70 percent in a mixture of Na^+
+ Ka^+, the amount of lysis is equal to that occurring in K^+
alone. Only when the proportion of Na^+ is greater than 70
percent does the amount of lysis begin to decrease to the
lower level incurred in osmotica of 100 percent Na^+. This
effect is observed equally in both freeze-thaw induced injury
or osmotically-induced injury. Thus, it appears that K^+ has
a greater affinity for the site and therefore the membrane is
in a more susceptible state even when suspended in osmotica
with relatively high $Na^+:K^+$ ratios.
    The above discussion implies that the effect of the ions
result from a binding to specific sites, but does not indicate
whether the ionic influence was associated with the site of
the contraction-induced alteration or the site where membrane
disruption occurs during expansion. The absolute surface
area increment that resulted in lysis was approximately 500
μm^2 in K^+ rather than 800 μm^2 as previously measured in Na^+.
From the preceeding discussion this would suggest that the
ionic influence on lysis is associated with a decrease in the

work required to disrupt the membrane and therefore alters
the intermolecular forces at the site where membrane disruption occurs.

B. pH

We initially set out to examine the pH dependence of osmotic lysis to determine whether there were any readily iden-
tifiable titratable groups associated with the site(s) of
injury. We found, however, that when protoplasts were suspended in either $CaCl_2$+NaCl or $CaCl_2$+KCl in a pH range of 3 to
9, osmotically-induced lysis gradually increased as the pH
decreased below pH 6.0. While the pH dependence of lysis
observed in the presence of salt is not inconsistent with the
presence of titratable groups associated with the site(s) of
injury, a simpler interpretation of the data is possible. H^+
can be characterized as an alkali cation and could therefore
be expected to act at the cation binding site discussed in the
preceeding section. It is not inconceivable that H^+ has a
much stronger affinity than even K^+ and thus this site becomes even more susceptible at relatively low pH's in the
presence of salts.

C. Ionic vs Non-Ionic Osmoticum

When protoplasts were frozen in solutions of varying ratios of $CaCl_2$+NaCl:sorbitol, an optimum in protoplast survival
was observed at a salt:sorbitol osmolar ratio of 1:1. In
other words, injury in a 1:1 mixture of salt:sorbitol was less
than injury in either osmoticum alone. Santarius (42)
observed an optimum in the photophosphorylation capacity of
spinach thylakoids frozen in various ratios of NaCl and Na
succinate. This phenomenon was interpreted as being due to
the existence of two sites or mechanisms of injury; at high
Cl^- concentrations injury occurred at a Cl^--sensitive site and
at high succinate concentrations injury occurred at a suc-
cinate-sensitive site and at intermediate concentrations of
both Cl^- and succinate, each component colligatively reduced
the final concentration of the other and thus, either component was concentrated sufficiently to be toxic. In the protoplast system a similar interpretation of the data can be made.
The existence of two sites is further supported by the fact
that lysis resulting from osmotic manipulation exhibited a pH
dependence when the osmoticum was ionic (Na^+ or K^+) but this
was not apparent in sorbitol. However, there is one disturb-
ing observation, when protoplasts were subjected to osmotic
manipulation in sorbitol, 50 percent lysis occurred at the

same absolute increment in surface area that resulted in 50 percent lysis in Na^+ (800 μm^2). This would infer that if two different sites of expansion-induced damage did occur they would be required to have the same magnitude of intermolecular forces. While this may be a fortuitous occurence, it is rather untenable and further work is required.

VI. SUMMARY

In summary, isolated protoplasts afford an excellent system in which to study the effects of freezing on plasma membrane structure and function. Several desirable attributes contribute to this desirability: Upon dehydration they contract as spheres facilitating accurate calculations of volume and surface area changes from diameter measurements; their size allows direct observation in a light microscope; and they can be isolated, rather easily, from tissues which can undergo cold acclimation. Furthermore, the amount of injury incurred during a freeze-thaw cycle can be quantitatively accounted for by injury that occurs when they are osmotically induced to contract and expand.

Freezing injury to protoplasts is the result of an alteration in the resilience of the protoplast. While the altered resilience is the result of an alteration in the plasma membrane that occurs during contraction it is not manifested until the protoplasts are induced to expand--either during osmotic dilution or thawing--when disruption of intermolecular forces in the membrane causes protoplast lysis. The extent of contraction primarily determines the alteration in the resilience, while several factors in the membrane environment can either mitigate or intensify the resilience.

REFERENCES

1. Bloj, G., and Zilversmit, D. B., *Biochemistry 15*, 1277 (1976).
2. Castellan, G. W., *in* "Physical Chemistry" (2nd ed.). Addison-Wesley, Reading, Mass. (1971).
3. Chandler, W. H., *Mo. Agr. Expt. Stat. Res. Bull. 8*, 141 (1913).
4. Cromer, A. H., *in* "Physics for the Life Sciences." McGraw-Hill, New York (1974).
5. de la Roche, I. A., Pomeroy, M. K., and Andrews, C. J., *Cryobiology 12*, 506 (1975).

6. de la Roche, I. A., Keller, W. A., Singh, J., and Siminovitch, D., *Can. J. Bot.* *55*, 1181 (1977).
7. Dodge, J. T., Mitchell, C., and Hanahan, D. J., *Arch. Biochem. Biophys.* *100*, 119 (1963).
8. Eisenman, G., *Biophysical J.* *2(Suppl.)*, 259 (1962).
9. Evans, E. A., and Leblond, P. F., *Biorheology 10*, 393 (1973).
10. Gorke, H., *Landw. Versuchs 65*, 149 (1906). (Cited by Chandler, 1913).
11. Garber, M. P., and Steponkus, P. L., *Plant Physiol.* *57*, 673 (1976).
12. Garber, M. P., and Steponkus, P. L., *Plant Physiol.* *57*, 681 (1976).
13. Hall, J. L., and Roberts, R. M., *Ann. Bot. 39*, 983 (1975).
14. Heber, U., *Planta 54*, 34 (1959).
15. Heber, U., *Plant Physiol. 42*, 1343 (1967).
16. Heber, U., *Cryobiology 5*, 188 (1968).
17. Heber, U., *in* "Ciba Found. Symp. on the Frozen Cell," p. 175. Churchill Ltd., London (1970).
18. Heber, U., and Ernst, R., *in* "Cellular Injury and Resistance in Freezing Organisms," Vol. II (E. Asahina, ed.), p. 63 (1967).
19. Heber, U. W., and Santarius, K. A., *Plant Physiol. 39*, 712 (1964).
20. Heber, U., and Santarius, K. A., *in* "The Cell and Environmental Temperature" (A. S. Troshin, ed.), p. 27. Pergamon Press, New York (1967).
21. Heber, U., and Santarius, K. A., *in* "Temperature and Life" (H. Precht, J. Christopherson, H. Heusel, and W. Larcher, eds.), p. 232 (1973).
22. Heber, U., Tyankova, L., and Santarius, K. A., *Biochim. Biophys. Acta 241*, 578 (1971).
23. Heber, U., Tyankova, L., and Santarius, K. A., *Biochim. Biophys. Acta 391*, 23 (1973).
24. Hodges, T. K., Leonard, R. T., Bracker, C. E., and Keenan, T. W., *Proc. Nat. Acad. Sci. (USA) 69*, 3307 (1972).
25. Kenefick, D. G., *Agr. Sci. Rev. 2*, 21 (1964).
26. Kenefick, D. G., Swanson, C. R., *Crop Sci. 3*, 202 (1963).
27. Levitt, J., "Responses of Plants to Environmental Stresses." Academic Press, New York (1972).
28. Levitt, J., and Dear, J., *in* "The Frozen Cell" (G. E. W. Wolstenholme and M. O'Conner, eds.). J. A. Churchill (1970).
29. Levitt, J., and Scarth, G. W., *Can. J. Res. C. 14*, 267 (1936).
30. Levitt, J., and Scarth, G. W., *Can. J. Res. C, 14*, 285 (1936).

31. Levitt, J., and Siminovitch, D., *Can. J. Res. C. 18*, 550 (1940).
32. Maximov, N. A., *Ber. der. Deutsch. Bot. Gesell.*, *30*, 52, 293, 504 (1912) (Cited by Chandler, 1913).
33. Mazur, P., *Ann. Rev. of Plant Physiol. 20*, 419 (1969).
34. Mazur, P., *Science 168*, 939 (1970).
35. Olien, C. R., *Ann. Rev. Plant Physiol. 18*, 387 (1967).
36. Olien, C. R., *USDA Tech. Bull. No. 1558* (1977).
37. Palta, J. P., Levitt, J., and Stadelmann, E. J., *Plant Physiol. 60*, 393 (1977).
38. Palta, J. P., Levitt, J., and Stadelmann, E. J., *Plant Physiol. 60*, 398 (1977).
39. Palta, J. P., Levitt, J., and Stadelmann, E. J., *Cryobiology 14*, 614 (1977).
40. Pomeroy, M. K., and Siminovitch, D., *Can. J. Botany 49*, 787 (1971).
41. Sakai, A., and Yoshida, S., *Cryobiology 5*, 160 (1968).
42. Santarius, K. A., *Plant Physiol. 48*, 156 (1971).
43. Santarius, K. A., *Biochim. Biophys. Acta 291*, 38 (1973).
44. Santarius, K. A., *Planta 113*, 105 (1973).
45. Santarius, K. A., and Heber, U., *Cryobiology 7*, 71 (1970).
46. Santarius, K., and Heber, U., *in* "Proc. Colloq. on the Winter Hardiness of Cereals" (S. Rajki, ed.), p. 7 (1972).
47. Scarth, G. W., *Trans. R. Soc. Can. Sect. 5, 30*, 1 (1936).
48. Scarth, G. W., and Levitt, J., *Plant Physiol. 12*, 51 (1937).
49. Scarth, G. W., Levitt, J., and Siminovitch, D., *Cold Spg. Harbor Sym. Quant. Biol. 8*, 102 (1940).
50. Schaffnit, E., *Mitt. Kaiser Wilhelm Inst. Landw. Bromberg 3*, 93 (1910) (Cited by Chancler, 1913).
51. Siminovitch, D., Gfeller, F., and Rheaume, B., *in* "Cellular Injury and Resistance in Living Organisms" (E. Asahina, ed.), p. 93. Inst. of Low Temp. Sci., Sapporo, Japan (1967).
52. Siminovitch, D., and Levitt, J., *Can. J. Res. C. 19*, 9 (1941).
53. Siminovitch, D., Rheaume, B., and Sacher, R., *in* "Molecular Mechanisms of Temperatures Adaptation" (C. L. Prosser, ed.), p. 3. Amer. Assoc. Adv. Sci., Washington, D. C. (1967).
54. Siminovitch, D., Rheaume, B., Pomeroy, K., and Lepage, M., *Cryobiology 5*, 202 (1968).
55. Siminovitch, D., and Scarth, G. W., *Can. J. Res. 16*, 467 (1938).
56. Siminovitch, D., Singh, J., and de la Roche, I. A., *Cryobiology 12*, 144 (1975).
57. Siminovitch, D., Singh, J., Keller, W. A., and de la Roche, I. A., *Cryobiology 13*, Abstract (1976).

58. Singh, J., de la Roche, I. A., and Siminovitch, D., *Nature 257*, 669 (1975).
59. Steponkus, P. L., *Plant Physiol. 47*, 175 (1971).
60. Steponkus, P. L., Garber, M. P., Myers, S. P., and Lineberger, R. D., *Cryobiology 14*, 303 (1977).
61. Stout, P. G., Cotts, R., and Steponkus, P. L., *Cryobiology 12*, 553, Abstract (1975).
62. Stout, D. G., Steponkus, P. L., and Cotts, R. M., *Plant Physiol. 60*, 374 (1977).
63. Thom, M., Laetsch, W. M., and Maretzki, A., *Plant Science Letters 5*, 245 (1973).
64. Wiest, S. C., and Steponkus, P. L., *Cryobiology 13*, Abstract (1976).
65. Wiest, S. C., and Steponkus, P. L., *J. Amer. Soc. Hort. Sci. 102*, 119 (1977).
66. Wiest, S. C., and Steponkus, P. L., *Plant Physiol.*, in press (1978).
67. Willemot, C., Hope, H. J., Williams, R. J., and Michaud, R., *Cryobiology 14*, 87 (1977).

CELL MEMBRANE PROPERTIES
IN RELATION TO FREEZING INJURY

J. P. Palta
P. H. Li

Laboratory of Plant Hardiness
Department of Horticultural Science
and Landscape Architecture
University of Minnesota
St. Paul, Minnesota

I. INTRODUCTION

Although frost injury to plants has been known for a long
time, its basic nature at cellular levels is not well understood. Slow cooling in nature produces extracellular freezing
which results in collapse of the cell walls (18, 23, 50).
This kind of freezing can occur with or without death of the
plant (51). Injury to plant cells due to freezing has been
thought for a long time to be located in the cell membranes
(24, 25, 30). In recent years, chloroplast membranes have
been reported to be injured by freezing (13, 15, 16), yet
there has been a lack of experimental evidence for cell membranes as primary sites of freezing injury.

One of the most common signs of freezing injury is the
infiltration of the tissue intercellular space with water.
This gives a soaked appearance to tissue as well as results
in loss of cell turgor many times. This has been reported
by many early workers (45, 57), which led them to believe that
cell rupture had occurred. Many workers have used this infiltration of the tissue as a criterion for evaluating injury
(5). Another common sign of freezing injury is the leakage
of ions from cells.

Biological membranes are differentially permeable (43) and
changes in the permeability properties of the protoplasmic
layer were proposed by Osterhout (34) as a sensitive and precise indicator of the vitality of the tissue. Based on this

93

leakage of ions, Dexter *et al.* (10) showed that measurement of the electrical conductivity of the effusate can provide a quantitative method for evaluating freezing injury.

Until very recently the efflux of ions from frozen and thawed tissue and its infiltration with water has been assumed to be due to break down of the semipermeable property of the cell membrane (23, 53, 54). In fact, the terms LT_{50} and LD_{50} have been used in evaluating freezing injury (31, 46). Here LT_{50} and LD_{50} being the freezing temperature at which the thawed tissue leaks out 50% of the total ions (50% conductivity in relation to killed tissue), presumably meaning 50% of cells killed and that tissue will never recover. This is probably based on the analogy with tissue cultures and cell suspensions (e.g. red blood cells or bacteria) in which actual counts of living and dead cells are made after freezing.

Recent work of Palta *et al.* (35-37) has shown that during progress of freezing, semipermeable properties remain intact whereas the active transport properties of these cell membranes are damaged. These studies provide a new direction in understanding the mechanism of freezing injury. This paper summarizes those recent findings and provides further experimental evidence in support of these results.

II. RECENT FINDINGS

Because of their large cell size and easy separation of inner epidermal layer, onion bulbs are very suitable material for cell physiological investigations (42). Moreover, it was found recently that they possess a high degree of freezing tolerance (killing point about -20° C) (37, 38). Therefore, onion bulbs are very useful material for investigating the physiology of freezing injury. Using this material Palta *et al.* (35-38) studied the various aspects of freezing injury. In these studies onion bulbs (*Allium cepa* L.) were frozen to various temperatures up to -20° C and kept frozen up to 12 days. After slow thawing, various measurements were made on the scale parenchyma as well as outer and inner epidermal cells. Results obtained were as follows.

A. *Infiltration of the Tissue and Conductivity of the Effusate in Relation to Cell Viability*

Although the scale tissue was completely infiltrated and up to 85% of total ions leaked out of the cells, still all of the cells plasmolysed in hypertonic mannitol solution

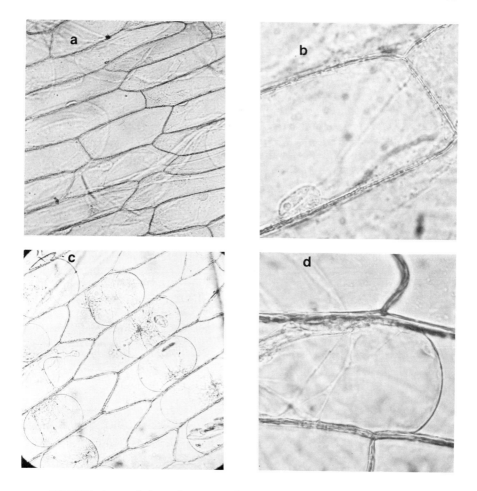

FIGURE 1. Light microscopic pictures of freeze injured onion epidermal cells. a--Immediately after thawing; b--close up of a; c--cells from a when transferred to 0.8 M mannitol solution; d--close up of c.

and exhibited protoplasmic streaming. An example of this is shown in Figure 1.

Freeze injured and subsequently thawed onion epidermal cells appear quite normal under light microscope. No separation of plasma membrane cell wall and coagulation of protoplasm was observed. This type of appearance of frost killed cell commonly called frost plasmolysis (26) was absent in these injured cells. On the contrary these cells exhibited protoplasmic streaming and protoplasmic strands could be easily observed just like in uninjured normal cells (Figure

1 a and b). In fact, no visual microscopic differences were found between injured and unfrozen control cells. These injured cells which appear throughly infiltered and are leaking out ions could be very easily plasmolysed in a hypertonic solution (about 0.8 M) of mannitol (Figure 1 c and d). Moreover, these cells stayed in plasmolysed condition for a long period of time (several days) just like uninjured control cells. This indicates that the semipermeable properties of these cell membranes are in fact intact and that leakage of ions from the cells contrary to the early reports is not due to breakdown of the semipermeable properties of these membranes.

It is important to point out here that there is some confusion in the literature regarding the differences between frost plasmolysis, normal plasmolysis and cell dehydration or cell contraction. In fact, many authors have referred to extracellular freezing causing plasmolysis (2, 31). This is, however, not correct for most plants or plant organs. Plasmolysis is defined as the separation of cell membrane from the cell wall when plant cells are exposed to hypertonic extracellular solution (Figure 1 c and d). During extracellular freezing, water is withdrawn from the cell in the presence of ice on the outside of the cell and in the absence of any hypertonic solution. Under such a case, the whole cell shrinks and collapse of the cell walls occurs. Some authors seem to have assured that ice penetrates through the wall and it proceeds between cell membrane and cell walls, thus causing separation of cell walls from the membrane (pseudo plasmolysis). This has never been shown to happen in intact plant tissues. Asahina (1) saw ice formation between walls and the plasmamembrane in hair cells which were removed from the leaf.

Things are quite different, however, when one is dealing with isolated protoplasts or cell suspensions. In these cases, protoplast shrinkage occurs when outer solution is concentrated due to formation of ice in the solution. Here cell shrinkage and plasmolysis may be used synonymously. Plasmolysis in natural conditions only occurs in the case of water plants, e.g. algal cells, etc. Here as the solution is concentrated during water or freezing stress beyond isotonic concentration, actual plasmolysis will occur (separation of cell membrane from the cell walls). If the injury to the cell is very severe so that the cell is killed, then on thawing the protoplast in the collapsed cell does not swell back whereas the cell wall, being elastic, comes back to original shape leaving the shrunken protoplast in the middle. This gives the condition of the protoplast which is commonly known as frost plasmolysis (26). This phenomenon was studied carefully by several early investigators (4, 32). As Levitt (23) has pointed out in his book, this term "frost plasmolysis" is an unfortunate one as it has no relationship with plasmolysis at all. Therefore, it may be better to drop this term to avoid confusion.

B. *Evaluation of the Conductivity Method*

It is quite clear from the above discussion that conductivity method does not measure dead cells in a tissue or loss of semipermeability of the cell membranes. The question arises then--what does this method measure? As it will be discussed in the following paragraphs, Palta *et al.* (35, 36) have shown that when the efflux of ions is from injured yet living cells, the conductivity method measures the average injury of the cells and therefore provides an insight into the mechanism of freeze injury. In spite of this, conductivity method has been used quite successfully for measuring frost hardiness and testing relative varietal resistance in plants. To be on the safe side, it is essential to relate the conductivity with other viability tests (41).

It is a common observation that even the unfrozen control tissue leaks out some ions when shaken with deionized water. Depending upon the time of shaking and temperature of shaking, as much as 30% of the total ions were found to leak out of onion scale tissue (35). Since injured tissue is usually infiltrated with ice water, to make a proper comparison unfrozen control tissue should also be vacuum infiltrated before shaking.

C. *Analysis of the Efflux from the Injured Cells*

The leakage of ions was found to be from injured yet living cells by Palta *et al.* (35). In order to study in detail the nature and cause of this efflux, they analysed it in detail and found the following:

(1) K^+ was the major cation present. Ca^{++} constituted only 0.1% of the K^+.

(2) K^+ content of the effusate plus its assumed counterion accounted for only 20% of the total solutes but for almost 100% of the conductivity. Correlation between K^+ content and conductivity was found to be 0.98. This was found to be true for all of the treatments.

(3) A large part of the nonelectrolytes in the remaining 80% of the solution were found to be sugars.

(4) Increased cell injury and infiltration with lower temperatures were paralleled by increases in conductivity, K^+ content, sugar content and pH of the effusate.

(5) The types of ions present were not analysed although large amounts of organic anions can be expected to be present in the efflux.

D. *Permeability of Cell Membranes to Water; No Change in Spite of Large Efflux of Ions and Sugars Following Freeze Injury*

It has been assumed that the efflux of ions following a freezing injury is due to the loss of semipermeability or increase in passive permeability of these killed cells (19, 23, 54). Since the first assumption was proven wrong (injured cells could be plasmolysed in hypertonic solutions), the second one could be correct. Although still alive, the cells may have suffered an injurious increase in permeability. In order to determine whether or not this occurred, direct measurements of the cell permeability to water were made by Palta *et al.* (35, 36). The onion epidermis was allowed to absorb tritiated water, then the rate of efflux into unlabeled water was measured using an accurate method (42). These measurements failed to detect any change in cell permeability to water. No measurements were made of ion permeability. Direct measurements on the passive permeability for K^+ of injured cells, however, must be made in order to test this possibility.

E. *Post-thaw Behavior of Injured Cells; Reversibility of Membrane Damage*

It has long been known that the extent of injury to plants kept at 0 to 5° C for 24 hours after freezing and thawing may be different than if immediately transferred to room temperatures (34). This is usually interpreted to mean that freeze-damaged but still living cells may die at higher temperatures but may recover at 0 to 5° C. This interpretation, however, has not been adequately investigated until recent studies by Palta *et al* (36). In this study onion bulbs were subjected for 12 days to either a moderate freeze (-4° C) or a severe freeze (-11° C). They were then thawed slowly over ice. During 7 to 12 days following the thaw, the injury progressed with time (irreversible injury) in the severely frozen bulbs, but appeared completely repaired (reversible injury) in the moderately frozen bulbs. This was shown by the following post-thawing changes.

Infiltration of the intercellular spaces increased from
80-90% to 100% after the severe freeze, decreased from 30-50%
to zero after the moderate freeze. All the cells were alive
imemdiately after thawing, whether the freeze was moderate
or sever. Corresponding to the infiltration results 7 to 12
days later, many to most were dead following the severe
freeze, all were alive following the moderate freeze.

The conductivity of the effusate from pieces of bulb
tissue increased after severe freezing, decreased after the
moderate freezing. The concentration of K^+, total solutes,
and sugars in the effusate paralleled the conductivity
changes. Neither the pH of the effusate, nor the permeability
of the cells (as long as the cells were living) to water were
changed following the severe to moderate freezes.

F. *Treatments of the Freeze-injured Tissue for Halting the
    Progress of Injury*

As pointed out earlier, a freeze-injured tissue frequently
becomes infiltrated with water on thawing and appears to have
lost turgor. This loss of turgor is the result of the high
concentration of solutes present in the extracellular space
due to leakage of ions and sugars from the vacuole. In these
studies with onion bulbs an attempt was made to halt the
progress of the injury by specific treatments after thawing.

It was found that removing the extracellular solutes
from the extracellular solution by washing the scale pieces
with distilled water, considerably improved the cell survival.
Treatments of the washed tissue with 0.01M $CaCl_2$ solution
further increased cell survival over the untreated cells.
The extent of the survival, however, depended upon the severity of the injury. Only when the injury was severe enough
but not too severe (most of the cells dying within a few days
after thawing) did this treatment improve cell survival.

G. *Primary Site of Freezing Injury Possibly to Active
    Transport System of the Cell Membranes*

These studies on onion bulbs by Palta *et al*. (35-37)
indicate that the semipermeable properties of the cell membranes are uninjured whereas the ion and sugar transport
mechanisms are damaged by freezing. It was pointed out that
the primary injury is most likely to the active transport
mechanism. This follows from the post-thaw recovery of
injury which led to disappearance of the infiltration of the
tissue and marked reduction in the efflux of ions and sugar
from these cells. The reason for this absorption of water by

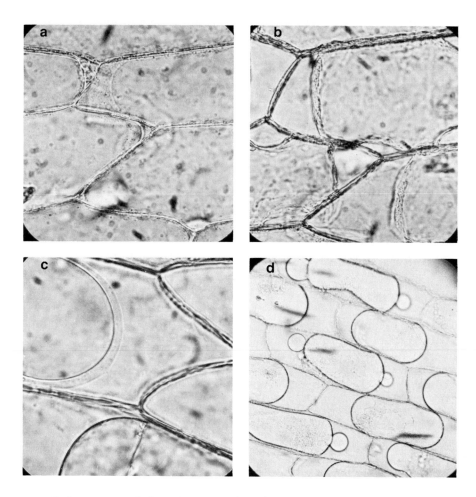

FIGURE 2.   Light microscopic pictures of freeze injured
onion epidermal cells.   Onions were slowly frozen to -10° C
and thawed over ice overnight.   a--Swelling of protoplasm on
thawing can be seen in some cells; b--same cells as in a
when transferred to 0.8 M mannitol for plasmolysis; c--cap
plasmolysis of swollen protoplast cells; d--separation of
plasma membrane observed when plasmolysed cells transferred
to equimolar KCl solution.

the cells is due to influx of ions and sugars from extracellular solution.   Since this influx is against the concentration
gradients it must be active in nature.

Membrane   located ATPases are now believed to constitute
ion pumps (47) and several such ATPases have been found involved   in ion uptake by plant cells (12, 17, 22, 28).   Heber

(15) reported that freezing of chloroplast membranes un-
couples photophosphorylation from electron transport and in-
activates light dependent ATPase.

Therefore, from the results presented above, the plasma
membrane ATPases and the mechanisms for active sugar transport
may conceivably be the loci of the initial freezing injury.
This hypothesis would also explain the high K^+ and sugar
conc in the effusate from the frozen and thawed, but still
living cells. If the freezing damaged the K^+ activated
ATPase and the mechanism for active sugar transport, this
would decrease the active influx of K^+ and sugars while permitting the passive efflux unaltered or enhanced (the damaged
ion pumps and mechanisms for active sugar transport could be
visualized as sites for facilitated effusion). This would
result in a net efflux of K^+ and sugars. Furthermore, the
repair of the damage can be visualized as a reversible injury
to these active transport pumps which temporarily get inactivated and following recovery are able to pumb back the K^+ and
sugar into the vacuole resulting in disappearance of the infiltration of the tissue, in the same way irreversible damage
to the pumps will lead to continued increased flux of ions
and sugars as was observed.

III.   FURTHER EVIDENCE IN FAVOUR OF DAMAGE TO ACTIVE TRANSPORT
       SYSTEM

Results discussed above are not unique to onion bulb
cells. These results have been confirmed by our recent work
with excised leaflets of several *Solanum* species (40). In
order to advance these results Palta and Li (41) conducted
various types of detailed studies with onion bulbs as well as
potato leaves. Their results further provide evidence that
initial freezing injury is located at the membrane associated
active transport systems. A summary of these results is
presented here.

A. Microscopic Observations; Swelling of Protoplasm

As indicated in the early discussion in spite of large
efflux of ions and sugars, infiltration of the tissue and many
times complete loss of turgor, the onion epidermal cells appear normal on thawing. (Figure 1, a and b). Furthermore,
these cells could be plasmolysed in a hypertonic mannitol
solution indicating that the semipermeable properties of these
membranes were intact (Figure 1, c and d). Many times with
higher degree of cell damage swelling of the protoplasm was

observed clearly (Figure 2, a). This was also true of many cells when the tissue with irreversible damage was showed increased in injury during post-thaw period. Apparently, the semipermeable properties of the cell membranes were still intact since these cells could be plasmolysed in a hypertonic mannitol solution (Figure 2, b) just like the other damaged cells which did not show swelling of protoplasm. When these cells were left for a few hours to equilibriate in plasmolysing solution of mannitol (about 0.8 M) a typical "cap" plasmolysis was observed (Figure 2, c).

Cap plasmolysis usually occurs when cells are exposed to hypertonic KCl or KNO_3 solutions for a long period of time (several hours). It is the result of the injury to plasma membrane produced by a high concentration of K^+ in the extracellular water. Due to this injury the possible permeability of K^+ through the plasma membrane is increased where the tonoplast remains relatively less permeable. This results in accumulation of K^+ in the protoplasm that leads to osmotic uptake of water giving rise to swollen protoplast, i.e., cap plasmolysis (Figure 2, c). One reasonable explanation for occurrence of swollen protoplasm in freezing injury is the presence of high amount of K^+ in the extracellular water that gives rise to high conductivity values for the effusate (35). This high amount of K^+ present in the extracellular water could very likely produce a similar swelling of protoplasm as if the cells were in hypertonic KCl or KNO_3 solution.

Freeze injured leaf parenchyma cells of *Solanum tuberosum* also showed swelling of protoplasm. Electron photomicrographs of such cells are shown in Figure 3, c and d. Here unfrozen control cells do not show any such swelling of protoplasm (Figure 3, a and b). Freezing injury was produced by cooling the excised leaflets at 1º/hr and ice nucleation was produced at -1º C (26). These were then thawed in ice and boxed overnight and 1 mm thick strips from the central portion of the leaflet were used for the electron microscopic studies. It is interesting to note that in spite of the swollen protoplast (clear separation of plasmalemma and tonoplast) for cells frozen to -2.5º C (very close to the killing temperature of the tissue) the chloroplasts and mitochondria were found to be quite normal (Figure 3, c and d). The grana stacks, outer chloroplast membrane and mitochondrial membranes were visibly intact. Only when the tonoplast and/or plasma membrane were found to be disrupted, was swelling of chloroplast and mitochondria found. This shows that tonoplast and/or plasma membrane are probably injured first resulting in leakage of K^+, leading to swelling of the protoplasm, whereas chloroplast and mitochondria are affected at a later stage.

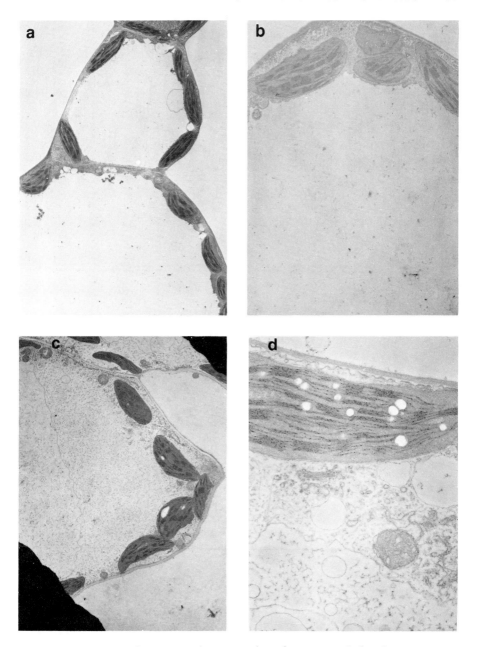

FIGURE 3.   Electron microscopic pictures of leaf paren-chyma cells of Solanum tuberosum.   a and b--Unfrozen control; c and d--freeze injured cells.

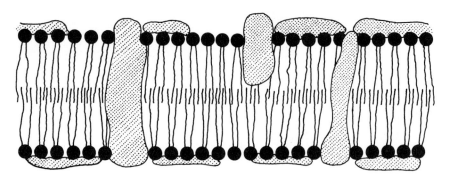

FIGURE 4. Diagram of latest membrane model. This is
based on Singer Nicholson's model (51). Membrane proteins
are of two types: intrinsic (embedded in the lipid bilayer)
and extrinsic (on the surface of the bilayer).

B.   Cell Membrane Transport Studies; Permeability to K^+ and
     Nonelectrolytes

From the preceeding discussion it is quite clear that
freezing injury results in alterations, of the cell membranes.
Permeability of these membranes to water, ions and nonelectrolytes is a very sensitive indicator of these alterations.
Since it has been assumed that freezing injury results in the
breakdown of the semipermeable properties of the cell membranes, no attempt was made to study the transport properties
of freeze injured cells. Palta et al. (35, 36) studied the
water transport properties of freeze injured thawed cells
using a radiotracer method (42). In spite of reversible or
irreversible freeze injury no change in the water permeability
of these cells was found. Water movement across cell membranes is best explained by the molecular theory of membrane
transport advanced by Traeuble (55). According to this theory
water moves through the imperfections in hydrocarbon tails
(so called "kinks"). Thus, no change in water permeability,
in spite of freeze injury, would seem to indicate that the
lipid portion of the membrane is unaltered. This also agrees
well with ability of these cells to plasmolyse in hypertonic
solutions of mannitol (a test of semipermeability).
For explaining the large efflux of ions and sugar from the
injured cells, Palta and Li (26) studied the transport of K^+,
urea and methyl urea across the injured cell membranes. A
plasmometric method described by Stadelmann (52) was used for
making these measurements. Freeze injured onion epidermal
cells were plasmolysed in hypertonic mannitol solution and
were then transferred to equimolar solution of substance for
which the permeability was to be measured. In response to

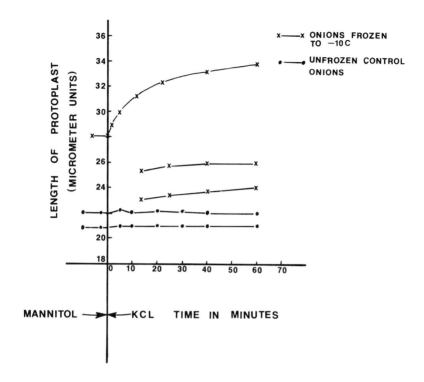

FIGURE 5.   Measurement of protoplast expansion in response to passive transport of KCl.   Plasmolysed cells in mannitol were transferred to equimolal KCl at zero time.

passive uptake of the substances (K^+, urea and methyl urea) a simultaneous transport of water takes place which leads to the expansion of the cell protoplast.   This protoplast expansion could be measured directly microscopically.   A typical result of such an experiment is shown in Figure 5.   The unfrozen control cells do not take up K^+ and remain plasmolysed to a constant volume (length of the protoplast remained constant, Figure 5).   On the other hand the injured cells (frozen to -10° C) started expanding in KCl solution and slowly deplasmolysed.   The extent of K^+ uptake, however, varied in injured cells.   This is expected since the injury in a tissue varies from cell to cell.

Permeability values for transport of urea and methyl urea across freeze injured and control cells are given in Table I.   Clearly, no significant difference could be detected between the injured and unfrozen control cells for both urea and methyl urea permeability constants.

TABLE I.   Urea and Methyl Urea Permeability Constants for Freeze Injured and Control Onion Epidermal Cells

	Permeability constant $(cm\ sec^{-1})$		
Substance	Control	$-10^{o}\ C$	$K_{oil}{}^{a}$
Urea	$(6.4 \pm 0.3)10^{-8}$	$(6.5 \pm 0.3)10^{-8}$	$15\ x\ 10^{-5}$
Methyl urea	$(2.9 \pm 0.4)10^{-7}$	$(2.8 \pm 0.8)10^{-7}$	$44\ x\ 10^{-5}$

$^{a}K_{oil}$ *is the oil/water distribution coefficient, a high value indicates high lipid solubility (8).*

In order to understand the significance of these results it is important to look at the membrane structure in relation to permeability of these substances. Singer and Nicolson (51) proposed a fluid mosaic model of membrane structure. This model is well accepted among the physiologists and appears to explain most of the observed experimental data. Similar to Davson and Danielli's (9) model, this model consists of two layers of lipids with polar ends sticking out. It assumes that proteins are dispersed throughout the membrane (Figure 4). The membrane proteins are supposed to be mainly two types, intrinsic and extrinsic. Some of the intrinsic membrane proteins are simply embedded on one side or the other while the rest pass entirely through the bilayer (Figure 4). Transport studies provide information to the site of the injury with the membrane. Two generalizations can be made from these studies.

1. Lipid Portion of the Membrane Seems Unaltered

All the experimental facts seem to indicate that the movement of nonelectrolytes across cell membrane is a simple diffusion process and follows Fick's law. Results of Collander and co-worker (6-8) showed that the permeability of a nonelectrolyte is a direct function of its lipid solubility - i.e., the oil/water distribution coefficient (K_{oil}, also known as oil/water partition coefficient). These results are now well established and accepted by membrane physiologists. Therefore, any changes in the nonelectrolyte permeability should reflect the alterations in the lipid portion of the membrane. The two nonelectrolytes selected for our studies varied greatly in their K_{oil} values; methyl urea being three

times more soluble in lipids than urea (Table I). This gave
us the opportunity to examine the alterations in the lipid
portion of the membrane of freeze injured cells. Since no
change in the permeability values for these two substances was
found (Table I) the damage or the alterations in the lipid
portion of the membrane by freezing injury can be ruled out.

2. *Intrinsic Membrane Proteins Seem to be the Sites of Injury*

In contrast to nonelectrolytes the molecular mechanisms
of ion permeation through the cell membrane are not as yet
well understood. The intrinsic membrane proteins that pass
entirely through the bilayer (Figure 4) have been proposed to
be the sites of active transport of ions (50). These built-in
proteins offer to the ion a hydrophilic pathway through the
apolar core of the membrane. The evidence for existence
of such proteins came from the studies with a simple channel
(ionic pathway) forming molecule gramicidin A. This molecule,
if introduced into biological membrane (44, 56) or into an
artificial lipid bilayer (20, 33), causes the membrane to
become cation permeable. There is a considerable amount of
evidence in the literature that membrane associated ATPases
are involved in ion transport (12, 17, 22, 28). These
ATPases are believed to constitute ion pumps (47). Therefore,
by study of the transport of K^+ across the cell membrane one
can get information about the intrinsic membrane proteins.

In our studies of K^+ transport across freeze injured cell
membrane, a sharp increase in its passive permeability was
observed (Figure 5). In injured cells, when plasmolysed in
mannitol solution and transferred to isotonic KCl solution
within a minute, a separation of the plasma membrane could be
observed, leading to formation of round caps on the surface
of the plasmolysed protoplasts (Figure 2, d). This was
found to be the case only in irreversibly injured cells. Unfrozen control cells on the other hand do not show such caps.
This sharp increase in K^+ permeability of the injured cells
supports the recent reports (35, 36) that membrane associated
ATPases are the sites of primary freeze injury. The intrinsic
membrane proteins as mentioned above serve as channels for
ion transport. It is possible, that these proteins, if denatured by freezing injury, can serve as channels of passive
ion transport. Large passive efflux of ions and sugars from
injured cells could also occur through these denatured
intrinsic membrane proteins in the direction of the concentration gradients (vacuole to the extracellular solution). Inactivation of the active transport system, due to changes in the
lipid-protein interaction, produced by the freezing cannot be
ruled out as possible injury.

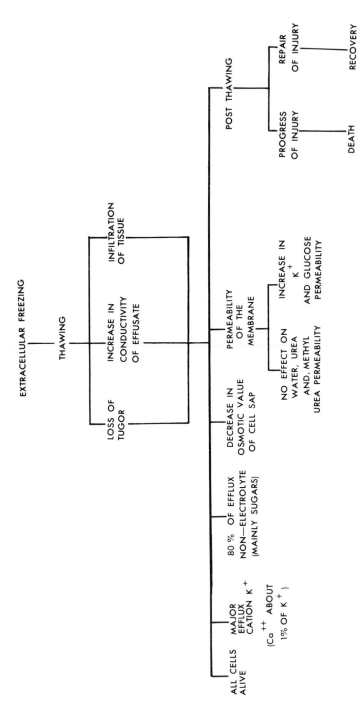

FIGURE 6. A summary of experimental results.

IV. SUMMARY OF EXPERIMENTAL RESULTS

Recent results, presented in this paper, of freeze injury studies with onion bulb cells (35-37) and potato leaf parenchyma (41) are summarized in Figure 6. First signs of injury, by extracellular freezing followed by slow thawing, are the loss of turgor, infiltration of the tissue with water and increase in the conductivity of the effusate. In spite of these signs of injury, all the cells were alive (for discussion of viability see Palta {39}) and no change in the passive permeability of the cell membranes to water, urea and methyl urea could be detected. Passive permeability of the injured cells to K^+, however, was markedly increased. Glucose permeability follows K^+ except that these results are only preliminary. In correspondence to the large efflux of ions and sugars there was a decrease in the osmotic value of the cell sap. Ions form only 20% of the total efflux; the rest was found to be of mainly sugars. Although there was some Ca^{++} present in the efflux, the major efflux was due to K^+ leakage from the cells. During the post-thaw period, depending on the degree of injury, either the cells completely recovered or eventually died. These results show that the semipermeability of the cell membrane is not broken down during initial freezing injury. Light and electron microscopic examinations of the freeze injured cells showed swelling of the protoplasm followed by swelling of chloroplast and mitochondria during progress of injury. When injury advanced further, breakdown of the tonoplast and plasmalemma occurred followed eventually by cell death (coagulation of the protoplasm).

V. PROPOSED HYPOTHESIS FOR A POSSIBLE SEQUENCE OF EVENTS
    LEADING TO CELL DEATH DURING FREEZING INJURY

Based on the results presented in this paper, we propose a possible sequence of events that lead to the inactivation of the active transport system and eventually cell death. This sequence is summarized in Figure 7 and involves the following steps.

(1) Various stresses (e.g., dehydration, mechanical, osmotic and low temperature itself) produced by extracellular freezing and subsequent thawing results in inactivation of the active transport system. This can be due to denaturation of the intrinsic cell membrane proteins and/or changes in the lipid protein interaction. As indicated in Figure 7 by

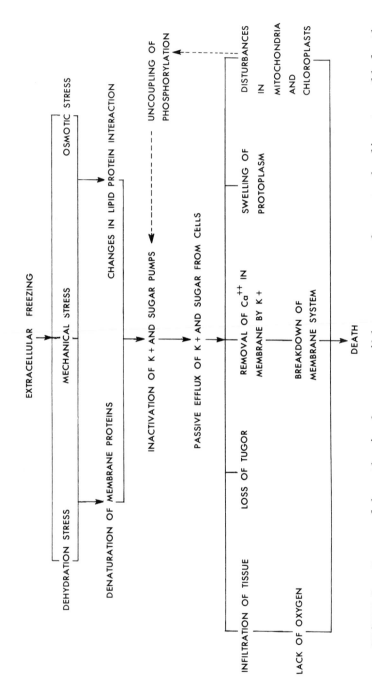

FIGURE 7. *Proposed hypothesis for a possible sequence of events leading to cell death during freezing injury.*

broken lines, disturbances in the mitochondria and chloroplasts can also result in inactivation of the active transport
system by shutting off the energy supply. Recent studies
have shown that chloroplast and mitochondria are damaged by
freezing (13, 15, 16). Our studies, however, show that swelling of the protoplasm and a large efflux of K^+ and sugar can
occur from the cells with normal looking chloroplasts and
mitochondria (Figure 3, c and d). It, therefore, appears that
disturbances in chloroplasts and mitochondria occur at a later
stage during the injury. More direct enzymatic studies are
needed to further this point.

(2) A large passive efflux of ions and sugars results
due to inactivation of active transport system. During thawing as ice melts in the extracellular space, ions and sugars
move down its concentration gradients (vacuole to extracellular solution) as the active transport system is unable to
pump them back into the vacuole. Cell is unable to absorb
water due to osmotic equilibrium between outside solution and
cell sap. This results in infiltration of the tissue with extracellular solution and loss of turgor--i.e., tissue becomes
flaccid. Although water permeability has been found to be insensitive to turgor changes (42) ion transport has been found
to be very sensitive to such changes in the plant cells (3).
Also, potato ATPase has been found to be sensitive to hydrostatic pressure (21). Since turgor loss during the freezing
injury is a result of the loss of ions and sugars from the
cell, its role in freezing injury is only secondary.

(3) A high concentration of K^+ in the extracellular solution leads to the swelling of the protoplasm followed by the
swelling of mitochondria and chloroplasts. This swelling of
the protoplasm can be produced in normal unfrozen plant cells
by exposing them to solutions of KCl or KNO_3. Under these
conditions cap plasmolysis (Figure 2, c) is formed. Swelling
of mitochondria has also been shown to result from the chilling injury (29).

(4) It has been long known that the presence of Ca^{++} is
essential for the stability of the natural membranes (14),
although small yet increasing efflux of Ca^{++} from freeze
injured cell has been found (35, 36). This is probably due to
removal of Ca^{++} from the cell membranes by high concentration
of K^+ present in the extracellular solution. Such a removal
of Ca^{++} causes instability of the membrane structure and finally results in breakdown of the membrane system. Furthermore, more infiltration of the tissue with water results in
lack of oxygen. All these together with lack of energy for
vital functions due to injury to the mitochondria and chloroplasts result in cell death. Electron microscopic observations of the potato leaf parenchyma shows that tonoplast probably breaks down before the plasmalemma does (41) yet it is

sometimes hard to assess the membrane damage from the electron microscopic pictures. Mostly ATPases involved in ion transport have been reported to be located in the plasmalemma (17). It would be more logical to think that active transport system located at the plasma membrane is damaged first. Yet, loading of the vacuole with ions and sugars is thought to involve tonoplast and recently some ATPases have been found to be located in the tonoplast (27). It is therefore possible that the injury is initiated at the tonoplast due to concentration of the solutes in the vacuole during freezing. Further work testing the activity of the enzymes, associated with individual membranes, during freezing injury will provide answers to these questions.

VI.   SUMMARY

The results presented in this paper can be summarized as follows:

(1) The semipermeability of the cell membranes remained intact in spite of freezing injury that resulted in a large efflux of ions and sugars from the cells.

(2) Depending upon the extent of freezing injury, cell damage was reversible or irreversible. During the post-thaw period reversibly damaged cells recovered completely whereas the irreversibly damaged cells ultimately died.

(3) A recovery in the post-thaw period corresponded to decrease in the efflux of ions and sugars and disappearance of infiltration resulted. This indicates the recovery of freezing injury is due to active absorption of ions and sugars by the cell from the extracellular water. An increase in efflux of ions and sugars was found for irreversibly damaged cells.

(4) During the progress of injury a swelling of the protoplasm was found. Similar condition of the protoplasm was found during the electron microscopic observation of freeze injured potato leaf parenchyma cells. It was shown that in spite of this condition of the protoplasm the chloroplast and mitochondria appear normal.

(5) No change in the permeability of the cell membranes to water, urea and methyl urea could be detected in spite of the freeze injury. However, passive transport of K^+ increased significantly through the injured cell membranes.

(6) Based on these experimental results, a hypothesis is proposed for the sequence of events leading to death during

freezing injury. These data can be best explained in terms of the active transport system as the primary site of freezing injury.

ACKNOWLEDGMENT

We would like to acknowledge the facilities provided by Professor E. J. Stadelmann for the light microscopic and membrane permeability work. We also wish to thank the International Potato Center, Lima, Peru and the Rockefeller Foundation for their support.

REFERENCES

1. Asahina, E., *Contrib. Inst. Low Temp. Sci. 10*, 83-126 (1956).
2. Becquerel, P., *Botaniste 34*, 57-74 (1949).
3. Bisson, M. A., and Gutknecht, J., *J. Membrane Biol. 24*, 183 (1975).
4. Buhlert, H., *Landwirt. Jahrb. 35*, 837 (1906).
5. Chen, P. M., Burke, M. J., and Li, P. H., *Bot. Gaz. 137*, 313-317 (1976).
6. Collander, R., *Acta Bot. Fenn. 6*, 1-20 (1930).
7. Collander, R., *Transact. Farad. Soc. 33*, 985-999 (1937).
8. Collander, R., and Bärlund, H., *Acta Bot. Fenn. 11*, 1-114 (1933).
9. Davson, H., and Danielli, J. E., "The Permeability of Natural Membranes." Cambridge Univ. Press, New York (1943).
10. Dexter, S. T., *Plant Physiol. 7*, 721-726 (1932).
11. Dexter, S. T., Tottingham, W. E., and Graber, L. F., *Plant Physiol. 7*, 63-78 (1932).
12. Edward, M. L., and Hall, J. L., *Protoplasma 78*, 321-338 (1973).
13. Garber, M. P., and Steponkus, P. L., *Plant Physiol. 57*, 673-680 (1976).
14. Hansteen-Cranner, B., *Meld. Norges Landbruksh. 2*, 1 (1922).
15. Heber, U., *Plant Physiol. 42*, 1343-1350 (1967).
16. Heber, U., Tyankova, L., and Santarius, K. A., *Biochim. et Biophys. Acta 291*, 23-37 (1973).
17. Hodges, T. K., and Leonard, R. T., *Methods Enzymol. 32*, 392-405 (1974).
18. Iljin, W. S., *Protoplasma 20*, 105-124 (1933).

19. Kacperska-Palacz, A., Dlugokecka, E., Breitenwald, J., and Wcislinska, B., *Biol. Plant. 19*, 10-17 (1977).
20. Kolb, H. A., Läuger, P., and Bamberg, E., *J. Membrane Biol. 20*, 133-154 (1975).
21. Kuiper, P. J. C., *Biochim. et Biophys. Acta 250*, 443-445 (1971).
22. Leigh, R. A., and Wyn-Jones, R. G., *J. Exp. Bot. 26*, 508-520 (1975).
23. Levitt, J., "Responses of Plants to Environmental Stresses." Academic Press, New York (1972).
24. Levitt, J., and Scarth, G. W., *Can. J. Res. C 14*, 267-284 (1936).
25. Levitt, J., and Siminovitch D., *Can. J. Res. C 18*, 550-561 (1940).
26. Li, P. H., and Palta, J., *In* "Recent Advances in Plant Cold Hardiness and Freezing Stress: Mechanisms and Crop Implications" (P. H. Li and A. Sakai, eds.). Academic Press, New York (1978).
27. Lin, W., Wagner, G. J., Siegelman, H. W., and Hind, J., *Biochim. Biophys. Acta 465*, 110-117 (1977).
28. Lindberg, S., *Physiol. Plant. 36*, 139-144 (1976).
29. Lyons, J. M., Wheaton, T. A., and Pratt, H. K., *Plant Physiol. 39*, 262-268 (1964).
30. Maximov, N. A., *Ber. Deutsch. Bot. Gesell. 30*, 52-65, 293-305, 504-516 (1912).
31. Mazur, P., *Ann. Rev. Plant Physiol. 20*, 419-448 (1969).
32. Molisch, H., "Untersuchungen über das Erfrieren der Pflanzen," pp. 1-73. Fisher, Jena (1896).
33. Mueller, P., and Rudin, D. O., *Biochem. Biophys. Res. Commun. 26*, 398-404 (1967).
34. Osterhout, W. J. V., *Science 40*, 488-491 (1914).
35. Palta, J. P., Levitt, J., and Stadelmann, E. J., *Plant Physiol. 60*, 393-397 (1977).
36. Palta, J. P., Levitt, J., and Stadelmann, E. J., *Plant Physiol. 60*, 398-401 (1977).
37. Palta, J. P., Levitt, J., and Stadelmann, E. J., *Cryobiol. 14*, 614-619 (1977).
38. Palta, J. P., Levitt, J., Stadelmann, E. J., and Burke, M. J., *Physiol. Plant. 41*, 273-279 (1977).
39. Palta, J. P., Levitt, J., and Stadelmann, E. J., *Cryobiol.*, in press (1978).
40. Palta, J. P., and Li, P. H., *Science*, in preparation (1978).
41. Palta, J. P., and Li, P. H., *Plant Physiol.*, in preparation (1978).
42. Palta, J. P., and Stadelmann, E. J., *J. Membrane Biol. 33*, 231-247 (1977).
43. Pfeffer, W. C., "Osmotische Untersuchungen. Studien zur Zellemechanik." Engelmann, Leipzig (1877).

44. Pressman, B. C., *Proc. Nat. Acad. Sci. (U.S.A.) 53*, 1076-1083 (1965).
45. Prillieux, E., *Ann. Sci. Nat. Paris (Ser. 5) 12*, 125-134 (1869).
46. Proebsting, E. L., Jr., *Proc. Amer. Soc. Hort. Sci. 74*, 144-153 (1959).
47. Racker, E., *Biochem. Soc. Trans. 3*, 785-802 (1975).
48. Scarth, G. W., Levitt, J., and Siminovitch, D., *Cold Spg. Harbor. Symp. Quant. Biol. 8*, 102-109 (1940).
49. Siminovitch, D., and Scarth, G. W., *Can. J. Res. 16*, 467-484 (1938).
50. Singer, S. J., "Cell Membranes," pp. 35-44. H. P. Publishing Co., New York (1975).
51. Singer, S. J., and Nicolson, G. L., *Science 175*, 720 (1972).
52. Stadelmann, E. J., *Methods in Cell Physiol. 2*, 143-216 (1966).
53. Steponkus, P. L., and Weist, S. C., *In* "Recent Advances in Plant Cold Hardiness and Freezing Stress: Mechanisms and Crop Implications" (P. H. Li and A. Sakai, eds.). Academic Press, New York (1978).
54. Sukumaran, N. P., and Weiser, C. J., *Plant Physiol. 50*, 564-567 (1972).
55. Traeuble, H., *J. Membrane Biol. 4*, 193-208 (1971).
56. Wenner, C. E., and Hackney, J. H., *Biochemistry 8*, 930-938 (1969).
57. Weigand, K. M., *Bot. Gaz. 41*, 373-424 (1906).

PHOSPHOLIPID DEGRADATION AND ITS CONTROL
DURING FREEZING OF PLANT CELLS

S. Yoshida

The Institute of Low Temperature Science
Hokkaido University
Sapporo

I. INTRODUCTION

One of the most remarkable characteristics of certain
perennial plants grown in cold climates is the ability to
survive a deep freezing in mid-winter. Twigs from some hardy
boreal trees such as willows, poplars, and white birch
survive freezing even at liquid nitrogen temperatures throughout winter. However, they may be severely injured by freezing
in a relatively higher temperature range when they are not in
a fully hardened state. Many questions, however, remain to
be solved as to the mechanisms of cold hardiness changes and
freezing injury of plant cells. One of the difficulties in
approaching these problems is probably due to the lack of
information as to any sort of biochemical changes to be
initiated in freezing cells at sublethal temperatures. Most
of the recent works in this area have been concerned with
the possible involvement of membrane damage in freezing
injury of plant cells (2, 3, 4, 11, 12, 18). In this study,
an enzymatic degradation of phospholipids during freezing of
plant cells at sublethal temperatures was mainly investigated
to see the primary biochemical events associated with freezing
injury. Special attention was focused on the possible involvement of an alteration of the regulatory system of the
phospholipid splitting enzyme due to freezing.

117

II.   MATERIALS AND METHODS

A.   *Plant Materials and Freezing Procedures*

As experimental materials, the living bark tissues of
poplar (*Poplus euramericana* cv. *gelrica*) and black locust
tree (*Robinia pseudoacacia*) were used.   The living bark tissues were cut into small pieces and then frozen in a test
tube (1.3 x 18 cm) at -5° C.   After exposure to -5° C for
2 hr, the frozen tissues were cooled in 5° C steps at hourly
intervals to successively lower temperatures to -30° C.   Some
tissues which were frozen to -30° C were further cooled to
50° C and then immersed in liquid nitrogen.   The frozen tissues were thawed in air at 0° C.   After standing at 0° C for
1 hr, they were placed at room temperature for more than 3
days.   Freezing injury was then evaluated visually.   Browning
of the tissues was used as the criterion for rating injury.
In some experiments injury was determined from the extent
of release of compounds reactive with $FeCl_3$ from the frozen
thawed tissues of poplar.   For this test, 1 g of the frozen
thawed tissue pieces was leached in 5 ml of distilled water
at 27° C for 4 hr.   Then 0.1 ml of 10% $FeCl_3$ solution was
added to the leaching solution.   Dark greenish color developed, depending on the degree of injury.   The absorbance was
read at 620 nm against the leaching solution from unfrozen
samples.   The results obtained by this method were found
to be comparable with those obtained by the visual browning
test.

B.   *Extraction and Analysis of Phospholipids*

Lipids were extracted both from frozen and thawed samples.
In the frozen samples, grinding was performed in a cold room
at -10° C.   To minimize the degradation of phospholipids by
phospholipases during grinding, isopropanol cooled to -10° C
was added.   After grinding, chloroform methanol (2:1, v/v)
cooled at -10° C was added and then lipids were extracted for
1 hr at room temperature.   Extraction was repeated twice more
with the same solvent, and the combined lipid extracts were
subjected fo Folch's procedure (1) to remove non-lipid contaminants.   Aliquots of the purified lipid samples dissolved
in a known amount of chloroform methanol (2:1, v/v) were
loaded on a Silica Gel H plate (5 x 20 cm) and were developed
with chloroform-methanol-acetic acid (65:25:8, v/v/v) at 25°
C.   After exposing the plate to iodine vapour, the areas corresponding to each standard phospholipid were scraped into

test tubes and directly heated with 0.5 ml of 70% perchloric acid. The lipid phosphorus was determined according to Marinetti's method.

C. *Preparation of Microsomal Fraction*

The living bark tissues of black locust tree were cut into small pieces and ground in a motor-driven mortar and pestle assembly for 2 min with washed sea sand at 0° C. The grinding medium consisted of 0.75 M sorbitol, 200 mM tris-HCl buffer solution, pH 7.8, 10 mM EGTA-K and 1% Ficoll. Five ml of the grinding medium, 0.3 g of Polyclar AT and 0.6 g of the abrasive were used per gram fresh weight of tissues. The brei was squeezed through one layer of gauze and Miracloth and then successively centrifuged at 200 g for 5 min, 1,500 g for 10 min, 14,000 for 15 min and 105,000 for 90 min. Microsomal pellets were resuspended in 2.5 mM tris-HCl buffer solution, pH 7.6, containing 1 mM EGTA and then recentrifuged. The EGTA-washed microsomal pellets were finally suspended in 2.5 mM tris-HCl buffer solution, pH 7.6. The protein concentration was adjusted to around 1.6 mg per ml.

D. *Procedure for the Determination of in situ Activity of Phospholipase D in Microsomes*

The reaction mixture for the determination of *in situ* activity of phospholipase D consisted of 0.5 ml of the EGTA-washed microsomal suspension, 0.3 ml of 200 mM of buffer solutions in different pH and divalent cations as required in a final volume of 1 ml. Incubation was carried out at 25° C for 5 or 10 min and then the reaction was terminated by addition of 0.2 ml of 1 N perchloric acid solution. The liberated choline in the supernatant was measured according to I_2-IK method (6).

III. RESULTS

A. *Phospholipid Degradation in vivo and Freezing Injury in Living Bark Tissues of Poplar*

In the less hardy tissues of poplar collected on October 4, no injury was observed after freezing at -5° C for 20 hr, whereas they suffered serious injury by freezing below -10° C.

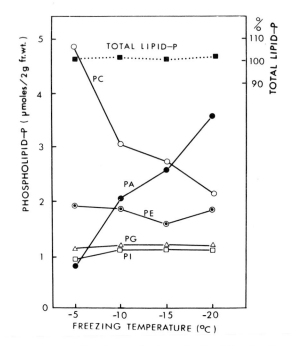

FIGURE 1.   Degradation of phospholipids in less hardy bark tissues of poplar during freezing.   PC--phosphatidyl-choline; PE--phosphatidylethanolamine; PG--phosphatidyl-glycerol; PI--phosphatidylinositol; PA--phosphatidic acid.

As presented in Figure 1, in the tissues frozen below -10° C for 20 hr, a remarkable decrease was observed in phosphatidyl-choline, while little or no change was observed in phospha-tidylinositol and glycerol. A slight decrease was observed in phosphatidylethanolamine at -15° C. The decrease in phosphatidylcholine was always balanced by the corresponding increase in phosphatidic acid. The total amount of phospho-lipids, however, showed no significant change during freezing at any temperature. In less hardy tissues from the twigs collected in late September, the freezing injury increased with the length of time held at -10° C. The time course of the degradation of phosphatidylcholine in the tissues held at -10° C is presented in Figure 2. Phosphatidylcholine was degraded with the length of time held at -10° C, which was also accompanied by a concomitant formation of phosphatidic acid. As can be clearly seen in Figure 2, the degradation of phosphatidylcholine in the early process of freezing was noted to be initially associated with the extent of freezing injury in the tissues. To strengthen this notion, experiments

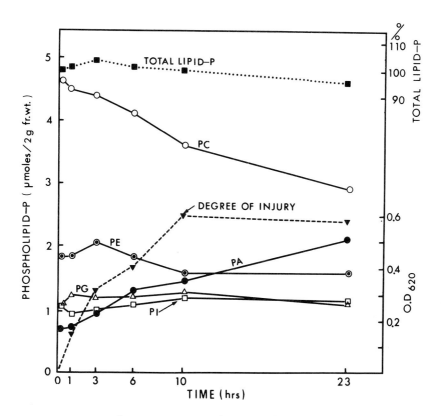

FIGURE 2. *Time course of degradation in phosphatidyl-choline in less hardy bark tissues of poplar during freezing at -10° C. Freezing injury was determined from the extent of release of compounds reactive with ferric chloride from the frozen-thawed tissues and expressed as absorbance at 260 nm.*

were performed with hardy tissues collected on November 6. The tissues survived slow freezing to any test temperature down to -30° C or even immersion in liquid nitrogen after prefreezing to -50° C. In these frozen tissues, little or no change in phospholipid components was detected. Even after incubation at 27° C for 2 hr following thawing, almost all of the phospholipid components still remained unchanged (Figure 3). On the other hand, the tissues frozen by a direct immersion in liquid nitrogen from room temperature (-196 RF in Figure 3) were killed and browning of the tissues was observed immediately after thawing. In these rapidly frozen tissues, however, a drastic degradation of phospholipids, including

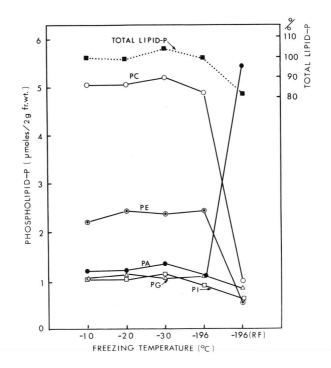

FIGURE 3. *Changes in phospholipids in hardy bark tissues of poplar after freeze-thawing. The tissues resisted slow freezing to the temperature of liquid nitrogen. -196(RF): Tissues were directly immersed in liquid nitrogen from room temperature.*

phosphatidylinositol and ethanolamine, and a slight decrease in the total amount of phospholipids was observed after thawing. To clarify the subfreezing temperatures under which phospholipid degradation proceeds, the tissues previously damaged by an immersion in liquid nitrogen from room temperatures were directly transferred to various temperatures ranging from 0 to 30° C following removal from liquid nitrogen. After standing at these temperatures for 20 hr, phospholipid components were analysed from frozen or thawed samples. As shown in Figure 4, a significant degradation of phospholipids was observed above -30° C. More than 80% of phosphatidylcholine and ethanolamine were degraded into phosphatidic acid below -10° C. These results indicate that phospholipid degradation may proceed even at subfreezing temperatures above -30° C when the tissues sustain serious injury, but such a degradation may be slowed down at below -30° C. Also, it was clearly demonstrated that a decrease in phospholipid

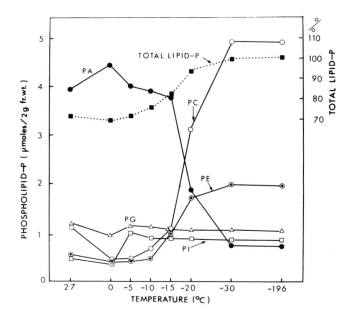

FIGURE 4. Temperatures causing phospholipid degradation in frozen poplar bark tissues. Tissues which resisted to -5° C were first killed by direct immersion in liquid nitrogen and were then transferred directly to various temperatures. After 20 hr, lipids were extracted from frozen tissues. Some of the frozen tissues were thawed at 0° C and then incubated at 27° C for one hr before being extracted..

components during freezing resulted in a concomitant increase in phosphatidic acid. Thus it may be postulated that the change in phospholipid components during freezing is mainly caused by phospholipid degradation catalysed by phospholipipase D.

B. *In situ Degradation of Phospholipids and its Control in Microsomal Membranes*

To better understand the mechanism of initiating *in vivo* reaction of phospholipase D caused by freezing at sublethal temperatures, several experiments were conducted with EGTA-washed microsomes, based on the *in situ* reaction of membrane-bound phospholipase D, endogenous phosphatidylcholine as the substrate. In these experiments, living bark tissues of black locust trees varied in cold hardiness were mainly used for the

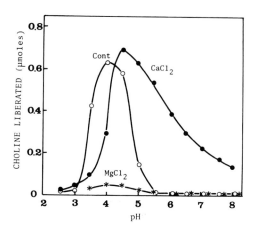

FIGURE 5.   pH-Activity profiles of in situ reaction of phospholipase D in microsome with presence and absence of divalent cations.   Microsome was prepared from hardy tissues. CaCl₂ or MgCl₂ was added in final concentration of 50 mM.

experimental materials.   The tissues were very easy to grind and depleted in phenolic compounds which sometimes made it difficult to prepare intact enzymes and cell organellas.

Figure 5 shows the effect of pH upon the *in situ* activity of phospholipase D in microsomes in the presence or absence of divalent cations.   In these experiments microsomes used were prepared from hardy tissues.   When pH was reduced to near 4, a remarkable degradation of endogenous phosphatidylcholine was observed without addition of $CaCl_2$.   However, little or no degradation was proceeded above pH 5.5.   The *in situ* reaction without addition of $CaCl_2$ was completely inhibited by the addition of EGTA or EDTA.   This fact may show that Ca is requisite to the *in situ* activity of phospholipase D in microsomes and the required amount of Ca may have been tightly bound to the enzyme or the EGTA-washed microsomes. According to atomic absorption spectrophotometric analysis of the EGTA-washed microsomes, nearly 10 nmoles of Ca was detected on mg basis.   When 50 mM of $CaCl_2$ was added to the reaction mixture, the optimal pH was shifted 0.5 unit toward the neutral pH and a remarkable increase in activity was also noted near neutral pH region.   Considerable reduction in activity was observed below pH 4.   In contrast with Ca, Mg showed a striking inhibition for the *in situ* reaction of phospholipase D in all pH tested.   Thus, there seems to exist a marked contrast in the effects on the *in situ* activity of phospholipase D between those cations.

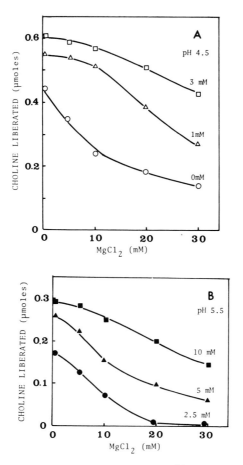

FIGURE 6.   Inhibitory effect of Mg^{2+} on in situ reaction
of phospholipase D depending on pH and Ca concentration.   The
reactions were performed at pH 4.5 (A) and pH 5.5 (B).

To clarify the interaction between these divalent cations
in the *in situ* reaction of phospholipase D, the inhibitory
effect of Mg was examined in the presence of varied amounts of
$CaCl_2$ as a function of pH.   The representative data are shown
in Figure 6.   The inhibitory effect of Mg was observed to be
much reduced either at lower pH or in the presence of a higher
amount of Ca.   The data represented in Figure 6 when plotted
in Lineweaver-burk double reciprocal plot are shown in Figure
7.   As clearly seen in this figure, the inhibition pattern
showed a mutual competitive inhibition between these cations.
Interestingly, nearly the same pattern was also observed
between pH and Ca just as observed between Mg and Ca.   From

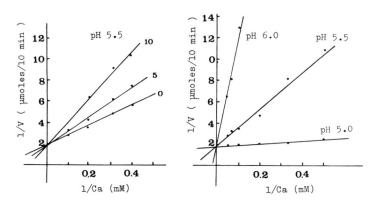

FIGURE 7.   Plot of the reciprocal of initial velocity of
in situ reaction of phospholipase D at pH 5.5 (V) versus the
reciprocal of Ca concentration as variable concentration of
Mg (left) and as variable pH (right).

these results, it is likely that Mg has a role on the conformation of phospholipase D just like increasing in pH.
The calculated apparent dissociation constants of a hypothetical enzyme-cation complexes as a function of pH are summarized in Table I.   The lower the pH, the smaller the dissociation constant for Ca was observed.   The inverse relationship was noted with Mg (Table II).   A higher concentration
of Ca produced a marked increase in the apparent dissociation
constant for Mg, when comparison was made at the same pH.
Microsomes from less hardy tissues (sampled in late spring)
gave much smaller dissociation constant for Ca and much
greater dissociation constant for Mg, when comparison was made
at the same pH.   Thus, it appeared that the affinity of phospholipase D in membranes for both cations was markedly dependent on the hardiness of the tissues from which microsomes were isolated.   Therefore, there seems to exist a great
difference in the properties of membrane-bound phospholipase
D between hardy and less hardy microsomes.
      The question now arises as to how freezing serves to alter
membrane properties and as to how the altered membrane properties confer the activation of membrane-bound phospholipase
D *in situ*.   To give some insights into this question, experiments were focused on the effects of freezing on properties
of membrane-bound phospholipase D in microsomes.   As seen in
Figure 8, freeze-thawing of less hardy microsomes at -30° C
gave a quite different mode of pH-activity profile in *in situ*
reaction of phospholipase D.   No great difference in the pH-
activity profile of *in situ* reaction of phospholipase D was

TABLE I.   *Apparent Dissociation Constants for* Ca^{2+}

pH	Ms from hardy tissues (mM)	Ms from less hardy tissues (mM)
4.5	0.17	–
5.0	2.04	–
5.5	13.50	0.19
6.0	70.00	–
7.0	–	4.78

TABLE II.   *Apparent Dissociation Constants for* Mg^{2+}

pH	Ca^{2+} conc. (mM)	Ms from hardy tissues (mM)	Ms from less hardy tissues (mM)
4.5	0	16.1	–
5.0	0	2.9	–
	1	8.3	–
	3	8.9	–
5.5	0	0.9	15.5
	2.5	4.6	–
	5.0	9.1	–
	10.0	21.8	–
6.0	0	$\overset{\cdot}{=}0$	–
	20	$\overset{\cdot\cdot}{3}0.8$	–
	50	140.0	–

observed in hardy microsomes. Freeze-thawing of less hardy
microsomes resulted in both shift of the optimal pH and much
increase in activity at neutral pH region.   Thus, it is
likely that microsomal membranes from different tissues varied in hardiness may respond to freezing stress in a different
manner.

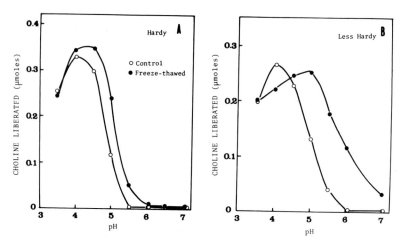

FIGURE 8.   Effect of freezing on pH dependency of in situ
activity of phospholipase D in microsomes. A--Microsome from
hardy tissues, B--Microsome from less hardy tissues. Microsomal suspensions were freeze-thawed at -30° C for 16 hr.

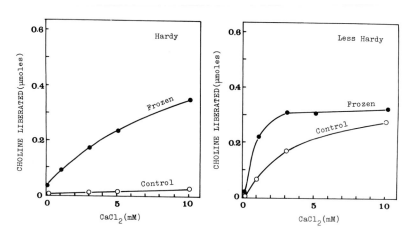

FIGURE 9.   Alteration of Ca dependency of in situ reaction
of phospholipase D as affected by freeze-thawing at -30° C for
16 hr. Left: Microsome from hardy tissues; Right: Microsome from less hardy tissues. Reactions were performed at
pH 7.

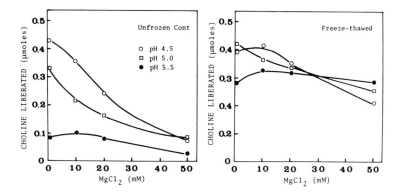

FIGURE 10   Alteration in the inhibitory effect of Mg upon in situ reaction of phospholipase D in microsomes from less hardy tissues as affected by freeze-thawing at -30° C.

Experiments were further conducted to clarify the changes in sensitivities of *in situ* activity of phospholipase D to divalent cations after freeze-thawing at -30° C by using both hardy and less hardy microsomes. As presented in Figure 9, freeze-thawing of microsomes at -30° C produced much increase in the Ca sensitivity in both samples at neutral pH.   The increase in the Ca sensitivity after freeze-thawing, however, was much greater in less hardy microsome than hardy microsome. In the case of freeze-thawed hardy microsome, the *in situ* activity of phospholipase D increased proportionally as increase in Ca concentration up to 10 mM, while in the case of freeze-thawed less hardy microsome the activity reach a plateau at 3 mM of Ca concentration.

Freeze-thawing of microsomes at -30° C also resulted in the decrease or loss of the inhibitory action of Mg upon the *in situ* reaction of phospholipase D depending on the hardiness of the samples. As presented in Figure 10, freeze-thawing of less hardy microsome resulted in a drastic loss of the inhibitory action of Mg independently of pH, whereas relatively small changes were observed in freeze-thawed hardy microsome. From these results, it is probable that the regulatory system operating in membrane-bound phospholipase D is susceptible to freeze-thawing, especially in less hardy microsome.

To check the effect of freezing temperatures on *in situ* reaction of phospholipase D, microsomal suspensions prepared from both hardy and less hardy tissues were frozen at various temperatures for 16 hr. Choline liberation was followed both immediately before thawing and after incubation at 25° C for 10 min in a buffer solution, pH 5.5, following thawing. Unexpectedly, when microsomal suspensions in 2.5 mM tris-HCl

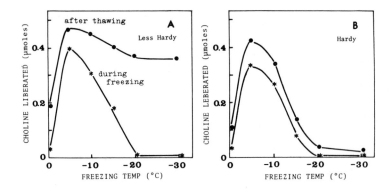

FIGURE 11. In situ degradation of phosphatidylcholine in microsomes during freezing and after thawing at different temperatures. Liberation of choline was followed immediately before thawing and after incubation at 25° C for 10 min in buffer solution, pH 5.5, following thawing. A (left)--Microsome from less hardy tissues; B (right)--Microsome from hardy tissues.

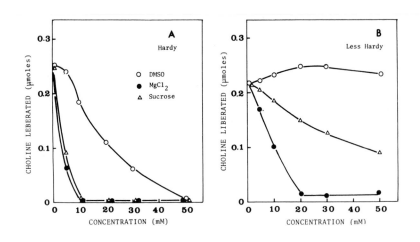

FIGURE 12. Protective effects of sucrose, DMSO and Mg on in situ degradation of phosphatidylcholine during freeze-thawing cycle. Microsomal suspensions from hardy (left) and less hardy (right) tissues were frozen at -10° C for 16 hr and incubated at 25° C for 10 min in buffer solution, pH 5.5, following thawing.

TABLE III.   *Effect of pH on Degradation of Phosphatidylcholine during Freezing at -10° C for 18 hr*

pH	Choline liberated (umoles)
7.8 (Tris-Hcl)	0.31
6.5 (Na-Acetate)	0.49
5.5 (Na-Acetate)	0.48
4.5 (Na-Acetate)	0.41

buffer solution, pH 7.6, were frozen at -5° C or -10° C for 16 hr, marked degradation of phosphatidylcholine occurred during freezing regardless of the hardiness of the materials (Figure 11). Below -20° C or at 0° C, however, little or no degradation was observed in both samples. When frozen specimens were further incubated at 25° C in a buffer solution, pH 5.5, following thawing, accelerated degradation was observed in less hardy microsome independently of the freezing temperatures. On the other hand, in hardy microsome, little or no accelerated degradation was observed during the post-incubation following thawing below -15° C. As seen in Table 3, the *in situ* degradation of phosphatidylcholine during freezing at -10° C was relatively less dependent on the pH of the microsomal suspension. Therefore, freezing of microsomes at relatively higher temperatures may have caused nearly the same conformational alteration on membrane-bound phospholipase D as decreasing in pH and increasing in Ca concentration.

In order to clarify the effect of protective substances upon *in situ* reaction of phospholipase D in microsomes, microsomal suspensions were frozen at -10° C for 16 hr with indicated concentration of protective substances and then incubated at 25° C in a buffer solution, pH 5.5, for 10 min after thawing. As indicated in Figure 12, very low concentration of sucrose as well as $MgCl_2$ exhibited a marked protection or an inhibition against *in situ* degradation of phosphatidylcholine during the freeze-thawing cycle. In hardy microsome, less than 10 mM of sucrose protected completely the *in situ* degradation of phosphatidyl choline. DMSO (dimethylsulfoxide), however, was less effective than sucrose. In less hardy microsome, however, the protective effects of sucrose and DMSO decreased remarkably. Even 50 mM of sucrose exhibited only 50% protection and DMSO showed little or no

protection in the whole range of concentrations tested.
The inhibitory effects of Mg upon *in situ* degradation of
phosphatidyl choline during freezing was also reduced in less
hardy microsome. From these results, it is reasonable to
think that freezing at relatively higher temperatures may
cause either an alteration of binding mode of the enzyme to
membranes or an alteration of enzyme conformation as well
through deleterious changes in membrane molecules. Sucrose
is thought to be capable of protecting against such a change
in membranes and/or a change in enzyme conformation caused by
freezing.

IV. DISCUSSION

     In less hardy living bark tissues of poplar as well as
black locust tree, a striking enzymatic degradation of phosphatidylcholine into phosphatidic acid was observed during
freezing at sublethal temperatures. On thawing, the enzymatic
reaction was drastically accelerated and resulted in degradation of the other phospholipids such as phosphatidylethanolamine and inositol. This degradation is clearly catalysed
by phospholipase D. This is the first evidence to suggest
that an irreversible biochemical change in membranes does
occur in frost susceptible plant cells during freezing at sublethal temperatures.
     Phospholipase D is well known to be widely distributed in
plants (7, 13, 15). However, questions remain to be solved
as to the physiological functions and as to the regulatory
system *in vivo*. Phospholipase D in living bark tissues of
black locust tree was found to be localized both in particu-
late and soluble cell fractions. Although the highest percentage of the total activity was distributed in soluble cell
fraction, the highest specific activity was associated with
microsomal fraction from wintering hardy tissues (unpublished
data). After linear sucrose density gradient of microsomal
fraction, phospholipase D was associated with several membranes, such as endoplasmic reticulum, Golgi apparatus and
other unidentified membranes with higher densities. Subjecting of microsomes to sonication, alternate freeze-thawing,
and washing in 0.5 M KCl solution, all failed to release
phospholipase D from microsomal membranes. In this regard,
the enzyme is likely to be tightly bound to the membranes.
     According to Matile *et al.* (9, 10) and other people (8)
provacuoles are known to be recovered together with mitochondria in differential centrifutation. Two distinct peaks
of phospholipase D activity were observed on linear sucrose

density gradient of mitochondrial fraction from wintering hardy tissues of black locust trees (unpublished data). One of the peaks with lower density (4, 8) was observed to coincide with the location profiles of NADH Cyt C reductase and acid phosphatase. Thus, the lighter membrane(s) may have been derived from vacuoles with smaller dimensions which are reported to exist in wintering hardy cells of black locust trees. Accordingly, soluble phospholipase D in living bark tissues of black locust trees is likely to be located in vacuoles.

In the present study, it was suggested that membrane-bound phospholipase D is regulated by binding either of Ca or Mg in a competitive manner. Mg has a strong inhibitory effect upon *in situ* reaction of phospholipase D in microsomal membranes from living bark tissues of black locust trees. The inhibitory action of Mg was also demonstrated in *in vitro* reaction system of phospholipase D with the presence of ethyl ether as the activator, when microsomal suspension and egg lecithin were used as enzyme source and exogenous substrate, respectively (unpublished data). On the other hand, little or no inhibitory action of Mg was observed in the *in vitro* reaction system, when 2 mM of SDS was replaced for ethyl ether as the activator (unpublished data). According to Quarles and Dawson (14), the conformation of phospholipase D is affected by anionic detergents such as SDS, resulting in marked shift of the optimal pH toward neutral region in *in vitro* reaction. Either the altered conformation caused by a strong absorption of enzyme on highly negatively charged surface of substrate or the dissociation of enzyme into subunits caused by the detergent may not allow the binding of Mg at the specific binding site(s) or otherwise nullify the comformational effect of Mg. In these regards, the regulatory effect of divalent cations observed in *in situ* reaction of phospholipase D might be ascibable to the direct interaction between those cations and enzyme.

According to Heller *et al.* (5), highly purified soluble phospholipase D from peanut seeds has a molecular weight of 200,000, which dissociates into smaller subunits with molecular weights of 50,000 under the presence of SDS or urea. From this fact, it may be speculated that phospholipase D consists both of catalytic and regulatory subunits. Binding of Ca at the regulatory site may be a prerequisite for the activity and a competitive binding of Ca and Mg at the regulatory subunit(s) may bring about a great change in the conformation of the enzyme molecule.

As the hardiness changes, the regulatory properties of membrane-bound phospholipase D were also changed. The apparent dissociation constant of hypothetical enzyme-Ca complex decreased as hardiness decreased, when a comparison was made at the same pH. The inverse relation was obtained in the apparent dissociation constant for Mg. Freeze-thawing of microsomes, especially in less hardy microsomes, caused a drastic increase in the Ca sensitivity and complete loss of the control action of Mg in *in situ* reaction of phospholipase D. Thus, the regulatory properties of the membrane-bound phospholipase D seem to be also dependent on the nature of the membranes and the interaction between the enzyme and membranes as well. In suggesting this notion, the *in situ* activity of phospholipase D in freeze-thawed microsomal membranes became more resistant to the digestive reaction by pronase and trypsin (unpublished data). This fact may imply that freezing of membranes causes a new interaction between phospholipase D and hydrophobic interior of the membrane architecture, resulting in a protection of enzyme against digestive reaction of proteases.

The alteration of the regulatory properties as affected by freezing was more effectively protected by lower concentration of sucrose in hardy microsome than in less hardy microsome. Accordingly, some qualitative changes in membranes affecting the regulatory properties of phospholipase D are likely to be involved in the mechanism of cold hardiness of plant cells, as demonstrated by Garber and Steponkus (3) in the thylakoid membranes from spinach chloroplasts.

According to our previous reports (16, 17, 19), changes in total phospholipid content *per se*, together with changes in lipid classes in membranes, such as endoplasmic reticulum, tonoplast and Golgi apparatus were suggested to have an important role in endowing resistance to cellular membranes against freezing stress. This fact, also, can be taken as evidence that membrane properties are altered as the cold hardiness changes. The regulatory system of phospholipase D may positively take part in the accumulation and depletion of phospholipids, mainly phosphatidylcholine and ethanolamine, per unit area of membranes.

Further experiments clearly are needed to understand entirely the regulatory system of phospholipase D *in vivo*. At the moment it seems possible to conclude that freezing of less hardy plant cells brings about primarily a drastic alteration of the regulatory properties of membrane-bound phospholipase D, and/or simultaneous disturbance of the intracellular compartmentations of Ca, pH and soluble phospholipase D. As a result, the deleterious degradation of phospholipids takes place in membranes. In this regard, it might be considered that the nature of tonoplast as well as the other membranes, such as plasma membranes has an important role for cold hardiness and freezing injury of plant cells.

REFERENCES

1. Folch, J., Lee, M., and Sloane-Stanley, G. H., *J. Biol. Chem. 26*, 497 (1957).
2. Garber, M. P., and Steponkus, P. L., *Plant Physiol. 57*, 673 (1976).
3. Garber, M. P., and Steponkus, P. L., *Plant Physiol. 57*. 681 (1976).
4. Heber, U. W., Tyankova, L., and Santarius, K. A., *Biochim. Biophys. Acta 291*, 23 (1973).
5. Heller, M., Mozes, N., Peri(Abramovitz), I., and Mozes, E., *Biochim. Biophys. Acta 369*, 397 (1974).
6. Kates, M., and Sastry, P. S., *Methods in Enzymology 14*, 197 (1969).
7. Kates, M., *Can. J. Biochem. Physiol. 32*, 571 (1954).
8. Madyastha, K. M., Ridgway, J. E., Dwyer, J. G., and Coscia, C. J., *J. Cell. Biol. 72*, 302 (1977).
9. Matile, P. H., and Moor, H., *Planta (Berl.) 80*, 159 (1968).
10. Matile, P. H., and Wiemken, A., *Methods in Enzymology 31*, 572 (1974).
11. Palta, J. P., Levitt, J., and Stadelmann, E. J., *Plant Physiol. 60*, 393 (1977).
12. Palta, J. P., Levitt, J., and Stadelmann, E. J., *Plant Physiol. 60*, 398 (1977).
13. Quarles, R. H., and Dawson, R. M. C., *Biochem. J. 112*, 787 (1969).
14. Quarles, R. H., and Dawson, R. M. C., *Biochem. J. 122*, 795 (1969).
15. Roughan, P. G., and Slack, C. R., *Biochim. Biophys. Acta 431*, 86 (1976).
16. Yoshida, S., *Plant Physiol. 57*, 710 (1976).
17. Yoshida, S., *Inst. Low Temp. Sci. B. 18*, 1 (1974).
18. Yoshida, S., and Sakai, A., *Plant Physiol. 53*, 509 (1974).
19. Yoshida, S., and Sakai, A., *Plant and Cell Physiol. 14*, 353 (1973).

Part III
Acclimation

MECHANISM OF COLD ACCLIMATION
IN HERBACEOUS PLANTS

A. Kacperska-Palacz

Institute of Botany
University of Warsaw
Warszawa

Herbaceous plants, contrary to the woody perennials,
do not develop a true dormancy during the late summer. Their
tissues maintain the growth capability throughout winter season, despite the environmental conditions imposing the cessation of growth rate. Simultaneously, the unfavourable environmental conditions bring a succession of strains into the
cell, e.g. water potential decrease, metabolic lesions, etc.
The questions arises of how these events correspond with the
frost tolerance changes and whether the term "cold acclimation" should be used in herbaceous plants for description of
the seasonal transition from tender to the frost hardy/
resistant conditions only.

It is commonly known that frost hardiness of herbaceous
plants, such as alfalfa, clover, rape, cabbage, wheat, and
turf grasses is limited; while not acclimated they can withstand only a few degrees of frost. In species capable of
acclimation the above-ground tissues may survive the temperature of about -20 to -25° C, if previously exposed to the
proper hardening conditions. There is a concordance in the
literature that low temperature and light are indispensable
factors in cold acclimation of herbaceous plants. On the
other hand, there is an inconsistency concerning temperature
and light requirements for elicitation of maximum hardines,
e.g. in some cases freezing temperature is used during the
hardening procedure, in other ones it is omitted. The inconsistency is often due to the variety of tissues as well as
to different experimental conditions and techniques used for
injury evaluation in different laboratories. Thus, characterization of factors influencing hardiness in one model
plant, description of its differentiated responses to the

hardening factors as well as comparison of the obtained data with the results gathered in the other laboratories for other plants, were thought to be a promising way for arriving at some general conclusions. The winter rape plants (*Brassica napus*, var. *oleifera*) were chosen by us as a model system for studies on mechanisms of cold acclimation in herbaceous plants.

I.   TEMPERATURE REQUIREMENTS

It has been shown (50) that frost hardening of winter rape plants occurs in three stages, related to the temperature conditions:

The first stage may be induced by lowering the temperature to 5° C or 2° C and results in the increase of frost tolerance by a few (4 or 5) degrees above the initial level which is -5° C for the winter rape leaves.

The second one relies on the appearance of subfreezing temperature (0° C to -2 or -3° C) and does not occur under field conditions if minimum air temperature does not fall below 0° C. This stage may result in the achievement of the maximum frost tolerance by the turgid tissue if the temperature conditions foregoing the hardening process were favourable for plant growth (50).

The third stage of frost hardening may be coincidental with the second one and depends on the occurence of prolonged frost inducing cell dehydration.

Similar requirements for temperature for hardening seem to be observed in alfalfa, red clover, and sweet clover plants (5), and also in cabbage leaves (32). Exposure to diurnal freezing temperature also increased the hardiness of winter wheat seedlings (3, 56). However, experiments of Grenier *et al*. (17) indicate that winter wheat plants may attain maximum hardiness under low (1° C) constant temperature.

II.   LIGHT REQUIREMENTS

It seems that the first stage of hardening may occur in darkness only when the cold-treated seedlings are supplied with sugar (sucrose) as an energy source (57) or they are not

depleted of the storage substance containing tissue--endo-
sperm (2) or cotyledons (Kacperska-Palacz and Forycka, unpublished).

   Shortening of photoperiod during the first stage of
hardening of winter rape plants at 5° C was found to decrease
the level of frost tolerance from $T_{k50} = -10^{o}$ C to $T_{k50} =$
-7.8^{o} C (30). A similar effect was observed in *Lolium*
perenne (36). These observations point to the photosynthetic
role of light in the first stage of frost hardening. On the
other hand, experiments performed on the winter rape seedlings (24) and on the winter wheat seedlings (58) indicate
that light may act in the process of hardening through its
morphogenetic influence. Phytochrome was demonstrated to be
involved in regulation of both processes--tissue elongation
and its frost tolerance (24)--light-induced inhibition of
hypocotyl growth corresponded to the increased seedling frost
tolerance. That effect was obtained without any cold treatment. At this point it should be kept in mind that cold
treatment always brings about cessation of tissue expansion
(12, 26). The cabbage leaves are the only case inconsistent
with that observation (8). The question arises whether the
requirement for cold during the first stage of hardening
would not be mainly related to the growth cessation and its
metabolic consequences.

III. FROST TOLERANCE IN RELATION TO THE GROWTH CESSATION

   If cessation of tissue expansion is correlated with the
increase of frost tolerance, then the other factors affecting
the rate of tissue growth should also bring the frost hardening effect. This is true for the light effect as mentioned
above. Growth retardants Alar, Phosfon-D, and AMO 1618 were
demonstrated to inhibit the elongation of winter rape hypocotyls and to increase their frost tolerance (25). Application of GA and ABA was found to change the pattern of growth
in alfalfa plants and to affect their frost tolerance (42).
Recent results of de la Roche (11) also indicate that the
turgid wheat seedlings developed a significant level of
freezing tolerance in darkness when subjected first to desic-
cative stress, which limited their growth during two-week
treatment.

   Thus, it is suggested that the first stage of frost
hardening of herbaceous plants is related to the cessation of
growth. Inhibition of cell expansion brings about the decreased tissue hydration (30). This may, in turn, limit the
probability of intracellular freezing and increase frost
tolerance even in the absence of the other hardening factors.

On the other hand, the important role of metabolic shift, caused by inhibition of cell expansion, should also be taken into consideration.

IV. METABOLIC CHANGES

During the first stage of hardening several metabolic events were shown to occur in the cabbage or winter rape plants, e.g. hydrolysis of starch and accumulation of reducing sugars (23, 30), accumulation of water soluble proteins (27, 31). These events are commonly known to occur in the other herbaceous plants treated with cold--in alfalfa (5, 6, 20, 22, 48), in cabbage (23, 32) and in wheat (59). On the other hand it has been demonstrated that light-induced inhibition of the winter rape hypocotyl growth was also coincidental with an accumulation of soluble proteins in the seedling tissue (24). Application of CCC into the older winter rape plants also increased the protein content in the leaves, even at higher temperature (30). The fact in many cases that the correlation between increased sugar or protein content and frost tolerance changes was rather poor indicates that these factors are not directly involved in the frost hardening process. There is a lack of data concerning the influence of factors other than cold on the content of starch, simple sugars, and water soluble proteins in relation to frost tolerance of herbaceous plants. However, it is interesting that exposure of dogwood plants to water stress brought about an inhibition of growth and simultaneously, the degradation of starch and the increase of sugar and protein contents (7).

These observations may indicate that the events mentioned above are related to the metabolic shift brought about by growth cessation, not necessarily by low temperature itself. However, when changes in specific sugars or proteins are discussed, the requirement for cold for their appearance seems to be undoubtful. Rafinose, sucrose, fructose, -D and -D glucose contents were found to increase in wheat under low temperature treatment (16). Some amino acids were observed to accumulate in the cold-treated winter rape leaves (28). The most striking difference was found in proline content which was markedly increased in the cold-acclimated tissues.

The specific water soluble protein fractions of low molecular weight, impoverished in proline and methionine residues and localized in the citosol subcellular fraction has been observed to accumulate in the winter rape leaves under cold treatment (28). A 24 hr exposure to 5° C conditions was demonstrated to induce in the cabbage leaves the formation of new protein (Fraction I) with a specific amino acid composition (49). Temperature conditions were also shown to affect the protein SH content in the cabbage leaves (32).

The other metabolic changes which were also shown to be a specific response of the plant tissue to lowering environmental temperature are lipid and phospholipid transformations. Low, but positive, temperature has been demonstrated to increase unsaturation of fatty acids (13, 51). The contents of phospholipids and some of the phospholipic fractions--phosphatidyl choline and phosphatidyl ethanolamine--were found to increase in the cold-hardened wheat (12, 18), alfalfa (35), and winter rape (51) tissues.

There is ambient evidence that accumulation of some compounds in cold-treated tissues is due not only to their decreased utilization for growth processes but also to the enhancement of their formation by low, but positive, temperature. Accumulation of water-soluble proteins in the cold-grown winter rape plants has been demonstrated to be prevented by cycloheximide application (27). Preferential synthesis of ferredoxin - $NADP^+$ reductase during hardening of wheat was suggested by Riov and Brown (43). Two particular proteins with a higher degree of hydrophily were shown to be synthesized during the cold hardening process of the hardy winter wheat variety (45). Increased unsaturation of phospholipid fatty acids under cold conditions has been demonstrated to be due to their cold-induced synthesis (11, 63). Stimulation of phospholipid biosynthesis during cold acclimation in winter wheat has been also observed (62) while such an effect of cold has not been found for winter rape leaves (Sikorska, 1976, unpublished).

All of these observations indicate that preferential synthesis of some metabolites occurs during the first stage of hardening of many herbaceous plants. On the other hand, several reports indicate that hydrolytic activities are enhanced by cold treatment. The temperature range which stimulates the activity of hydrolytic enzymes seems to be of importance. Hydrolysis of starch is stimulated by low, but positive, temperature (15). Starch-degradative enzymes such as α-amylase, β-amylase and glucan phosphorylase showed the greatest activity at 15^0 C and 10^0 C respectively (14).

However, some other reports indicate that light frosts are needed to stimulate hydrolysis of oligosaccharides to mono- and disaccharides in winter wheat (55, 56). Degradation of phospholipids under freezing conditions was also noted in the winter rape leaves (50) and in soybean cotyledons (64) and also in woody plants (65). This is probably due to the increased phospholipase -D activity at freezing, caused by dispersal of the enzyme during cell disruption.

Considering that subfreezing temperature is needed to induce the second stage of hardening in many plants, it is tempting to assume that the role of low, negative temperature in the process of hardening should not be attributed only to its physical (dehydrative) influence. It seems to be also related to some metabolic events occuring at that temperature. Our recent results (50) indicate phospholipid transformations to be directly affected by subfreezing temperature.

In front of the ample evidence of the intensive metabolism involved in the process of hardening, at least two further questions are immediately raised:

(1) What are the driving forces for the metabolic shifts occuring during the first and second stages of hardening?

(2) Which of the metabolic events registered at low, positive temperature are directly related to the frost resistance development and which are connected with functioning of tissue at low temperature?

V. DRIVING FORCES IN THE COLD ACCLIMATION PROCESS

Since growth cessation seems to be an indispensable condition for the achievement of the first stage of hardening, the metabolic shifts observed during that stage may be due to the hormonal balance change and/or energy availability.

Our preliminary (unpublished) studies on the changes of the hormonal balance in the winter rape leaves indicate that a decrease of the gibberellin-like substance content in the leaves corresponds to the inhibition of the leaf growth. On the other hand, changes in the content of growth inhibitors were correlated with the frost tolerance changes; the activity of inhibitors increased markedly during the autumn and winter months and reached a maximum in January, at the time of the highest (frost-induced) leaf frost tolerance. Brief exposure of winter rape seedlings to light and/or to cold (5° C) also increased significantly the inhibitor (possibly ABA) content in the tissue while it did not affect so much the gibberellin-like activity (40). Studies by Rikin *et al.* (42) and Waldman *et al.* (60) also point at the involvement of ABA/GA balance in monitoring capacity of two alfalfa cultivars to becoming cold-acclimated when exposed to low temperature.

Changes in hormonal balance may certainly account for modification of protein synthesis and/or for specific enzyme activation in the cold-acclimated tissue. Hormones may also affect directly the properties of cellular membranes; ABA is known to increase the water permeability of membranes while

kinetin exerts an opposite effect (35). These findings point
at the important role of hormones as a driving force in cold
acclimation of herbaceous plants, as it was already proposed
for woody plants (61). It seems very probable that low
temperature, which brings about the enhancement of growth
inhibitors' activity and the inhibition of cell expansion,
affects also the properties of hormone receptor(s). That,
in turn, may change the response of the tissue to the hormone action and direct the cell metabolism into the specific,
cold-affected pathways.

The significant increase of ATP and of E.C. (energy
charge) observed in the winter rape leaves treated with cold
(2° C) may indicate the existence of other than hormonal,
driving forces in leaf metabolism. It has been shown that
E.C. increase in the cold-treated leaves occurred not only in
light but also in darkness (52). This indicates that not
only photosynthesis (37) but also some respiratory systems
may contribute to the increased energy availability under cold
conditions. Since changes in energy charge are an important
factor in control of some metabolic pathways (4), one may predict that they can be involved in protein and lipid preferential synthesis under cold conditions (when acetyl-coenzyme A,
not utilized for growth processes is also available).

Recent studies on the influence of cold on photosynthetic
carbon metabolism in the winter rape leaves (41, 53) showed
that incorporation of ^{14}C into alanine, aspartate, glutamate
and malate was increased in the cold-grown winter rape plants
despite the decrease in the total $^{14}CO_2$ assimilation and in
the ^{14}C incorporation into carbohydrates. These findings may
confirm Levitt's supposition that cold-induced changes in
$NADPH_2/NADP$ ratio may contribute substantially to the cold
acclimation events, through its influence on the photosyn-
thate production. Cold affected formation of these products
may in turn affect the synthesis of other metabolites (e.g.,
proteins) or it may exert the direct effect on the stability
of proteins and membranes under stress conditions.

What are the driving forces in the second stage of frost-
hardening which seems to rely on the occurrence of subfreezing
temperature? There is much less experimental data allowing
us to draw more general conclusions. In the winter rape
plants phospholipid transient degradation at freezing was observed when leaf tissue showed "incipient" degree of injury
(5). The studies indicated also that there was some "thresh-
hold" value for the phospholipid content in the cells which
enabled cells to tolerate the freezing evoked strain. It was
a prerequisite for the sufficient phospholipid reconstruction
after the frost-induced stress was removed. It was also
shown that the subfreezing-dependent increase of leaf frost
tolerance to the second level, was coincidental with the

phospholipid increase.   It seems that membrane rapid alterations might be the main factor in the induction of the second
stage of hardening,   since rapid wilting of cabbage leaves
was shown to induce a rapid and high degree of freezing
resistance in darkness (9).   Studies on the main effect of
sudden wilting on phospholipid content and frost tolerance
might provide some evidence in favour of this supposition.
So far, we have already stated that the sudden desiccation
of winter rapd hypocotyls to a certain degree (WSD higher
than 60%) increased desiccation tolerance of that tissue.

VI.   ACCLIMATION EVENTS LEADING TO INCREASED FROST TOLERANCE

Taking into consideration that extracellular freezing
prevents intracellular ice formation but imposes desiccation
stress on the macromolecular cellular structure, it may be
predicted that frost hardening process involves some mechanisms that:

(1) Facilitate water movement across the membranes, out
of a cell.
(2) Protect the cell constituents against desiccation
effects.

Facilitation of water movement across the membranes may
be assured by increased unsaturation of lipids. However, a
recent observation of Smoleńska and Kuiper (51) that the
degree of unsaturation increases both in the tissues both
capable and uncapable of frost hardening (winter rape leaves
and roots, respectively), and most recent findings of de la
Roche (11), that increased frost tolerance of wheat seedlings
was accomplished without changes in the lipid unsaturation,
as well as the fact that acclimated parenchyma cells of woody
tissue do not show increased lipid unsaturation indicate this
phenomenon to be related to the cell functioning at low temperature rather than to frost tolerance mechanisms.   That
supposition would be consistent with the observation that
chilling resistant but frost sensitive plants also contain
more unsaturated fatty acids in membranes than the chilling
sensitive ones (38).
Changes in phospholipid content and in proportion of
phosphatidyl choline and phosphatidyl ethanolamine in the
lipids undoubtedly affect the membrane properties and increase
their permeability to water. Phospholipid transient decrease
at freezing may also have important bearing on the water re-
lations in the cell.   It certainly brings about the increased

water flow into the intercellular spaces at freezing and protects the cell against intracellular freezing. Thus, phospholipid changes seem to be directly involved in frost
hardening mechanisms (50).

The frost-induced desiccation injuries in the cold acclimated tissue may be prevented in the two following ways:

(1) Cell constituents undergo the structural or conformational changes during the process of acclimation. This
phenomenon may involve either a reorientation of macromolecules into the stable forms which resist severe dehydration
and/or cold-induced synthesis of specific macromolecules
resistant to desiccation.

(2) Cell constituents are protected against desiccation
by specific interactions with low molecular substances.

It remains obscure whether and how the phospholipid
changes may increase the membrane resistance to freezing and
desiccation injuries. There is also rather insufficient
experimental evidence on protein molecule reorientation or
transformations which would be responsible for its increased
stress resistance. While most of the studies indicate
changes in the enzyme activities in cold acclimated herbaceous plants (1, 29, 33, 44) which might be related to cell
functioning at low temperature, only recent results of Huner
and MacDowall (21) show that cold adaptation of wheat plants
increased stability of the enzyme (RuDP carboxylase) under
stress conditions. Increased heat stability of glucose-6-
phosphate dehydrogenase in the cold-hardened wheat seedlings
was also demonstrated by Shcherbakova (47). Studies performed on spinach thylakoids (54) seem to confirm the supposition
that cold acclimation alters membrane structure per se and
increases the membrane stability during freezing. Obviously,
more studies are needed on the acclimation events which may
directly affect the macromolecule resistance to stress
conditions.

The protection of cell membranes by low molecular substances such as sucrose, rafinose, and some organic and amino
acids has been demonstrated by Heber and Ernst (19), Santarius (46) and recently by Steponkus *et al*. (54). It has
been shown that spinach thylakoid membranes are protected
against freezing and desiccation injuries by specific
interactions of the low molecular substances with membrane
component or by colligative reduction of salt concentration in
the membrane environment. Thus, preferential formation of
these compounds, through its cold-induced synthesis or hydrolysis may be directly involved in the frost hardening process.

VII. ACCLIMATION EVENTS ENABLING CELLS TO FUNCTION AT LOW
      TEMPERATURES

Increased unsaturation of membrane lipids assures higher membrane fluidity during cold (39) and protects cells against
metabolic lesions. On the other hand, it brings about increased water permeability of membranes and in turn, may expose plant tissue to the secondary strain of dehydration.
Such a strain may be especially dangerous for plants growing
under natural environmental conditions with limited water
supply (26). It seems that an accumulation of water-soluble
proteins, starting from the beginning of the cold treatment
(27), is one of the mechanisms protecting plants against the
secondary effects of the low-temperature induced changes in
the membrane permeability. A similar role may be ascribed to
the accumulation of osmotically active substances such as
simple sugars, organic acids and others. These mechanisms,
assuring the abundance of water for life processes in cells,
operate parallel to the frost hardening mechanisms and may
partially counteract their effects by increasing the amount
of freezable water in the tissues of herbaceous plants.

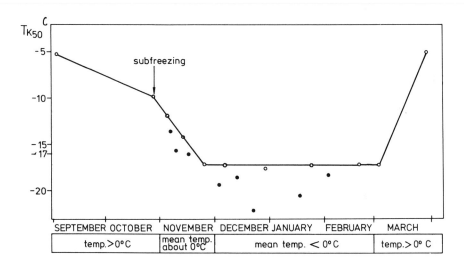

FIGURE 1. Three stages of frost acclimation in the winter
rape plants: open and full dots mean frost tolerance of the
turgid and frost-desiccated tissues, respectively.

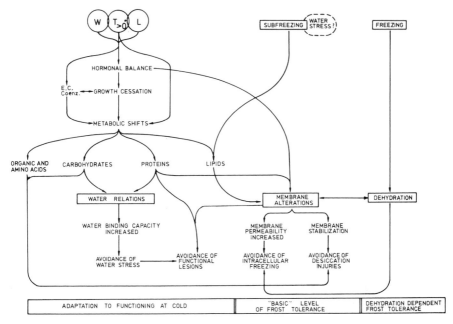

FIGURE 2.   Scheme of cold acclimation events in herbaceous plants W, T, and L mean: water stress, temperature (° C) and light, respectively.

VIII.   SUMMARY

In Figure 2, I have presented my views on the sequence of events which may occur during acclimation of herbaceous plants to cold.   The figure summarizes the preceding discusion and points out different aspects of the acclimation process.

REFERENCES

1.   Alden, J., and Hermann, R. K., *Bot. Rev. 37*, 1 (1971).
2.   Andrews, J. E., *Can. J. Plant Sci. 38*, 1 (1958).
3.   Andrews, C. J., Pomeroy, M. K., and de la Roche, I. A., *Can. J. Bot. 52*, 2539 (1974).
4.   Atkinson, D. E., *Ann. Rev. Microbiol. 23*, 47 (1969).
5.   Bulka, R. J., and Smith, D., *Agron. Jour. 46*, 397 (1954).
6.   Bula, R. J., Smith, D., and Hodgson, H. J., *Agron. Jour. 48*, 153 (1956).

7. Chen, P. M., and Li, P. H., *Plant Physiol. 59*, 236 (1977).
8. Cox, W., and Levitt, J., *Plant Physiol. 47*, 98 (1969).
9. Cox, W., and Levitt, J., *Plant Physiol. 57*, 553 (1976).
10. Dear, J., *Cryobiology 10*, 78 (1973).
11. de la Roche, I. A., *Plant Physiol. (Suppl.) 59*, 198 (1977).
12. de la Roche, I. A., Andrews, C. J., Pomeroy, M. K., Weinberger, P., and Kates, M., *Can. J. Bot. 50*, 2401 (1972).
13. de la Roche, I. A., Pomeroy, M. K., and Andrews, C. J., *Cryobiology 12*, 506 (1975).
14. Glier, J., and Caruso, J. L., *Biochem. Biophys. Res. Comm. 58*, 573 (1974).
15. Glier, J. H., and Caruso, J. L., *Cryobiology 10*, 328 (1973).
16. Green, D. G., and Ratzlaff, C. D., *Can. J. Bot. 53*, 2198 (1975).
17. Grenier, G., Tremolieres, A., Therrien, H. P., and Willemot, C., *Can. J. Bot. 50*, 1681 (1972).
18. Grenier, G., and Willemot, C., *Cryobiology 11*, 324 (1974).
19. Heber, U., and Ernst, R., *in* "Cellular Injury and Resistance in Freezing Organisms," p. 63. Intl. Conf. on Low Temp. Sci., Conf. on Cryobiol., Hokkaido Univ,, Sapporo, (1967).
20. Hodgson, H. J., and Bula, R. J., *Agron. Jour. 48*, 157 (1956).
21. Huner, N. P. A., and MacDowall, F. D. H., *Biochem. Biophys. Res. Comm. 73*, 411 (1976).
22. Jung, G. A., Shih, S. C., and Shelton, D. C., *Cryobiology 4*, 11 (1967).
23. Kacperska-Palacz, A., Blaziak, M., and Wcislinska, B., *Bot. Gaz. 130*, 213 (1969).
24. Kacperska-Palacz, A., Debska, Z., and Jakubowska, A., *Bot. Gaz. 136*, 137 (1975).
25. Kacperska-Palacz, A., and Dlugokecka, E., *Bull. de l' Academic Polonaise des Sciences (Serie des sciences biologique, CL. II) 19*, 537 (1971).
26. Kacperska-Palacz, A., and Egierszdorff, S. T., *Bot. Gaz. 133*, 355 (1972).
27. Kacperska-Palacz, A., Dlugokecka, E., Breintenwald, J., and Wcislinska, B., *Biol. Plant. 19*, 10 (1977).
28. Kacperska-Palacz, A., Jasinska, M., Sobczyk, E. A., and Wcislinska, B., *Biol. Plant. 19*, 18 (1977).
29. Kacperska-Palacz, A., and Uliasz, M., *Physiol. Vég. 12*, 561 (1974).
30. Kacperska-Palacz, A., and Wcislinska, B., *Biol. Plant. 14*, 39 (1972).

31. Kacperska-Palacz, A., and Wciślińska, B., *Physiol. Vég.* *10*, 19 (1972).
32. Kohn, H., and Levitt, J., *Plant Physiol.* *40*, 467 (1965).
33. Krasnuk, M., Jung, G. A., and Witham, F. H., *Cryobiology* *12*, 62 (1975).
34. Kuiper, P. J. C., *Plant Physiol.* *45*, 684 (1970).
35. Kuiper, P. J. C., *Plant Physiol.* *23*, 157 (1972).
36. Lawrence, T., Cooper, J. P., and Breese, E. L., *J. Agric. Sci.* *80*, 341 (1973).
37. Levitt, J., "Responses of Plants to Environmental Stress," 697 pp. Academic Press, New York (1972).
38. Lyons, J. M., *Ann. Rev. Plant Physiol.* *24*, 445 (1973).
39. Lyons, J. M., and Asmundson, C. M., *Jour. Am. Oil Chemists Soc.* *42*, 1056 (1965).
40. Lopka, D., and Kacperska-Palacz, A., *in* "Proc. IV Symp. on Plant Growth Regulators", p. 161. Toruń 1974, Acta Universitatis Nicolai Copernici, Biologia (1976).
41. Maciejewska, U., Maleszewski, St., and Kacperska-Palacz, A., *Bull. de l'Academie Polonaise des Sciences 18*, 513 (1974).
42. Rikin, A., Waldman, M., Richmond, A. E., and Dovrat, A., *J. Exper. Bot. 26*, 175 (1975).
43. Riov, J., and Brown, G. N., *Can. J. Bot. 54*, 1896 (1976).
44. Roberts, D. W. A., *Can. J. Bot. 45*, 1347 (1967).
45. Rochat, E., and Therrien, H. P., *Can. J. Bot. 53*, 2411 (1975).
46. Santarius, K. A., *Planta. 113*, 105 (1973).
47. Shcherbakova, A. M., *Dokl. Acad. Nauk SSSR (Russ.) 205*, 993 (1972).
48. Shih, S. C., Jung, G. A., and Shelton, D. C., *Crop Sci. 7*, 385 (1967).
49. Shomer-Ilan, A., and Waisel, Y., *Physiol. Plant. 34*, 90 (1975).
50. Sikorska, E., and Kacperska-Palacz, A., Submitted for Publication.
51. Smoleńska, G., and Kuiper, P. J. C., *Physiol. Plant. 41*, 29 (1977).
52. Sobczyk, E. A., and Kacperska-Palacz, A., *in* "Abstracts XII Internatl. Botanical Congress," Leningrad (1975).
53. Sosińska, M., Maleszewski, St., and Kacperska-Palacz, A., *Z. Pflanzenphysiol. Bd. 83*, 285 (1977).
54. Steponkus, P. L., Garber, M. L., Myers, S. P., and Lineberger, R. D., *Cryobiology 14*, 303 (1977).
55. Trunova, T. I., *Sov. Plant Physiol. 10*, 140 (1963).
56. Trunova, T. I., *Sov. Plant Physiol. 12*, 70 (1965).
57. Tumanov, I. I., and Trunova, T. I., *Sov. Plant Physiol. 4*, 379 (1957).

58. Tumanov, I. I., Trunova, T. I., Smirnova, N. A., and Zvereva, G. N., *Sov. Plant Physiol.* *22*, 1232 (1975).
59. Vasil'yev, I. M., Lebedeva, L. A., and Rafikova, F. M., *Sov. Plant Physiol.* *11*, 761 (1964).
60. Waldman, M., Rikin, A., Dovrat, A., and Richmond, A. E., *J. Exper. Bot.* *26*, 853 (1975).
61. Weiser, C. J., *Science 169*, 1269 (1970).
62. Willemot, C., *Plant Physiol.* *55*, 356 (1975).
63. Willemot, C., *Plant Physiol.* *60*, 1 (1977).
64. Wilson, R. F., and Rinne, R. W., *Plant Physiol.* *57*, 270 (1976).
65. Yoshida, S., and Sakai, A., *Plant Physiol.* *53*, 509 (1974).

PROTEIN SYNTHESIS MECHANISMS
RELATIVE TO COLD HARDINESS

G. N. Brown

Department of Forestry
Iowa State University
Ames, Iowa

I. INTRODUCTION

Protein synthesis, playing a major role in cell regulation, logically should be involved in a metabolic process such
as development of cold hardiness in living plant cells. Numerous observations have been reported relative to changes in
protein concentrations, patterns, syntheses and degradation
during cold hardening. These numerous observations will be
discussed and cited in the following pages. One major question arises relative to the work reported to date on protein
synthesis during cold hardening, *i.e.* cause and effect. Since
induction of cold hardiness often is accompanied by a reduction in growth rates, changes in protein synthesis might be
expected to reflect growth rates, and not be directly related
to cold hardening. On the other hand, protein synthesis, including transcription, processing and translation, is central-
ly involved in regulation of cell metabolism as influenced by
growth hormones, environmental factors and other cell regulators. Thus, one might logically expect at least some stages
of protein synthesis to be involved in regulation of cold
hardening. Proteins have been implicated in many roles during
cold hardening; protective, enzymatic, membrane structural
and soluble characteristics.
The following pages will summarize information relative to
protein synthesis and degradation, protein and enzyme profiles,
and protein structures and characteristics during cold harden-
int, cool temperature treatment and dehardening. Each of
these changes could reflect regulation via protein synthesis,

but also could reflect protein structural changes or direct
activation or release from bound or inactive forms, concepts
going beyond the scope of this paper.

Levitt (44) provided a complete disucssion of literature
through 1969 relative to protein trends, nucleic acid trends
and enzymatic changes during cold hardening. Therefore, this
paper will focus primarily on literature since 1969. Chilling
studies will be presented as they apply to plants capable of
cold hardening under chilling conditions. Chilling in itself
at low positive temperatures may stimulate early processes of
cold hardening, but only if applied to plants capable of developing cold resistance.

Information relative to protein synthesis and cold treatment or acclimation in bacteria, fungi, yeast and animals is
relevant to the scope of this paper, but will not be included
for the sake of brevity.

II.   HISTORICAL

As early as 1972, Doyle and Clinch (19) suggested protein
synthesis involvement in cold hardening of conifer needles
through observations of changes in major hydrolytic enzyme
activities. Sometime later, in 1949, Siminovitch and Briggs
(72) observed increases in soluble protein concentrations
during cold hardening of black locust (*Robinia pseudoacacia*
L.). This trend has since been observed for many species during development of cold hardiness (8, 20-23, 26, 31, 34, 35,
38, 51, 77). Siminovitch (71) provided the early evidence
that protein synthetic processes were directly involved.
During subsequent years, many workers reported nucleic acid
changes (10, 34, 45, 53, 54, 67, 69, 73, 83) and total protein
changes (8-10, 42, 45, 47, 48, 74) during hardening which
likely were related to protein synthesis. In contrast, a lack
of association between cold hardiness and protein concentration also has been observed (50, 93).

III.   SOLUBLE AND MEMBRANE PROTEINS

Changes in soluble protein concentrations during cold
hardening could reflect either changes in soluble protein
synthesis and degradation, release from membrane or other
bound forms (35) or changes in protein characteristics during
extraction (21, 38). In any event, for many species examined,
soluble protein increases have been associated with cold hardening. The specific significance of these observations has

only recently started to unfold. Glycoproteins with water-binding capacity have been implicated (8, 85, 86), various protective proteins have been characterized and described (30) and some specific soluble enzymes have been identified (discussed later), all of which fit into various cold hardening models.

Insoluble or membrane proteins have been less exhaustively studied during cold hardening than have been soluble proteins. Disagreement exists on quantitative shifts in membrane protein during hardening (8, 26, 35, 93). Hatano *et al.* (29) have reported that hardened *Chlorella ellipsoidea* are more susceptible to detergents than are the non-hardened algal cells. These observations suggest a change in membrane lipids and/or proteins. Since membrane proteins may be either structural or enzymatic, and since most of the integral membrane proteins have not been defined in higher plants, little can be suggested at this time relative to the role of membrane proteins in hardening. However, numerous roles of membranes in cold hardening, and the major roles of proteins in membranes certainly provide for interesting speculation as to how this group of proteins and their turnover could influence development of cold hardiness.

Recently, chloroplast membrane systems have been examined for protein compositional changes relative to cold hardening. Some conflict as to quantitative trends and qualitative patterns in total chloroplast membrane protein during hardening exists (16, 31). DeYoe (16) has found changes in major photosynthetic proteins in chloroplast thylakoid membranes of eastern white pine (*Pinus strobus* L.) during hardening. Steponkus *et al.* (76) have reported relationships between coupling factor, ATPase activity, plastocyanin, membrane semipermability and proton gradient formation during freezing and freeze acclimation in spinach (*Spinacia oleracea* L.). Even though the plasma membrane is likely the major regulatory membrane for the cell, one must consider that in tissue which remains chlorophyllous throughout the winter that once the plasma membrane acclimates, other subcellular membranes then would become limiting to survival without acclimation.

Characteristics of soluble proteins have been described as demonstrating some degree of correlation with freezing and acclimation. Hydrophobicity (11, 60, 70), amino acid composition (11, 70), sulfhydryl content (28, 51) and molecular weight distribution (12) are included among these characteristics. Sycheva and Drozdov (78) suggest that tissue pH can influence protein characteristics which, in turn, can influence development of cold hardiness. Many hypotheses provide roles by which these characteristics could fit into proposed schemes for cold hardening.

Qualitative shifts in soluble protein patterns during cold hardening have been observed in several studies (12, 13, 20, 21, 31, 37, 60, 83). While these suggest shifting roles of specific proteins, for the most part these proteins have not been identified and speculation must be restricted regarding relationship to cold hardening.

IV. NUCLEIC ACIDS

While qualitative and quantitative changes in proteins may imply protein synthesis mechanisms, they don't in themselves provide evidence for changes in synthetic mechanisms. Protein synthesis should be examined from evidence related to the major stages of the protein synthesis mechanism. The first stage, transcription, which involves RNA synthesis from the DNA template, has not received much attention relative to cold hardening. Deoxyribonucleic acid concentrations have been found to remain relatively stable (26, 45). Of course numerous types of DNA exist and numerous mechanisms regulate DNA activity. Regulation of DNA activity involves nuclear proteins such as histones, and chromatin preparations serve as a crude source of both DNA and nuclear proteins. Schnare (66) has found a reduction in chromatin's ability to support RNA synthesis during chilling in soybean (*Glycine max* L.), even in the presence of an exogenous source of RNA polymerase, an enzyme catalysing RNA synthesis. His data suggest a reduction in DNA template availability during chilling. However, chromatin contains many regulatory sites, and specificity relative to cold hardiness remains obscure.

Total RNA concentrations have been observed to increase during cold hardening (10, 34, 45, 53, 54, 69, 73, 83) with some contradictory evidence (9, 23, 43, 67). Observations of changes in base compositions of total RNA have suggested changes in specific types of RNA during cold hardening (33), while other observations suggest no change (53). Using total RNA fractionation procedures, patterns in specific types of RNA have been followed during cold hardening. These other RNA types are involved in a different major stage of protein synthesis called translation. Relative increases in ribosomal RNA (rRNA) concentrations have been observed (9, 15, 26, 46, 63, 84). Polysome profiles during cold hardening suggest stability of the polysome component relative to monomer units (2, 6, 36, 83, 84). However, the presence of polysomes doesn't insure active protein synthesis. A change in ribosomal structure has been reported during cold hardening in black locust (3). Ribosomal proteins also have been followed during hardening (3) and specific changes have been observed, even though these specific proteins were not identified. Transfer (tRNA) {or soluble RNA (sRNA), which includes components in addition to tRNA} concentrations also have been found to increase during cold hardening (63), while other workers have found tRNA or sRNA concentrations to be relatively stable during hardening (9, 26). Changes in specific tRNA activities have been identified during chilling in soybean (91), possibly suggesting a mechanism of regulation at the site of translation. The specific enzymes responsible for binding specific amino acids to specific tRNA's, *i.e.* aminoacyl-tRNA synthetases, have been observed to increase in activity during cold treatment in some studies (79) and to lose activity in other studies (52), again suggesting translation regulation.

V. PROTEIN SYNTHESIS

Studies of actual protein synthesis using techniques such as radioisotope-tagged amino acid incorporation into new protein have been followed during cold hardening (55, 71) and during low temperatures (67). During hardening, actual amino acid incorporation into protein increases (55, 71), while a reduction in protein synthesis was observed using tissue culture cells at low temperatures (67).

Processing of transcribed RNA's, a third major stage in protein synthesis, has not been examined relative to cold hardening. Likewise, messenger RNA's have not been followed during hardening.

Many tools currently are available for studying specific sites of protein synthesis and regulation during cold hardening. However, many of these have not been applied to study cold hardening mechanisms because these tools require tissue which can be easily manipulated such as isolated tissue cultures. If tissue cultures or isolated organ cultures can be repeatably cold hardened, new levels of information should be attained in the relationship between protein synthesis and cold hardening.

Studies incorporating specific inhibitors of RNA and protein synthesis provide one means of identifying specific aspects of protein synthesis as being directly related to induction of cold hardiness. Hatano *et al.* (29) have found that specific inhibitors of RNA and protein synthesis do inhibit the development of cold hardiness in *Chlorella ellipsoidea* suggesting direct involvement of at least cytoplasmic protein synthesis.

VI.   ENZYMES

Hydrolytic enzyme systems such as RNase are directly related to RNA and protein turnover rates. Various studies have
observed changes in RNase activity to be related to cold hardening (7, 26, 36). Generally, RNase activity appears to decrease during cold hardening (7, 26) which would be consistent
with an accumulation of RNA and protein discussed earlier.
Two specific RNases with two different temperature optima have
been reported in winter wheat seedlings (64), opening the possibility that low temperatures required during cold acclimation may activate or deactivate different RNase systems,
thereby regulating some aspects of RNA activity and protein
synthesis.
Smith and Powell (74), using cotton (*Gossypium hirsutum*
L.), observed protein breakdown during chill damage. If hydrolytic enzymes do increase in activity during stress {in-
cluding drought (5, 18) and low temperature (26, 74)}, a
decrease in their activities during hardening would seem appropriate for development of resistance to stress.
Numerous reports of shifts in enzyme activities during the
cold acclimation process or during cold temperatures have been
reported (1, 4, 14, 17, 22, 24, 25, 32, 38-41, 49, 56, 57, 59,
61, 62, 65, 68, 75, 80, 81, 87-89). These shifts in activities could reflect changes in enzymatic structures and characteristics {*e.g.* kinetics or activation energies (58, 88,
90)}, direct activation from inactive forms or new enzyme
synthesis. Unless these shifts relate to new enzyme synthesis
or hydrolytic enzymes involved in RNA and protein synthesis,
shifts in their activities go beyond the scope of this paper.
At least one study has implicated new indoleacetic acid oxi-
dase synthesis during cold treatment of winter wheat (*Triticum
vulgare* L.) (4).

VII.   DEHARDENING

At the other end of the seasonal spectrum, relatively less
work has been done on physiological mechanisms during dehardening. Contradictory evidence exists relative to protein
(42, 69, 82, 92) and nucleic acid (26, 46, 69) concentration
trends during dehardening. Evidence is accumulating relative
to shifts in specific enzyme activities during dehardening
(27, 38, 40). In some instances, these processes are the
reverse of processes observed during dehardening, but in other
instances they reflect entirely new synthetic or activation
processes. Again, further examination of these dehardening

processes could aid in pinpointing those processes during
cold hardening which are causal in relationship and those
which are effect or only correlative.

VIII. CONCLUSION

Most work relating protein turnover and characteristics to
cold stress or acclimation has only shown correlation to date,
and very little definitive results confirm cause of direct
regulation of cold hardening. Newly developed techniques,
when applied to biological material capable of developing cold
hardiness and also capable of being manipulated in the laboratory, should sort those aspects of protein metabolism which
are only correlated with cold hardening and those which are
causal.

Much of the published work to date relative to cold hardiness with perennial plants has failed to draw a distinction
between processes involved with winter dormancy and those involved with cold hardiness. These two phenomena of course
differ in their regulatory sites, but often are examined and
described as a single phenomenon. Future attempts must distinguish these two phenomena.

REFERENCES

1. Bervaes, J. C. A. M., and Kylin, A., *Physiol. Plant 27*,
    178 (1972).
2. Bewley, J. D., *Plant Sci. Letters 1*, 303 (1973).
3. Bixby, J. A., and Brown, G. N., *Plant Physiol. 56*, 617
    (1975).
4. Bolduc, R. J., Cherry, J. H., and Blair, B. O., *Plant
    Physiol. 45*, 461 (1970).
5. Brandle, J. R., Hinckley, T. M., and Brown, G. N.,
    Physiol. Plant. 40, 1 (1977).
6. Brown, G. N., *Plant & Cell Physiol. 13*, 345 (1972).
7. Brown, G. N., and Bixby, J. A., *Cryobiology 10*, 152
    (1973).
8. Brown, G. N., and Bixby, J. A., *Physiol. Plant. 34*, 187
    (1975).
9. Brown, G. N., and Sasaki, S., *J. Amer. Soc. Hort. Sci.
    97*, 299 (1972).
10. Chen, P. M., and Li, P. H., *Plant Physiol. 59*, 240 (1977).
11. Chou, J. C., and Levitt, J., *Cryobiology 9*, 268 (1972).
12. Coleman, E. A., Bula, R. J., and Davis, R. L., *Plant
    Physiol. 41*, 1681 (1966).

13. Craker, L. E., Gusta, L. V., and Weiser, C. J., *Can. J. Plant Sci. 49*, 279 (1969).
14. DeJong, D. W., *Amer. J. Bot. 60*, 846 (1973).
15. DeVay, M., and Paldi, I., *Plant Sci. Letters 8*, 191 (1977).
16. DeYoe, D. R., *Ph.D. thesis*, University of Missouri, Columbia (1977).
17. Domanskaya, E. N., and Strekozova, V. I., *Byull. Glavn. Bot. Sada 81*, 92 (1971).
18. Dove, L. A., *Plant Physiol. 42*, 1176 (1967).
19. Doyle, J., and Clinch, P., *Proc. Roy. Irish Acad. Sect. B37*, 373 (1927).
20. Faw, W. F., and Jung, G. A., *Cryobiology 9*, 548 (1972).
21. Faw, W. F., Shih, S. C., and Jung, G. A., *Plant Physiol. 57*, 720 (1976).
22. Gerloff, E. D., Stahmann, M. A., and Smith, D., *Plant Physiol. 42*, 895 (1967).
23. Ghazaleh, M. Z. S., and Hendershott, C. H., *Proc. Amer. Soc. Hort. Sci. 90*, 93 (1967).
24. Glier, J. H., and Caruso, J. L., *Biochem. Biophys. Res. Commun. 58*, 573 (1974).
25. Glier, J. H., and Caruso, J. L., *Cryobiology 14*, 121 (1977).
26. Gusta, L. V., and Weiser, C. J., *Plant Physiol. 49*, 91 (1972).
27. Hall, T. C., McLeester, R. C., McCown, B. H., and Beck, G. E., *Cryobiology 7*, 130 (1970).
28. Hashizume, K., Kakiuchi, K., Koyama, E., and Watanabe, T., *Agr. Biol. Chem. 35*, 4 (1971).
29. Hatano, S., Sadakane, H., Tutumi, M., and Watanabe, T., *Plant & Cell Physiol. 17*, 643 (1976).
30. Heber, U., and Ernst, R., *Proc. Intl. Conf. Low Temp. Sci., 1966, Low Temp. Sci., Hokkaido Univ., Sapporo 2*, 63 (1967).
31. Huner, N., and Macdowall, F., *Plant Physiol. (Supp.) 56*, 80 (1975).
32. Huner, N. P. A., and Macdowall, F. D. H., *Biochim. Biophys. Res. Commun. 73*, 411 (1976).
33. Jung, G. A., Shih, S. C., and Martin, W. G., *Cryobiology 11*, 269 (1974).
34. Jung, G. A., Shih, S. C., and Shelton, D. C., *Plant Physiol. 42*, 1653 (1967).
35. Kacperska-Palacz, A., and Wcislinska, B., *Biol. Plant. 14*, 39 (1972).
36. Kenefick, D. G., Johnson, C., and Whitehead, E. I., *Mechanisms of Regulation of Plant Growth. Bull. 12*, 505 (1974).

37. Khokhlova, L. P., Eliseeva, N. S., Stupishina, E. A., Bondar, I. G., and Suleimanov, I. G., *Sov. Plant Physiol. 22*, 723.

38. Krasnuk, M., Jung, G. A., and Witham, F. H., *Cryobiology 12*, 62 (1975).

39. Krasnuk, M., Jung, G. A., and Witham, F. H., *Cryobiology 13*, 375 (1976).

40. Krasnuk, M., Witham, F. H., and Jung, G. A., *Cryobiology 13*, 225 (1976).

41. Kuiper, P. J. C., *Physiol. Plant 26*, 200 (1972).

42. Lasheen, A. M., and Chaplin, C. E., *J. Amer. Soc. Hort. Sci. 96*, 154 (1971).

43. Lebedev, S. I., and Komarnitskii, P. A., *Sov. Plant Physiol. 18*, 151 (1971).

44. Levitt, J., *in* "Responses of Plants to Environmental Stresses." Academic Press, New York (1972).

45. Li, P. H., and Weiser, C. J., *Proc. Amer. Soc. Hort. Sci. 91*, 716 (1967).

46. Li, P. H., and Weiser, C. J., *Plant & Cell Physiol. 10*, 21 (1969).

47. Li, P. H., Weiser, C. J., and van Huystee, R., *Proc. Amer. Soc. Hort. Sci. 86*, 723 (1965).

48. Li, P. H., Weiser, C. J., and van Huystee, R., *Plant & Cell Physiol. 7*, 475 (1966).

49. McCown, B. H., Hall, T. C., and Beck, G. E., *Plant Physiol. 44*, 210 (1969).

50. Mitchell, H. L., and Toman, F. R., *Phytochem. 7*, 365 (1968).

51. Morton, W. M., *Plant Physiol. 44*, 168 (1969).

52. Norris, R. D., and Fowden, L., *Phytochem. 13*, 1677 (1974).

53. Oslund, C. R., and Li, P. H., *Plant & Cell Physiol. 13*, 201 (1972).

54. Oslund, C. R., Li, P. H., and Weiser, C. J., *J. Amer. Soc. Hort. Sci. 97*, 93 (1972).

55. Pomeroy, M. K., Siminovitch, D., and Wightman, F., *Can. J. Bot. 48*, 953 (1970).

56. Riov, J., and Brown, G. N., *Can. J. Bot. 54*, 1896 (1976).

57. Riov, J., and Brown, G. N., *Cryobiology*, (in press) (1977).

58. Roberts, D. W. A., *Can. J. Bot. 45*, 1347 (1967).

59. Roberts, D. W. A., *Can. J. Bot. 53*, 1333 (1975).

60. Rochat, E., and Therrien, H. P., *Can. J. Bot. 53*, 2411 (1975).

61. Sagisaka, S., *Plant Physiol. 50*, 750 (1972).

62. Sagisaka, S., *Plant Physiol. 54*, 544 (1974).

63. Sarhan, R., and D'Aoust, M. J., *Physiol. Plant 35*, 62 (1975).

64. Sasaki, K., and Sasaki, S., *Biochem. Biophys. Res. Commun. 72*, 850 (1976).
65. Sawada, S., Matsushima, H., and Miyachi, S., *Plant & Cell Physiol. 15*, 239 (1974).
66. Schnare, P. D., *Ph.D. thesis*, Univ. of Missouri, Columbia (1974).
67. Scholtissek, C., *Biochim. Biophys. Acta 145*, 228 (1967).
68. Sekiya, J., Kajiwara, T., and Hatanaka, A., *Plant & Cell Phhysiol 18*, 283 (1977).
69. Shih, S. C., and Jung, G. A., *Cryobiology 7*, 200 (1970).
70. Shomer-Ilan, A., and Waisel, Y., *Physiol. Plant 34*, 90 (1975).
71. Siminovitch, D., *Can. J. Bot. 41*, 1301 (1963).
72. Siminovitch, D., and Briggs, D. R., *Arch. Biochem. Biophys. 23*, 8 (1949).
73. Siminovitch, D., Rheaume, B., Pomeroy, K., and Lepage, M., *Cryobiology 5*, 202 (1968).
74. Smith, C. W., and Powell, R. D., *Plant Physiol. (Supp.) 57*, 34 (1976).
75. Srivastava, G. C., and Fowden, L., *J. Exp. Bot. 23*, 921 (1972).
76. Steponkus, P. L., Garber, M. P., Myers, S. P., and Lineberger, R. D., *Cryobiology 14*, 303 (1977).
77. Svec, L. V., and Hodges, H. F., *Can. J. Plant Sci. 53*, 165 (1972).
78. Sycheva, Z. F., and Drozdov, S. N., *Doklady Akademii Nauk SSR 195*, 1466 (1970).
79. Tao, K. L., and Khan, A. A., *Biochem. Biophys. Res. Commun. 59*, 764 (1974).
80. Taylor, A. O., Slack, C. R., and McPherson, H. C., *Mechanisms of Regulation of Plant Growth Bull. 12*, 519 (1974).
81. Taylor, A. O., Slack, C. R., and McPherson, H. G., *Plant Physiol. 54*, 696 (1974).
82. Tromp, J., and Ovaa, J. C., *Physiol. Plant 29*, 1 (1973).
83. Vigue, J. T., and Li, P. H., *Plant Physiol. (Supp.) 53*, 12 (1974).
84. Vigue, J., Li, P. H., and Oslund, C. R., *Plant & Cell Physiol. 15*, 1055 (1974).
85. Williams, R. J., *Cryobiology 9*, 313 (1972).
86. Williams, R. J., *Cryobiology 10*, 530 (1973).
87. Wilson, R. F., and Rinne, R. W., *Plant Physiol. 57*, 270 (1976).
88. Wolfe, J. K., and Gray, I., *Cryobiology 10*, 295 (1973).
89. Wu, H. B., Yu, T. T., and Liou, T. D., *Mechanisms of Regulation of Plant Growth Bull. 12*, 483 (1974).
90. Yamaki, S., and Uritani, I., *Plant & Cell Physiol. 15*, 669 (1974).

91. Yang, J. S., and Brown, G. N., *Plant Physiol.* *53*, 694
     (1974).
92. Yoshida, S., *Plant Physiol.* *57*, 710 (1976).
93. Young, R., *J. Amer. Soc. Hort. Sci.* *94*, 252 (1969).

THE ROLE OF WATER IN COLD HARDINESS
OF WINTER CEREALS

P. Chen
L. V. Gusta

Crop Development Centre
University of Saskatchewan
Saskatoon, Saskatchewan

I. INTRODUCTION

Cold hardiness of winter cereals is an inducible character, the expression of which is dependent upon a source of
energy and temperatures below 10° C. Photoperiod has no inducing effect and is only important in supplying energy
through photosynthesis. The hardiest tissue of winter cereals
is the crown which generates shoots and roots. The roots are
the least hardy and the herbage somewhat more hardy. During
cold acclimation certain biochemical and biophysical alterations allow the cells to tolerate extracellular ice and dehydration. The amount of energy available for crystallization
of water, the water content and the distribution of water and
ice throughout the tissue determine the degree of injury.
Injury is thought to originate from three types of stresses,
nonequilibrium and equilibrium freezing and frost desiccation.
The duration of the frost required to produce injury is only
seconds for nonequilibrium freezing, minutes for equilibrium
freezing and hours to days for desiccation injury (17, 34).

II. WATER CONTENT

One of the first measurable events to occur during cold
acclimation of winter cereals is a reduction in crown moisture content. Within a cereal species there is a positive
correlation (0.9 or better) between crown water content and

165

cold hardiness as measured by artificial freeze tests. However, between species this correlation does not hold. For example, Kharkov winter wheat (*Triticum aestivum*) and Frontier rye (*Secale cereale*), when fully acclimated, may have the same moisture content, but there is a 10° C difference in cold hardiness.

The importance of moisture regulation to the freezing process and cold hardiness in herbaceous tissue has been noted by many researchers (7, 21, 29, 30, 13). Metcalf *et al.* (30) found that a moisture content of 65% in winter cereal crowns was optimum for the maximum cold expression. In tissue with a higher moisture content most of the water freezes upon nucleation causing mechanical damage (32). In tissue with a lower moisture content (less than 60%) isolated pockets of water may supercool and then freeze intracellularly (33).

Attempts to increase the hardiness of plants by regulating the cell water content through induced water stress have had varying degrees of success. Chen *et al.* (5), working with the woody species *Cornus stolonifera*, reported that the hardiness of stems was increased from -3° C to -11° C by slowly withholding water. Cox and Levitt (11) reported that rapid wilting of cabbage leaves (*Brassica capitata*) resulted in a high degree of freezing resistance. Gusta (unpublished data) found that benzyl adenine ($10^{-5}M$) or polyethylene glycol 6000, added to the nutrient solution in which winter wheat plants were grown resulted in a two to three fold reduction in crown water content. However, there was little or no increase in the cold hardiness of crowns frozen under equilibrium conditions.

Depending upon the cultivar of winter wheat, the water content may decrease from 6g H_2O/g dry weight in the tender state to 2g H_2O/g dry weight in the hardened state. However, the water potential of the crown remains relatively constant throughout the hardening period. This suggests that under natural conditions the reduction in crown water content is not merely due to a restriction in water uptake, but may be due to an exclusion in water in the cell by the accumulation of dry matter.

Cell aggregates of Kharkov winter wheat, grown either on agar or in a liquid media (1-B5), readily cold acclimated upon exposure to a cold hardening environment. Fully acclimated cells frozen slowly (2° C/hr) can tolerate -18° C without the use of cryoprotectants. Yet, during this period there is only a slight reduction in moisture content (13g H_2O/g dry weight in the tender state to 11 g H_2O/g dry weight in the hardened state). Since cell aggregates can acclimate to nearly the same level as crowns, a decrease in water content is not the key to cell survival.

Water content of intact crowns may be more critical for survival due to possible mechanical damage by the growing ice crystal (34) or under conditions which promote intracellular freezing. Under conditions of equilibrium freezing ice crystals grow extracellularly at preferred sites which can accomodate their growth (12, 42). Water content of the tissue is crucial under conditions of fast freezing and rapid removal of latent heat. Fully hardened winter wheat crowns submerged in ice water for 24 hr and frozen slowly (2° C/hr) on a heat sink (aluminum plate) could tolerate -12° C (Gusta, unpublished data). Similarly frozen crowns, but not submerged in water, could tolerate -19° C. The difference in water content between the treated and untreated crowns was approximately 8%. Thus, as shown by Metcalf *et al.* (*30*), water content can have an effect on survival, depending on the freezing conditions, but merely reducing the water content does not have an effect on the cold hardiness of winter cereals.

III. WATER DISTRIBUTION

In differentiated plant tissue, water is distributed between the intracellular and extracellular spaces. Within the cell at least two populations of water exist which differ greatly in their properties. The population of water associated with hydrophilic compounds is termed water of hydration or "bound" water. This fraction of water comprises about 0.2 to 0.4g H_2O/g dry tissue (4, 16). Bound water is water closely associated to macromolecules and is not available for freezing. The bulk of the intracellular water is considered to have properties similar to those of a dilute salt solution (4, 16, 37).

Through nuclear magnetic resonance (NMR) analysis, it has been possible to quantitate the amount of water in the intracellular and extracellular spaces. Approximately 80% of tender winter wheat crown tissue water is found in the intracellular spaces. During cold acclimation, depending upon the cultivar, there is a reduction in the intracellular water.

Burke *et al.* (4), Stout (37, 38) and Chen *et al.* (5) concluded that the long relaxation times for both (T_1 and T_2) were associated with the intracellular water and the short relaxation times with the extracellular water. Since the two relaxation rates differ greatly, it may be assumed that intracellular and extracellular water do not exchange rapidly. The rate of water exchange between the two fractions will depend upon the permeability of the plasma membrane. The mathematics of the effect of water exchange between two fractions of water with different intrinsic relaxation times on the relaxation times and fraction sizes have been given by Hazlewood *et al.* (20).

To determine if freezing injury was associated with plasma mebrane damage, Chen *et al.* (6) froze tender, semi-hardened and hardened Kharkov winter wheat crowns slowly (2° C/hr) to a lethal temperature or quickly by immersion in liquid N_2. Relaxation times were determined as soon as the crowns thawed. There was a significant decrease in the long relaxation times associated with intracellular water in tender or semi-hardened Kharkov winter wheat crowns following either a slow freeze (2° C/hr) to a lethal temperature or following direct immersion in liquid N_2. This suggests that freezing to a lethal temperature altered the permeability properties of the plasma membrane. In the case of fully hardened crowns there were no significant differences in the long relaxation times of crowns not frozen or frozen slowly to a lethal temperature. However, there was a significant difference in the distribution of water between the intracellular and extracellular spaces. Direct immersion of fully hardened crowns into liquid N_2 had a significant shortening effect on the long relaxation times. Again, this would suggest fast freezing, and thus freezing before some of the intracellular water can diffuse out of the cell, results in marked destruction of the plasma membranes, possibly due to mechanical damage.

Since factors other than membrane permeability may result in relaxation time changes (3, 20), futher testing was necessary to prove that membrane permeability was increased during freezing. The plasma membrane is the main barrier to Mn^{2+} diffusion into the cell (8, 37). If the permeability of the plasma membrane was increased as a result of freezing, exogenously applied Mn^{2+} would diffuse into the protoplast at a faster rate. Paramagnetic cations such as Mn^{2+} cause protons to relax in a short time, and thus the long T_2 from intracellular water would be decreased by the Mn^{2+}. Using the Carr-Purcell-Beiboom-Gill (CP/MG) pulse sequence method for determining T_2 with the pulse sequence time between the 90° pulse and the first 180° pulse of radio frequency energy (t_{cp}) set at 6 msec, water under the influence of Mn^{2+} would be undetectable in the NMR spectrometer.

Freezing hardened Kharkov crowns to a non-lethal temperature did not affect the diffusion of Mn^{2+} across the plasma membrane. However, the plasma membrane of crowns frozen slowly (2° C/hr) to a lethal temperature were more permeable to Mn^{2+} than plasma membranes of non-injured crowns. The entire integrity of the plasma membrane was not destroyed, since a long T_2 was still detectable after 5 hr incubation in

Mn^{2+}. A long T_2 could not be detected in crown tissue killed in liquid N_2 after 2 hr of incubation in Mn^{2+} indicating the plasma membrane was no longer a major barrier to Mn^{2+} diffusion into the cell.

The difference in relaxation times between slow-killed and fast-killed tissues exposed to Mn^{2+} suggest that the type and/or degree of injury to the plasma membrane is different. Slow freezing injury may result from the freeze induced dehydration stress, causing certain biophysical alterations in the membrane which prevent the cell from functioning.

IV. FREEZING OF EXTRACELLULAR AND INTRACELLULAR WATER

Under conditions which lead to extracellular freezing, winter cereals will initially supercool. The temperature at which the tissue freezes depends upon the solute concentration of tissue water. Plant tissues will generally supercool 3 to 8° C before spontaneous nucleation occurs. Under conditions of slow freezing and a few degrees of supercooling, ice formation in hardened tissue occurs extracellularly. Ice nucleation may occur intracellularly in a few cells and then nucleate extracellular water (22). If freezing occurs extracellularly, intracellular supercooled water will move out of the cell to the growing ice crystal due to its higher vapor pressure. This results in a concentration of the intracellular solutes, a freezing point depression of the intracellular water. Therefore, the amount of cell dehydration depends upon both temperature and the initial solute concentration of the intracellular water.

Scarth (35) originally proposed that plasma membrane permeability increased with cold hardiness, thus permitting rapid escape of water to extracellular sites of ice nucleation. Olien (32) proposed that with increasing hardiness the plasma membrane became more elastic and more stable to the presence of ice. Both Stout *et al.* (39) and Gusta (unpublished observation) found that if the membranes in plant tissues were first injured by rapid freezing in liquid nitrogen and then refrozen the rate of ice formation was significantly faster than the control. Under conditions of equilibrium freezing the rate of ice formation at -3° C in hardened winter cereal crowns was faster than in tender crowns. The difference in the rate of ice formation may be due to an increase in water permeability of the plasma membrane or due to the fact tender crowns have a higher water content than hardened crowns. If the latter situation was the case, then the rate of heat removal would determine the rate of ice formation.

Hardened winter wheat crowns of hardy cultivars held at
-3° C for 16 h and then slowly frozen at 2° C/hr can readily
tolerate -18° C to -20° C (Gusta, unpublished data). If the
crowns are frozen at a rate of 15° C/hr the survival temperature was reduced from -12° C to -13° C. However, if the
crowns are frozen to -6° C at 2° C/hr first and then frozen
there on at a rate of 15° C/hr the survival temperature is
identical to crowns frozen at 2° C/hr. Within the first 6° C
of freezing, approximately 30% of the tissues' total water
freezes (16). However, it can not be determined from this
study whether injury is due to the rate of water efflux or
due to the rate of water migration to sites which can accomodate the growing ice crystal.

V.   PROPERTIES OF INTRACELLULAR WATER

In 1936 Stark proposed that cold hardiness of apple
(*Malus pumila*) shoots was dependent upon the ability of the
cell to retain water. This involved what came to be known
as the "bound" water concept to explain cold hardiness. An
increase in bound water would result in a decrease in the
freeze induced dehydration stress. Water may be bound in
the cell due to the colligative properties of solutes and by
the hydrophilic surfaces of macromolecules (27).
Several authors have reported an increase in bound water
during cold acclimation (24, 26, 28). In contrast, Burke *et
al.* (4) and Gusta *et al.* (16) could not detect an increase in
bound water in *Cornus stolonifera* and winter cereals,
respectively.
Johansson (23, 24), using calorimetry to study freezing of
water in winter cereals of varying hardiness, concluded that
increases in hardiness resulted from an increase in cell sap
concentration which decreased cell dehydration at low temperatures. Regardless of hardiness the plants were killed when
more than 85% of the freezable water was frozen. MacDowall
and Buchanan (28), using continuous wave NMR, reported that
cold hardened winter wheat leaves and the expressed sap from
these leaves had more unfrozen water at -25° C than tender
winter wheat leaves. Stout (37), using NMR to study freezing
of water in *Hedra helix*, found at any given freezing temperature there was more liquid water present in hardened stems
than in tender stems.
Burke *et al.* (4), Gusta *et al.* (16) and Stout (37) agree
that when the quantities of unfrozen water of cold acclimated
tissue and tender tissue are compared at the tissue killing
temperature, cold acclimated tissue tolerates more cell dehydration than tender tissue. Also there was no difference in

the quantity of water frozen whether the tissue was living or dead. Thus, it appears no specific amount of liquid water is associated with the living state as suggested by several hypotheses (40, 41).

Two schools of thought have evolved on the properties of intracellular water. One hypothesis is that the bulk of the intracellular water has properties; viscosity, degree of hydrogen bonding, etc. similar to those of a dilute salt solution and thus a minor fraction of the water (bound water) has altered properties because of its strong interaction with macromolecules (25). This hypothesis was supported by the findings that self-diffusion of intracellular water in several plant and animal tissues is similar to that of bulk water (1, 18). Cooke and Wien (9) using NMR data on muscle tissue, also concluded that the bulk of the intracellular water has a property similar to free water and less than 4 to 5% of the total water has altered properties.

A second hypothesis suggests that the bulk of the intracellular water is in a more highly structured phase or phases than water in a dilute salt solution. This hypothesis is based on steady state NMR data indicating that water protons in muscle have a wider line width than free water (19). Also pulsed NMR studies have shown that water in muscle and brain tissue has a shorter relaxation time than free water (10). This hypothesis suggests that when water molecules interact with charged molecules, they line up with the positive ends of the water molecule to the negative sites on the macromolecules. Another layer of water molecules lines up to the initial layer of water molecules and so forth, resulting in concentric layers of oriented water molecules around the molecule. As the distance from the charge molecule decreases, the order in the water layers decreases.

Burke *et al.* (4) observed a 250-fold increase in the linewidth of the peak for water during cold acclimation of *Cornus stolonifera* stems, but did not attribute the increase in linewidth to an increase in bound water. MacDowall and Buchanan (28) and Gusta (unpublished observation), using continuous wave NMR, observed that the water absorption peak in winter wheat increased during cold acclimation. Burke *et al.* (4) and Chen *et al.* (5), using pulsed NMR, reported that the transverse relaxation time (T_2) of *Cornus stolonifera* and winter wheat crowns decreased with cold acclimation. Stout (37) could not detect any significant effect of cold acclimation on the relaxation times in *Hedra helix* bark. This difference may be related to the dramatic changes in water content during cold acclimation of *Cornus stolonifera* bark (80 to 50%) and winter wheats (58 to 65%) as compared to *Hedra helix* bark where the water content changes very little. Recently, a number of studies have shown that the relaxation

times and line width of water are related to the total
water content of the tissue (14, 15, 31). However, relaxation
times may be altered without a significant change in the
water content (2). Neville *et al.* (31) concluded that the
one possible reason for the shorter relaxation times in bio-
lgoical tissue may be due to the rapid exchange between the
water of hydration and the bulk water giving rise to an
observed relaxation time intermediate between these two phases. This, in conjuction with a reduction in water content
and in increase in hydrophilic compounds, would also give
rise to shorter relaxation times with cold acclimation of
cereals. Reduction in relaxation times during cold acclimation of winter cereals does not provide evidence for ordering
of the bulk of the cell water.

VI. SUMMARY

Winter cereals are injured from the presence of extracellular ice or from freeze induced dehydration or a combination of both. The quantity of liquid water at the killing
temperature is not associated with survival. Thus, cold acclimation is associated with an increased tolerance of the
protoplasm to dehydration. The structured mobility of the
bulk of the cell water is not greatly altered during cold
acclimation.

Injury from slow freezing is due to a lesion created in
the plasma membrane which alters the functional properties
of the cell, whereas there is a marked alteration in the
permeability properties of cells killed by fast freezing.
The rate of freezing (the rate of heat removal) during the
initial few degrees of freezing has a significant effect on
the temperature that the tissue can tolerate.

REFERENCES

1. Abetsedarskaya, L. A., Miftakhutdinova, F. C., and
   Fedotov, V. D., *Biofizika 4*, 630 (1968).
2. Beall, P. T., Hazlewood, C. F., and Rao, P. H., *Science
   192*, 904 (1976).
3. Belton, P. S., and Packer, K. J., *Biochim. Biophys. Acta
   354*, 305 (1974).
4. Burke, M. J., Bryant, R. G., and Weiser, C. J., *Plant
   Physiol. 54*, 392 (1974).
5. Chen, P. M., Li, P. H., and Burke, M. J., *Plant Physiol.
   59*, 236 (1977).

6. Chen, P. M., Gusta, L. V., and Stout, D. G., *Plant Physiol. (submitted)*, (1977).

7. Clausen, E., *Bryologist 67*, 411 (1964).

8. Conlon, T., and Outhred, R., *Biochim. Biophys. Acta 288*, 354 (1972).

9. Cooke, R., and Wien, R., *Biophys. J. 2*, 1002 (1971).

10. Cope, F. W., *Biophys. J. 9*, 303 (1969).

11. Cox, W., and Levitt, J., *Plant Physiol. 57*, 553 (1976).

12. Dennis, F. G. Jr., Lumis, G. D., and Olien, C. R., *Plant Physiol. 50*, 527 (1972).

13. Fowler, D. B., and Gusta, L. V., *Can. J. Plant Sci. 57*, 751 (1977).

14. Fung, B. M., *Biochim. Biophys. Acta 497*, 317 (1977).

15. Fung, B. M., Durham, D. L., and Wassil, D. A., *Biochim. Biophys. Acta 399*, 191 (1975).

16. Gusta, L. V., Burke, M. J., and Kapoor, A. C., *Plant Physiol. 56*, 707 (1975).

17. Gusta, L. V., and Fowler, D. B., *Can. J. Plant Sci. 57*, 213 (1977).

18. Hansen, J. R., *Biochim. Biophys. Acta 230*, 482 (1971).

19. Hazlewood, C. F., Nichols, B. L., and Chamberlain, N. F., *Nature 222*, 747 (1969).

20. Hazlewood, C. F., Chang, D. C., Nichols, B. L., and Woessner, D. E., *Biophys. J. 14*, 583 (1974).

21. Hudson, M. A., and Idle, D. B., *Planta 57*, 718 (1962).

22. Idle, D. B., *Science J. 4*, 59 (1968).

23. Johansson, N.-O., *Nat. Swed. Inst. Plant Protection Contrib. 14*, 364 (1970).

24. Johansson, N.-O., and Krull, E., *Nat. Swed. Inst. Plant Protection Contrib. 14*, 343 (1970).

25. Kuntz, I. D., Brassfield, T. S., Law, G. D., and Purcell, G. V., *Science 163*, 1329 (1969).

26. Levitt, J., *Plant Physiol. 14*, 93 (1939).

27. Levitt, J., *Plant Physiol. 33*, 674 (1959).

28. MacDowall, F. D. H., and Buchanan, G. W., *Can. J. Biochem. 52*, 652 (1974).

29. Mayland, H. F., and Cary, J. W., *Adv. Agron. 22*, 203 (1970). Academic Press, New York.

30. Metcalf, E. L., Cress, C. E., Olien, C. R., and Everson, E. H., *Crop Sci. 10*, 362 (1970).

31. Neville, M. C., Paterson, C. A., Rae, J. L., and Woessner, D. E., *Science 184*, 1072 (1974).

32. Olien, C. R., *Ann. Rev. Plant Physiol. 18*, 387 (1967).

33. Olien, C. R., *Barley Genetics 2*, 356 (1969).

34. Olien, C. R., *USDA Tech. Bul. No. 1558*, 1 (1976).

35. Scarth, G. W., *Roy. Soc. Can. Sect. 5, 30*, 1 (1936).

36. Stark, A. L., *Plant Physiol. 11*, 689 (1936).

37. Stout, D. G., *Ph.D. thesis*, Cornell University, Cornell, New York (1976).

38. Stout, D. G., Cotts, R. M., and Steponkus, P. L., *Can. J. Bot. 55*, 1623 (1977).
39. Stout, D. G., Steponkus, P. L., and Cotts, R. M., *Plant Physiol. 60*, 374 (1977).
40. Tumanov, I. I., *Izu. Acad. Nauk USSR Sev. Biol. 4*, 469 (1969).
41. Weiser, C. J., *Science 169*, 1269 (1970).
42. Wiegand, K. M., *Bot. Gaz. 41*, 373 (1906).

Plant Cold Hardiness and Freezing Stress

STUDIES ON FROST HARDINESS
IN CHLORELLA ELLIPSOIDEA:
EFFECTS OF ANTIMETABOLITES,
SURFACTANTS, HORMONES, AND SUGARS
ON THE HARDENING PROCESS
IN THE LIGHT AND DARK

S. Hatano

Department of Food Science and Technology
Faculty of Agriculture
Kyushu University
Fukuoka

SUMMARY

Cells of *Chlorella ellipsoidea* Gerneck (IAM C-27) were synchronously grown under a 28-hr light/14-hr dark regime at 25° C. Hardened at 3° C for 48 hrs, the cells at the L_2 stage (ripening phase) showed the maximum survival rate (70%). Effects of various antimetabolites, surfactants, hormones, and sugars on the hardening process in L_2 cells were examined under different light-period conditions. The results of these studies were as follows:

(1) RNA and protein synthesis and lipid changes are involved in the hardening process and the main hardening process in *Chlorella* seems to be similar to that in higher plants.
(2) The various plant hormones tested had little or no effect on the hardiness increase.
(3) Mitochondria and chloroplasts closely interact at low temperature, and the former plays a principal role in the hardening process and the latter serves as substrate-donor in the light.
(4) Mitochondria change their membranes into a structure much hardier than chloroplasts and function for the repair of freeze-injured chloroplasts.

175

(5) Almost all ATP and $NADPH_2$ produced by chloroplasts are used up for carbon-assimilation, and ATP for the hardiness increase is mainly provided by mitochondria and NADP is reduced via pentose-phosphate cycle.

(6) Glucose enhances the algal hardiness at a very low concentration of 10^{-3} M in the dark, and glucose is not only transformed into energy for the hardiness increase but also accumulated in the form of sucrose.

(7) The hardening process can be divided into an arranging process and a constructing process, and coefficient of utilization of light energy and glucose for the development of hardiness increases with progress of the arranging process.

(8) Protein synthesis on cytoplasmic ribosomes in the arranging process is a prerequisite to the hardiness increase and that on chloroplastic ribosomes is involved in the hardening process at early stages in the light.

(9) The structural changes in *Chlorella* during hardening are different in some respects from those in woody plants.

I.   INTRODUCTION

The world-wide scarcity of food has been predicted to result from the increasing population and unusual atmospheric phenomena, and accordingly, the research to endow plants with chilling and freezing resistance has been increasingly important for food production and preservation. The success of physiological approaches to increasing the hardiness of plants will highly depend on basic research to understand the mechanisms of cold injury and resistance in plants. To investigate the mechanisms more effectively, the plant to be tested should be one which can be readily grown and hardened, and many environmental conditions of the plant should be easily controllable.

We attempted to harden unicellular green algae in several different species of *Chlorella* and *Schenedesmus*, and then found three strains which were different in susceptibility to low temperature. These are *Chlorella ellipsoidea* Gerneck (IAM C-27), (IAM C-102), and *Chlorella ellipsoidea* from the Tokugawa Institute. C-27 strain became hardy during low temperature treatment. C-102 strain was injured by exposure to low temperature. This chilling injury took place only in the light and it was stimulated by oxygen. The last strain was found to be insensitive to low temperature. The hardened cells were able to survive slow freezing down to extremely low temperature. The purpose of the present study is to investigate the mechanisms of frost hardening and survival in *Chlorella ellipsoidea* Gerneck (IAM C-27).

II. ABBREVIATIONS

CAP: chloramphenicol;   CHI: cycloheximide;   DCMU: 3-(3,4-dichlorophenyl)-1,1-dimethylurea;   OGM: oligomycin.

III. MATERIALS AND METHODS

A. *Plant Materials*

Chlorella ellipsoidea Gerneck (IAM C-27) and (IAM C-102) were obtained from the Algal Culture Collection of the Institute of Applied Microbiology, University of Tokyo, Tokyo, Japan.

B. *Synchronous Culture*

Chlorella ellipsoidea was grown synchronously by the method of Tamiya *et al.* (16, 17) with some modifications (3). All cultures for obtaining synchrony of algal growth were run aseptically at 25° C under 9 to 10 kilolux with constant bubbling of filtered air containing about 1% CO_2. The culture for the first cycle was subjected to a 40-hr light/20-hr dark regime. Good synchronization of algal cells was obtained after a second or third cycle of 28-hr light/14-hr dark regime.

C. *Hardening*

Algal cells synchronized at 25° C were directly hardened at 3° C for 48 hrs. During treatment, the culture was kept under light intensity of 9 to 10 kilolux and aerated with air containing about 1% CO_2, unless otherwise stated. To examine the effects of antimetabolites, sugars and so forth, such a substance was added to the culture during hardening and removed by centrifugation.

D. *Freezing and Thawing*

Five milliliters of the culture in the sterilized test tube, which was set aslant to decrease supercooling, was cooled in an air blast freezer at -20° C for 20 hrs, unless otherwise specified. The frozen specimen was transferred to a

bath kept at 25° C and thawed rapidly. The cooling and thawing rates, represented by the time required to change the
temperature between 10° C and -10° C, were about 41 min and
about 57 sec, respectively.

E. Determination of Viability

The viability of algal cells was determined with the
growth curve on the basis of OD_{420}. Previous studies (3, 4)
have demonstrated that the viability determined with the
growth curve agreed with that determined by both colony
count and packed cell volume.

F. Measurement of O_2 Uptake and Evolution

O_2 uptake of algal cells was first measured in the dark,
and then O_2 evolution of the same cells was measured in the
light (5 kilolux). They were measured at 25° C with a Clark-
type oxygen electrode inserted into a 2.3 ml reaction chamber
(made of acryl resin). The culturing medium for algal cells
was used as the reaction medium. Algal suspension (0.2 ml)
containing about 2×10^7 cells was added into the reaction
medium in the chamber. When the effects of oligomycin (OGM)
and DCMU on O_2 uptake and evolution were examined, 5 µl of
their concentrated solutions was injected into the chamber.
Cell number in the reaction medium was counted (with a Thoma
hemacytometer) for calculating the rate of O_2 uptake and evolution per algal cells.

G. Analysis of Glucides

The algal cells were extracted four times with boiling
80% ethanol. The combined extracts were evaporated on the
water bath to dryness. After four washings with ether, the
final precipitate was dissolved with water. This solution
was analyzed by phenol-sulfuric acid method (1) for total
sugars, by Somogyi method (14) for glucose, and by Roe's
method (11) for fructose and sucrose.

The residue obtained after extraction with 80% ethanol
was extracted three times with boiling water. The combined
extracts were analyzed by phenol-sulfuric acid method for
dextrin. The residue remaining after extraction with boiling
water was hydrolyzed with 2% HCl in a boiling water bath for

2 hrs. After the hydrolyzate was centrifuged, the residual material was washed twice with hot water. The supernatant and washings were combined and analyzed by Somogyi method for starch.

IV. RESULTS AND DISCUSSION

A. *Synchronous Culture*

Figure 1 shows the schematic representation of changes of cell size and nuclear pattern at different stages and phases during the life cycle of *Chlorella ellipsoidea* Gerneck (IAM C-27). A mother cell divides into four daughter cells, and a daughter cell grows up through six stages, Da, D-L, L_1, L_2, Lr, and Lf. Of course, the susceptibility of *Chlorella* cells to low temperature varies with the stage. Therefore, we cultured these cells synchronously. Cultured at a 28-hr light/ 14-hr dark regime, C-27's were found to synchronize well after the second or third cycle of the regime (3). To synchronize C-102's, the culture was subjected to a 14-hr light/8-hr dark regime. C-27's were most hardened at the L_2 stage, while C-102's were most injured at the D-L stage. Therefore, we used C-27's at the L_2 stage for frost hardiness.

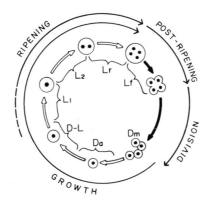

FIGURE 1. *Schematic representation of the changes of cell size and nuclear pattern during the life cycle of Chlorella ellipsoidea (IAM C-27).*

TABLE I.   *Effects of antimetabolites and Surfactants on the Process of Hardening in the Light*

Compounds	Effects
RNA-synthesis inhibitors	
Actinomycin C	Complete inhibition
Rifampicin	No effect
5-fluolouracil	Inhibition
Protein-synthesis inhibitors	
Cycloheximide	Complete inhibition
Chloramphenicol	Little or no inhibition
Energy-synthesis inhibitors	
Oligomycin	Inhibition
DCMU	Inhibition
Surfactants	
Triton X-100	Inhibition
Sodium dodecyl sulfate	Inhibition
Cholate, Deoxycholate	No effect

B.   *Effects of Antimetabolites and Surfactants on the Hardening Process*

Effects were examined of various antimetabolites and sur-factants on the hardening process (4).   The results are summarized in Table I.   Actinomycin D, which binds to DNA and inhibits RNA synthesis, completely inhibited the development of frost hardiness.   5-fluorouracil also remarkably inhibited the hardening process.   Cycloheximide (CHI) and chloramphen-icol (CAP) have been reported to inhibit protein synthesis on cytoplasmic ribosomes (8) and chloroplastic ribosomes (18), respectively.   CHI inhibed the hardening process while CAP did not.   It has been reported that oligomycin (OBM) specific-ally inhibits the energy coupling mechanism in mitochondria and DCMU inhibits the electron transport system in chloro-plasts (5, 6).   OGM and DCMY also remarkably inhibited the de-velopment of frost hardiness.   As the examples of inhibition, the experimental data obtained with OGM and DCMU were shown in Figures 2 and 3.

Triton X-100 and sodium dodecyl sulfate suppressed the hardiness increase but sodium cholate did not.   Furthermore, we observed differences between hardened and unhardened cells in susceptibility to these surfactants (4).

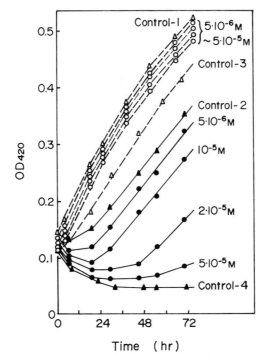

FIGURE 2.   Effect of OGM on the process of hardening.
L_2 cells were hardened in the presence of OGM at the different
concentrations indicated (open and solid circles).   Controls
(open and solid triangles) represent the absence of OGM.
After removal of OGM from the culture, the viability was de-
termined with the growth curve.   Open Symbols: unfrozen
cells; solid symbols: frozen and thawed cells.   Control 1:
hardened and unfrozen cells; Control 2:   hardened and frozen-
thawed cells; Control 3:   unhardened and unfrozen cells;
Control 4:   unhardened and frozen-thawed cells.

These results indicate that RNA and protein synthesis and
lipid changes are involved in the hardening process.   This
fact suggests that the main hardening process in *Chlorella* is
similar to that in higher plants.

C.   Effects of Plant Hormones on the Hardiness Increase

The tests made for L_2 cells showed that the following hormones exert no effect on the development of frost hardiness
at various concentrations tested (10^{-5} to 10^{-3} M):   abscisic
acid; colchicine; kinetin; indole acetic acid; 2,4-D;

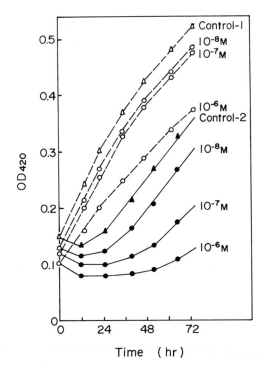

FIGURE 3.   *Effect of DCMU on the process of hardening.*
Meanings of the symbols are the same as in Figure 2.   Controls
3 and 4 were not shown.

gibberellin A_3; maleic hydrazide.   However, abscisic acid
$(10^{-3}$ M) and colchicine (5×10^{-3} M) only slightly enhanced
the hardiness of D-L cells.   It is necessary to reexamine the
experimental conditions and it is interesting to study the
combined effects of the hormones by twos.

D.   *The Mode of Involvements of Mitochondria and Chloroplasts*
     in the Hardening Process

As shown in Figures 2 and 3, OGM and DCMU remarkably inhibited the development of frost hardines.   These results indicate that both mitochondria and chloroplasts are involved in
the process of hardening, because OGM inhibited only the O_2-
uptake activity by mitochondria and DCMU inhibited only the
O_2-evolution activity as shown in Table II.   In *Chlorella*
also, OGM and DCMU have the unique specificity of inhibitory
effect.

TABLE II.   *Effects of OGM and DCMU on the O_2 Uptake and Evolution in Unhardened Cells at 25° C*

Inhibitor	Conc. (M)	O_2 uptake (µ moles $O_2/10^7$ cells·hr)	O_2 evolution
Control	0	60 (100)[a]	316 (100)
OGM	2 x 10⁻⁵	29 (47)	305 (97)
OGM	5 x 10⁻⁵	30 (50)	310 (98)
DCMU	1 x 10⁻⁷	59 (99)	149 (47)
DCMU	1 x 10⁻⁶	59 (99)	-59 (--)[b]

[a]*The figure in parenthesis shows the percentage of activity to that of control.*
[b]*In the presence of 10^{-6} M DCMU, O_2 evolution did not occur and O_2 uptake was detected regardless of illumination.*

An interesting question is: Which organelle is more directly involved in the process of hardening, mitochondrion or chloroplast? To clarify this question, some experiments were carried out. Effects were examined of OGM and DCMU on the changes in the O_2-uptake and the O_2-evolution activity during hardening (Figure 4). During normal hardening (without inhibitor in the light), the O_2-uptake activity increased, while the O_2-evolution activity decreased. OGM, which is to inhibit only the O_2-uptake activity (Table II), stimulated the decrease of O_2 evolution during hardening, and increased the increase of O_2 uptake in an early period of hardening but decreased it later. DCMU, whose nature is to inhibit only the O_2-evolution activity (Table II), had no effect on the decrease of O_2 evolution but completely inhibited the increase of O_2 uptake during hardening. Under dark conditions, neither O_2 uptake nor O_2 evolution substantially changed regardless of the presence of inhibitor. The algal hardiness have been reported to be very limited in the dark (3). These results suggest that mitochondria and chloroplasts closely interact on each other at low temperature, that is, chloroplasts function for mitochondria to maintain the increase of O_2 uptake and mitochondria function for chloroplasts to minimize the decrease of O_2 evolution.

As mentioned above, the O_2 uptake increased during normal hardening while the O_2 evolution decreased. The increase of O_2 uptake was suppressed by both OGM and DCMU. These results suggest that mitochondria are more directly involved than chloroplasts in the process of hardening.

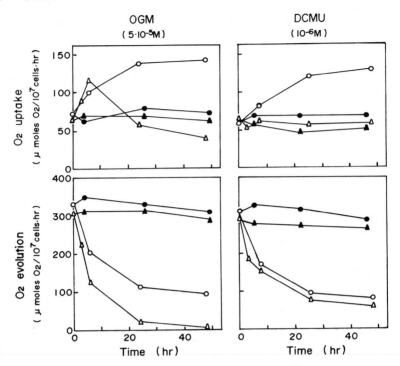

FIGURE 4. *Time-course of O_2-uptake and O_2-evolution activity in Chlorella cells during hardening with OGM or DCMU. L_2 cells were hardened under each condition and harvested at suitable intervals. After removal of the inhibitor, the activities were measured at 25° C. Open circles: cells hardened without inhibitor in the light; open triangles: cells hardened with inhibitor in the light; solid circles: cells hardened without inhibitor in the dark; solid traingles: cells hardened with inhibitor in the dark.*

Furthermore, to understand the mode of involvement of chloroplasts in the hardening process, the algal cells were hardened in the dark by adding glucose into the culture. As shown in Figure 5, the addition of glucose in the dark resulted in the development of frost hardiness which was not all inferior to that in the light. The addition of glucose in the light had no effect on the hardiness. These results suggest that chloroplasts do not involve in the hardening process with glucose in the dark. DCMU did not inhibit the hardiness increase resulting from the addition of glucose in the dark.

We also followed the changes in O_2 uptake and evolution during incubation of frozen-thawed L_2 cells after or without hardening. As Table III shows, the freeze-thaw treatment

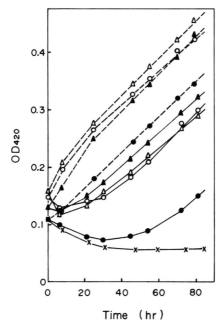

FIGURE 5. Development of frost hardiness by addition of glucose in the dark. L_2 cells were hardened in the presence or absence of glucose (0.1%) and in the light or dark. After removal of glucose from the culture, the viability of cells was determined with the growth curve. Solid and broken curves show, respectively, the growth curves of frozen-thawed cells and unfrozen cells. Open circles: cells hardened without glucose in the light; open triangles: cells hardened with glucose in the light; solid circles: cells hardened without glucose in the dark; solid traingles: cells hardened with glucose in the dark; crosses: unhardened cells. Glucose was added to the culture at the final concentration of 1×10^{-3} M.

completely inhibited the O_2-uptake and the O_2-evolution activity in unhardened cells. The increased O_2-uptake activity in hardened cells was decreased by freeze-thaw treatment to a 70% survival rate. This rate is approximately equal to the survival rate of hardened and frozen-thawed cells. Therefore, it seems that the O_2-uptake activity in freeze-survived cells was hardly inactivated by freeze-thaw treatment. The O_2 uptake in hardened and frozen-thawed cells further decreased during incubation nearly to the same level as before hardening. The decreased O_2-evolution activity in hardened cells further decreased by freeze-thaw treatment. However, this

TABLE III.   *Changes in O_2 Uptake and Evolution During Incubation of Frozen-thawed Cells after or without Hardening*

	Incubation time (hr)	Unhardened cells		Hardened cells	
		O_2 uptake	O_2 evolution	O_2 uptake	O_2 evolution
		μ moles O_2 / 10^7 cells·hr		μ moles O_2 / 10^7 cells·hr	
Unfrozen cells	–	60	356	148	98
Frozen-thawed cells[a]	0	0	0	103	19
	1	–	–	98	32
	2	–	–	87	63
	4	0	0	78	150
	6	–	–	76	256
	8	–	–	70	362
	12	0	0	70	391
	16	–	–	63	384
	24	–	–	30	140

[a]*Frozen-thawed cells were incubated in the same conditions as in synchronous culture.*

decreased evolution was repaired during incubation to the level before hardening. Cell division was observed in 12 hrs after the cells started being incubated. These results suggest that mitochondria change their membranes into a structure much hardier than chloroplasts and that mitochondrial functions are essential for the repair of chloroplasts.

E.   *Metabolism of Glucose and Related Compounds during Hardening in the Dark*

Since glucose remarkably stimulated the development of frost hardiness in the dark, effects were examined of glucose and related compounds on the hardiness increase in the dark. As Table IV shows, glucose was most effective on the hardiness increase in the dark. Sucrose, galactose, and fructose were

TABLE IV. *Development of Frost Hardiness in the Dark by Addition of Sugars and their Related Compounds (5 x 10⁻² M)*

		Unfrozen cells	Frozen-thawed cells
		Increment of OD_{420}	Increment of OD_{420}
None	Light	0.280	0.150
None	Dark	0.215	0.025
Glucose	Dark	0.290	0.175
Sucrose	Dark	0.300	0.140
Galactose	Dark	0.270	0.100
Fructose	Dark	0.250	0.065
Mannose	Dark	0.220	0.030
Pyruvate	Dark	0.215	0.025
Succinate	Dark	0.210	0.025
Citrate	Dark	0.210	0.020
Lactate	Dark	0.195	0.020

The viability of cells was represented by the increment of OD_{420} on the growth curve in the first 72 hrs of incubation.

also capable of increasing hardiness, but pyruvate, succinate, and citrate were incapable. Mannose and lactate were not oxidized by *Chlorella*.

Levitt (7) has reported that ATP and NADPH₂ are necessary to hardening plants. Sagisaka (12, 13) has demonstrated the definite increase in enzymatic activities of pentose-phosphate cycle in poplar xylem during wintering, and has also demonstrated the distinct metabolic shift of glucose 6-phosphate from pentose-phosphate cycle predominance to glycolysis predominance at the time of budding. The sugars tested, which increased the algal hardiness, can be easily metabolized through pentose-phosphate cycle and glycolysis pathway. However, the acids tested, which did not increase the algal hardiness, are difficult to metabolize through pentose-phosphate cycle. The results suggest that the hardening process in *Chlorella* cells also has a high level of pentose-phosphate cycle to reduce NADP.

From these and other results, it appears that, in *Chlorella*, almost all ATP and NADPH₂ produced by chloroplasts are used up for carbon-assimilation, and that ATP for the

TABLE V.   Development of Frost Hardiness in the Dark by
Addition of Glucose at Different Concentrations

Concentration	Unfrozen cells Increment of OD_{420}	Frozen-thawed cells Increment of OD_{420}
5×10^{-2} M	0.300	0.185
2×10^{-2} M	0.305	0.190
1×10^{-2} M	0.300	0.190
5×10^{-3} M	0.290	0.185
2×10^{-3} M	0.285	0.185
1×10^{-3} M	0.285	0.190
5×10^{-4} M	0.270	0.135
2×10^{-4} M	0.250	0.095
1×10^{-4} M	0.235	0.025
0	0.210	0.015

hardiness increase is provided by mitochondria and NADP is
reduced via pentose-phosphate cycle in protoplasm. From the
fact that oligomycin did not completely inhibit the hardiness
increase, it is inferred that ATP in chloroplasts is directly
used for the frost hardiness through a little in its amount
at the highest estimate, but there also remains a possibility
that ATP in chloroplasts is not directly used at all for the
frost hardiness.

As Table V shows, glucose enhanced the algal hardiness at
a very low concentration of 10^{-3} M. Accordingly, there is a
good reason to believe that the hardiness increase by glucose
does not result from its function as a cryoprotective agent.
However, Steponkus (2, 15) has contended that cold acclimation
involves both alterations in membranes and an accumulation of
sucrose. Therefore, the contents of glucides were measured
and the results were shown in Figure 6. The contents of
monosaccharide and dextrin did not change during hardening.
Starch and sucrose remarkably increased in the light. Sucrose
considerably increased in the dark and in the presence of
glucose. These results supported the contention by Steponkus.

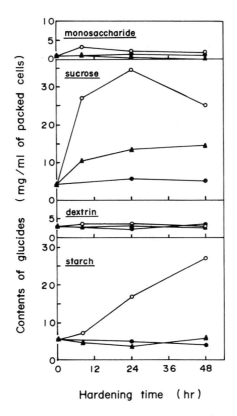

FIGURE 6. Changes of glucides during hardening. Open circles: cells hardened in the light; solid circles: cells hardened in the dark; triangles: cells hardened in the dark with glucose.

F. *An Arranging and a Constructing Process in the Hardening Process*

The hardening process is complicated. If the process can be divided into some parts, it will be easier to study the hardening process.

A 12-hr light period was placed at different stages during the 48-hr hardening. As shown in Figure 7, the illumination was slightly effective on the hardiness increase during the first 24 hrs of hardening. But, at later stages, the illumination became more effective. The cells illuminated for the last 12 hrs were hardier than those continuously illuminated during hardening.

Figure 8 shows the results obtained in a 6-hr light period. A 6-hr light period produced similar results to those

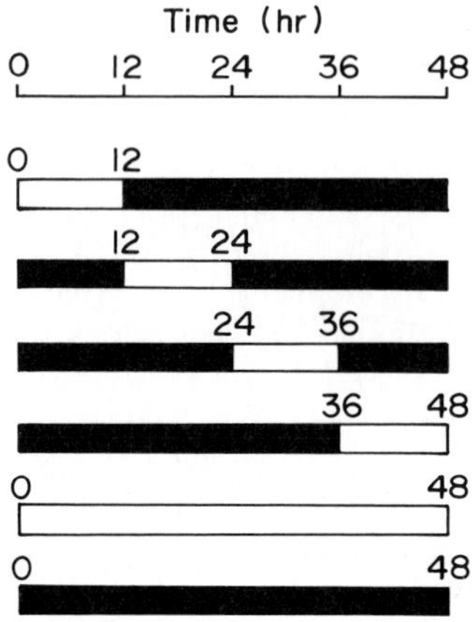

Time (hr)	Unfrozen cells	Frozen-thawed cells
0 12 24 36 48	Increment of OD$_{420}$	Increment of OD$_{420}$
0 12	0.240	0.080
12 24	0.235	0.075
24 36	0.245	0.155
36 48	0.250	0.175
0 48	0.260	0.150
0 48	0.215	0.020

FIGURE 7. *Effect of a 12-hr light period on the hardiness increase. Cultures of L₂ cells were illuminated for 12 hrs at different stages during hardening at 3° C for 48 hrs. Light conditions: white part, light period; black part, dark period.*

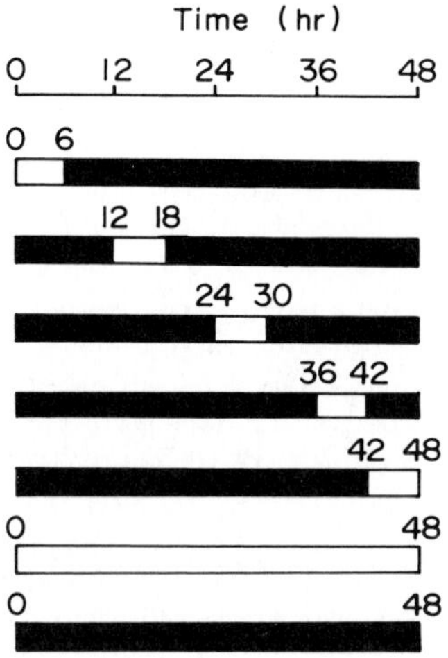

Time (hr)	Unfrozen cells	Frozen-thawed cells
0 12 24 36 48	Increment of OD$_{420}$	Increment of OD$_{420}$
0 6	0.220	0.030
12 18	0.215	0.025
24 30	0.230	0.090
36 42	0.240	0.140
42 48	0.245	0.160
0 48	0.260	0.145
0 48	0.215	0.020

FIGURE 8. *Effect of a 6-hr light period on the hardiness increase. Meanings of the symbols are the same as in Figure 7.*

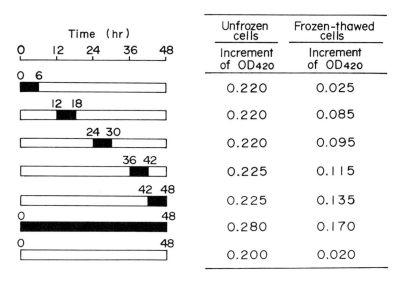

Time (hr)	Unfrozen cells	Frozen-thawed cells
	Increment of OD420	Increment of OD420
0 6	0.220	0.025
12 18	0.220	0.085
24 30	0.220	0.095
36 42	0.225	0.115
42 48	0.225	0.135
0 48	0.280	0.170
0 48	0.200	0.020

FIGURE 9.   Effect of glucose added for 6 hrs at different hardening stages upon the hardiness increase in the dark. Black part: presence of glucose.

obtained in a 12-hr light period. Hardiness did not increase during the first 24 hrs of hardening and the last 6-hr illumination was more effective than the continuous 48-hr illumination.

Figure 9 shows the results obtained with a 6-hr addition of glucose to the cultures. Glucose was added for 6 hrs at different stages in the dark. Glucose was also more effective at later stages. However, in contrast with illumination, glucose was considerably effective for the first 24 hrs of hardening and the last 6-hr addition was less effective than continuous 48-hr addition.

From these results, it appears that the hardening process can be divided into an arranging process and a constructing process. The arranging process proceeds both in the light and in the dark, while the constructing process is endergonic. Coefficient of utilization of light energy and glucose for the development of hardiness increases with progress of the arranging process.

We further examined the effects of CHI and CAP on the hardiness increase in a 12-hr light period. As Figure 10 shows, CHI added in the dark period remarkably inhibited the increase of hardiness regardless of hardening stage, whereas CHI added in the light period slightly inhibited the hardiness increase.

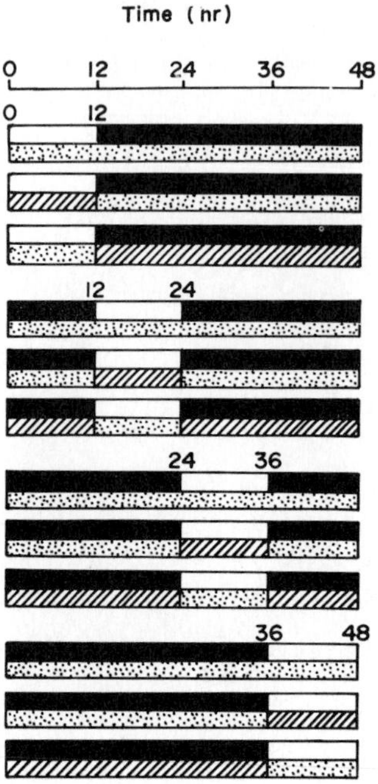

Time (hr)	Unfrozen cells	Frozen-thawed cells
	Increment of OD420	Increment of OD420
	0.285	0.090
	0.270	0.065
	0.255	0.035
	0.280	0.090
	0.270	0.065
	0.260	0.030
	0.285	0.145
	0.285	0.135
	0.250	0.035
	0.295	0.185
	0.300	0.180
	0.270	0.040

FIGURE 10. Effect of CHI added in the light and dark upon the hardiness increase under a 12-hr light period at different hardening stages. Cultures of L_2 cells were illuminated for 12 hrs at different hardening stages. CHI was added to the culture in the light or the dark period at the final concentration of 1×10^{-4} M. To remove CHI, the cells were quickly washed with fresh medium by centrifugation. After or without freeze thawing, their viabilities were determined. Hardening conditions: white part, light period; black part, dark period; hatched part, with CHI; dotted part, without CHI.

These results show that the protein synthesis on cytoplasmic ribosomes in the arranging process is a prerequisite to hardiness increase.

Figure 11 shows the data obtained with CAP. CAP added in a 12-hr light period during the first 24 hrs of hardening caused a slight but definite inhibition while the inhibitory effect on CAP was not observed after 24 hrs of hardening. As mentioned before, CAP had little or no effect on the hardiness increase when it was continuously added during the 48-hr hardening. The results suggest that the protein synthesis on chloroplastic ribosomes was also involved in the hardening

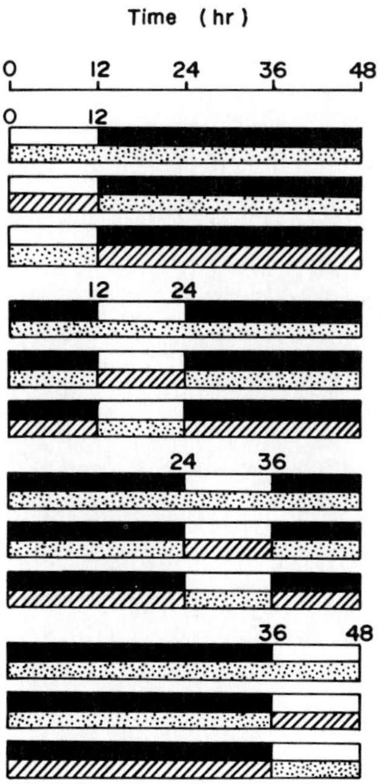

Time (hr)	Unfrozen cells	Frozen-thawed cells
0 12 24 36 48 0 12	Increment of OD_{420}	Increment of OD_{420}
0—12 row 1	0.265	0.085
row 2	0.255	0.025
row 3	0.255	0.080
12 24 row 1	0.260	0.080
row 2	0.250	0.025
row 3	0.250	0.075
24 36 row 1	0.270	0.140
row 2	0.265	0.145
row 3	0.265	0.135
36 48 row 1	0.290	0.180
row 2	0.280	0.180
row 3	0.285	0.180

FIGURE 11. Effect of CAP added in the light and dark upon the hardiness increase under a 12-hr light period at three different hardening stages. Hardening conditions: hatched part, with CAP; dotted part, without CAP. Meanings of other symbols are the same as in Figure 10.

process at early stages in the light. After completion of the protein synthesis, light energy was efficiently utilized for the development of frost hardiness.

G. Structural Changes in Chlorella during Hardening

The structural changes observed by electron microscopy are briefly described. As shown in Figures 12 and 13, the structural changes in *Chlorella* during hardening were different in some respects from those in woody plants (9, 10). Large vacuoles divided into smaller ones also in *Chlorella*, but the infolding of plasmalemma was not apparent. Starch remarkably accumulated during hardening. Swelling of stromata and meandering of chloroplast membranes were observed in hardened cells. As Figure 14 shows, numerous lipid particles were observed along the cell wall. These particles were

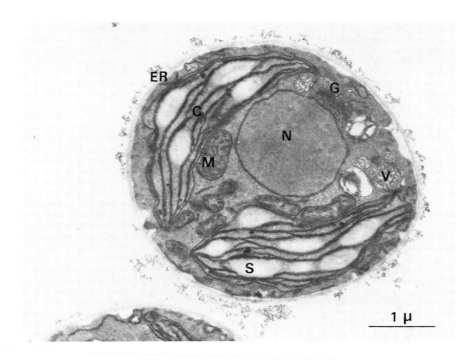

*FIGURE 12.   A hardened cell fixed in glutaraldehyde and
potassium permanganate solutions.   C: chloroplast; G:
Golgi complex; ER: endoplasmic reticulum; M: mitochondrion;
N: nucleus; S: starch granule; V: vacuole.   X 20,000.*

closely related to the development of frost hardiness.   When
the development of frost hardiness was inhibited by various
antimetabolites, the formation of the particles was inhibited,
too.   These particles seem to show one spot on thin-layer
chromatogram and this spot could be divided into four peaks on
gas-chromatogram.

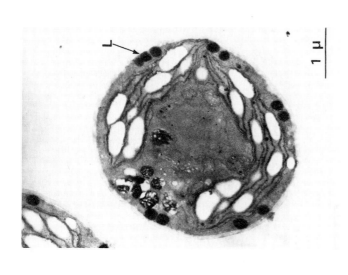

FIGURE 14. A hardened cell fixed in glutaraldehyde and osmic acid solutions. L: lipid body. X 15,000.

1 μ

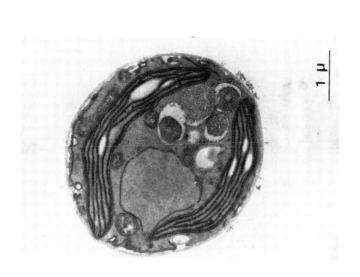

FIGURE 13. An unhardened cell fixed in glutaraldehyde and potassium permanganate solutions. X 15,000.

1 μ

195

REFERENCES

1. Dubios, M., Gilles, K. A., Hamilton, J. K., Rebers, P. A., and Smith, F., *Anal. Chem. 28*, 350 (1956).
2. Garber, M. P., and Steponkus, P. L., *Plant Physiol. 57*, 681 (1976).
3. Hatano, S., Sadakane, H., Tutumi, M., and Watanabe, T., *Plant & Cell Physiol. 17*, 451 (1976).
4. Hatano, S., Sadakane, H., Tutumi, M., and Watanabe, T., *Plant & Cell Physiol. 17*, 643 (1976).
5. Katoh, S., and San Pietro, A., *Arch. Biochem. Biophys. 122*, 144 (1967).
6. Lee, C., and Ernster, L., *Biochem. Biophys. Res. Commun. 18*, 523 (1965).
7. Levitt, J., *in* "Cellular Injury and Resistance in Freezing Organisms" (E. Asahina, ed.), p. 56. Inst. Low Temp. Sci., Sapporo (1967).
8. McKeehn, W., and Hardesty, B., *Biochem. Biophys. Res. Commun. 36*, 625 (1969).
9. Otsuka, K., *Low Temp. Sci., Ser. B 30*, 33 (1972).
10. Pomeroy, M. K., and Siminovitch, D., *Can. J. Bot. 49*, 787 (1971).
11. Roe, J. H., *J. Biol. Chem. 107*, 15 (1934).
12. Sagisaka, S., *Low Temp. Sci., Ser. B 28*, 43 (1970).
13. Sagisaka, S., *Plant Physiol. 50*, 750 (1972).
14. Somogyi, M., *J. Biol. Chem. 160*, 61 (1945).
15. Steponkus, P. L., *Plant Physiol. 47*, 175 (1971).
16. Tamiya, H., Morimura, Y., Yokota, M., and Kunieda, R., *Plant & Cell Physiol. 2*, 383 (1961).
17. Tamiya, H., Morimura, Y., *in* "Synchrony in Cell Division and Growth" (E. Zeuthen, ed.), p. 565. John Wiley & Sons, Inc. (1964).
18. Wolfe, A. D., and Hahn, F. E., *Biochim. Biophys. Acta 95*, 146 (1965).

COLD ACCLIMATION OF CALLUS CULTURES
OF JERUSALEM ARTICHOKE

Y. Sugawara
A. Sakai

The Institute of Low Temperature Science
Hokkaido University
Sapporo

I. INTRODUCTION

Hardy plants characteristically undergo a series of changes in the autumn which enable them to withstand freezing stress. It has been suggested that cold acclimation involves a sequence of complicated processes which are mutually dependent (29). To understand these processes, extensive studies have been performed using intact plants under natural or artificially controlled conditions (10). However, such materials are sometimes unsuitable for studying the cellular mechanism of cold acclimation in plant due to the complex interrelationships between different tissues and organs, and the difficulty of isolation of cell organellles, especially cell membrane under mild conditions. The use of tissue culture technique may reduce these troubles and also enable us to work with more homogeneous and large masses of tissue on the sterile and chemically defined medium under easily controlled conditions the whole year round. From such a system we can easily elicit a clue of gaining an insight into the mechanism of cold acclimation in plants.

Tumanov *et al*. (24) first demonstrated that the callus derived from cherry twigs became hardy to -20 or -30° C when hardened at 2° C and subsequently subjected to sub-freezing temperatures for 20 days. Bannier and Steponkus found that callus tissues of *Chrysanthemum* survived freezing to -16° C after hardening (1). We (15) demonstrated that callus tissues from poplar twigs which were frozen to the temperature of liquid nitrogen proliferated vigorously when cultured at 26°C.

Recently, Ogolevets performed several studies to induce frost hardiness in callus tissues from several fruit twigs (13).

In the present report we describe some characteristic features of cold acclimation processes in callus cultures of Jerusalem artichoke.

II. MATERIALS AND METHODS

In this experiment tubers of Jerusalem artichoke (*Helianthus tuberosus* L.) were used as materials. Tubers which had been harvested in late November were stored at $0°$ C in the dark and mainly used within six months, since the hardening of the callus from tubers stored over six months became less effective.

Tubers were sterilized with about 4% sodium hyperchloric acid for 15 min. Cylinders with 1 cm diameter were isolated from the surface-sterilized tuber and disks with 1 mm thickness were prepared. The slices were inoculated on the modified Linsmaier-Skoog's medium (11) (with only NO_3 as nitrogen source) containing 2.4-D as auxin and incubated at $26°$ C in the dark. The procedures for tissue culture were done according to the methods described by Yasuda *et al.* (25). Under this condition, the first DNA synthesis for callus formation occurs synchronously in the least 60% of cells in the disk within 36 hrs (25). Cold acclimation experiments were done with the callus which had been grown at 26° C for 8 days in the dark.

Flasks containing callus on agar medium were transferred to the chamber of 8 or $0°$ C and incubated there for different lengths of time in the dark or light to induce hardiness of callus. After being removed from agar medium and blotted briefly, acclimated and non-acclimated calluses were frozen in petri dish (9 cm in diameter) or test tube (10 cm long, 1 cm diameter) by transferring these containers to the temperature-controlled boxes of -5, -10, -15 and -20° C, successively, at 2 hr intervals. Frozen calluses were held at each test temperature for 18 hrs. They were then thawed slowly in air at 0 to 3° C.

Survival rate of callus after freeze-thawing was mainly determined by TTC reductions test as described by Steponkus and Lanphear (20). In some experiments callus survival was estimated by the amino acid release test. Released amino acid was determined by the method of Cocking and Yemm (4).

Freeze-fracturing of callus was done essentially according to the procedure of Moor and Muhlethaler (12). Prior to freeze-fracturing callus tissue was fixed with 3%

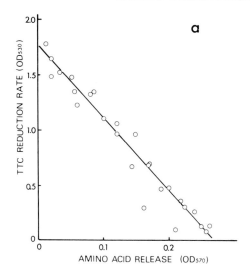

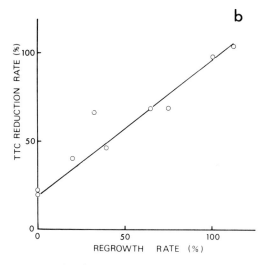

FIGURE 1.   Correlations between TTC reduction rate and amino acid release (a), regrowth rate (b) of callus after freeze-thawing. Calluses hardened at 0° C for 18 days were frozen to the temperatures ranging from -3 to -20° C and then thawed in air at 3° C. For determination of released amino acid, frozen-thawed calluses were shaken in the fresh medium at 0° C for 2 hrs. Amino acid released in the medium was determined by ninhydrin reaction (10). Thawed calluses were trnasplanted to the medium on which they were grown just before freezing (conditioned medium), and cultured at 26° C for 14 days before determination of regrowth rate.

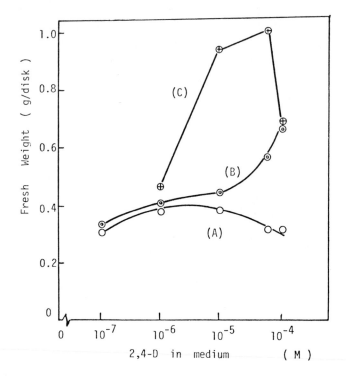

FIGURE 2.   Effect of 2,4-D on the growth rate of callus during growing and hardening.   (A), 26° C for 8 days; (B), (A) plus 8° C for 10 days and 0° C for 8 days; (C), (A) plus 8° C for 10 days and 0° C for 28 days.   Each value was represented as average of seven to twelve calluses.

glutaraldehyde in 0.05M phosphate buffer, pH 7.0 containing 0.25M sucrose for 3 hrs at 1° C and then treated with 30 or 40% glycerol solution.   Specimen was fractured and minimum etching allowed at -96° C.

III.   RESULTS

First we examined the validity of three methods used for determination of the survival rate of frozen-thawed callus. Results are shown in Figure 1.   A reciprocal correlation was observed between TTC reduction rate and amino acid release in the frozen-thawed calluses (a), and a parallel correlation between TTC reduction rate and regrowth rate was also observed

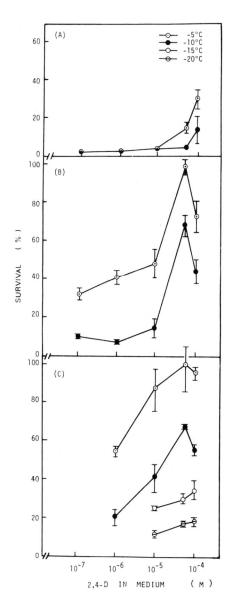

FIGURE 3. *Survival of the calluses grown on the media with various concentrations of 2,4-D before and during hardening. (A), before hardening; (B), after hardening at 8° C for 10 days and then at 0° C for 8 days; (C), after hardening at 8° C for 10 days and then at 0° C for 28 days.*

(b). These good correlations suggest that these methods can
be utilized as reliable criteria for determination of callus
survival after freeze-thawing.

The callus formation and growth rate of artichoke tuber
firmly depend on the concentration of exogeneously supplied
auxin (26). Thus, the effects of 2,4-D concentration in the
culture medium on the callus formation, growth rate, and the
cold acclimation of callus were determined. As shown in
Figure 2(A), the optimal concentration of callus formation and
growth of callus was in the range of 10^{-5} to 10^{-6}M as reported
by several workers (25, 26). Artichoke calluses were precon-
ditioned at 8° C for 10 days before being hardened at 0° C
for 8 or 28 days. Results were shown in Figures 2 and 3. The
calluses grown on the medium containing lower concentrations
of 2,4-D than 10^{-5}M showed little or no growth during hardening at 0° C for 8 days (Figure 2B), while calluses on the
medium containing 2,4-D higher than 10^{-5} grew vigorously during hardening at 0° C for 28 days (Figure 2 B and C).

Increase in the hardiness of calluses grown on the media
containing various concentrations of 2,4-D during hardening
were summarized in Figure 3. Before hardening nearly all
of the calluses were killed by freezing at -5° C. A rapid
increase in the hardiness of the callus acclimated at 8° C
for 10 days, and subsequently at 0° C for 8 days was observed
in the callus grown on the medium with a high concentration
of 2,4-D (Figure 3 B). Concentration of 2,4-D, 5 x 10^{-5}M was
most effective in increasing hardiness of artichoke callus.
Further hardening, however, produced only a slight increase
in hardiness (Figure 3 C). These results suggest that the
artichoke callus grown on the medium with 5 x 10^{-5}M had already hardened to near the maximal level after treatment at
8° C for 10 days and then at 0° C for 8 days. Figure 4
shows the time course in increase of hardiness of artichoke
callus exposed to 0° C. The hardiness reached to a maximal
level at the 10th day. At this time, the survival rate
amounted to about 70% to -10° C. Previously, we reported that
the survival rate of artichoke tuber wintering in our campus
was about 40% at -10° C when determined by TTC reduction rate
(22). This indicates that the hardiness of callus hardened
fully is slightly higher than that of the tuber.

In the following experiments, the calluses grown on the
medium with 5 x 10^{-5}M 2,4-D were mainly used and they were
hardened at 0° C for the time longer than 10 days to induce
the maximal hardiness.

In our previous study, a positive effect of light illumination for increasing hardiness of poplar callus was observed
(15). As shown in Table I, illumination also stimulated
hardiness increase of artichoke callus. Illumination during
both growning at 26° C and hardening at 0° C was most

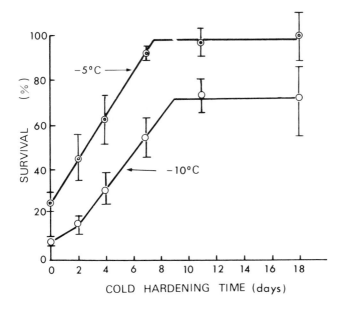

FIGURE 4.   *Increase in hardiness of callus during harden-ing at 0° C.   Calluses grown at 26° C for 8 days were hardened at 0° C for different lengths of time.*

TABLE I.   *Effect of Light on the Development of Cold Hardiness of Jerusalem Artichoke Callus*

| | | Survival (%)[a] | | |
| | Cold | | | |
Growing	hardening	-5° C	-10° C	-15° C
Light	Light	100 ± 4.0	94.6 ± 6.3	44.8 ± 1.0
	Dark	100 ± 8.1	90.2 ± 13.2	34.4 ± 1.6
Dark	Light	100 ± 7.0	86.7 ± 17.0	16.4 ± 2.7
	Dark	100 ± 8.8	47.9 ± 2.3	9.5 ± 1.0

Calluses were grown at 26° C for 8 days in the dark or light at 3,800 lux. and then hardened at 0° C for 10 days in the dark or light at 2,600 lux.

[a]*Average ± SE of four values.*

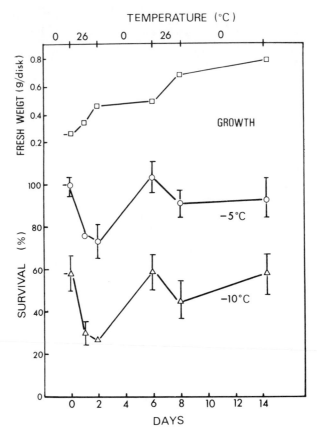

FIGURE 5.　Changes in fresh weight and hardiness of callus during hardening and dehardening cycle.　Calluses previously hardened at 0° C for 14 days were dehardened at 26° C for 3 days and then rehardened at 0° C.

effective for enhancing callus hardiness.　Also, calluses illuminated only during either growing or hardening were observed to be considerably effective.

　　In the next experiment, we examined hardiness change of callus which was hardened and dehardened repeatedly.　The callus hardened at 0° C for 18 days was dehardened at 26° C for 2 days in the dark (Figure 5), resulting in a great decrease in the survival rate to 70% at -5° C.　The hardiness decrease of dehardened callus was restored to the previous hardy level by rehardening at 0° C for 4 days.　However, during subsequent dehardening hardiness change of callus was little.

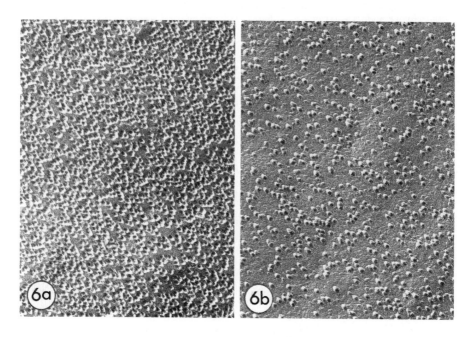

FIGURE 6.   Fracture faces of plasma membrane of callus
grown at 26° C for 8 days.   a, Fracture face of P; b, fracture
face E.   X100,000.

Many studies suggest that some alterations occur in cell
membrane during cold acclimation (6, 9, 17, 18, 27, 28).
Thus, plasma membrane of unhardened and hardened calluses was
examined by freeze-fracturing method. Many studies revealed
that little or no difference was observed in the figures of
fractured planes whether specimen had been fixed with glutaraldehyde prior to freeze-fracturing (2, 3).   Therefore, the
callus tissues were fixed with 3% glutaraldehyde at 1° C
prior to freeze-fracturing. Electron microscopical views of
the fracture faces of plasma membrane of the unhardened callus
are shown in Figure 6. The face of half-membrane left frozen
to the cytoplasm (fracture face P) has a higher concentration
of intra-membrane particles than that of another half-membrane
left frozen to the cell wall (fracture face E).   The fracture
faces of plasma membrane of the acclimated callus are shown
in Figure 7. In comparing those two calluses, the most striking difference between them is less particle concentration on
the fracture face D of hardened callus than that of the unhardened one (Figures 6-8). Determination of particle concentration showed that the particle concentration of fracture
face E of the unhardened callus was 370 $\pm$ 55 particles/μm^2,

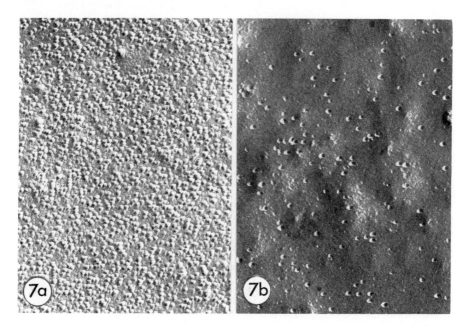

FIGURE 7.   Fracture faces of the plasma membrane of hard-ened callus.   Callus previously grown at 26⁰ C for 8 days was hardened at 0⁰ C for 12 days.   a, Fracture face of P; b, frac-ture face E.   X100,000.

whereas that of hardened one was 115 ± 29 particles/μm^2. However, in the particle concentration of fracture face P, little or no significant difference was observed between unhardened and hardened calluses (Figure 8).

The decreased concentration of intramembrane particles on the fracture face E was restored to the initial level by incubating the hardened callus at 26° C for 3 days (Figure 8 C). When hardened callus was rehardened at 0° C for 4 days, the particle concentration on the fracture face E of plasma membrane decreased to the fully hardened level (Figure 8 D). During this procedure (alternative temperature treatments), the particle concentration of the fracture face P of plasma membrane showed a tendency to decrease gradually, probably due to the proceeding of callus aging.

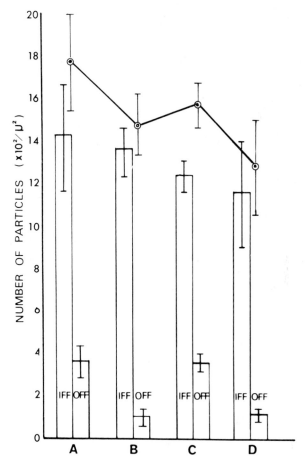

FIGURE 8.   Changes in particle concentration on the fracture faces of plasma membrane of callus during hardening and dehardening cycle. A, before being hardened; B, hardened at 0° C for 14 days; C, dehardened at 26° C for 4 days; D, dehardened at 26° C for 3 days and then rehardened at 0° C for 4 days. Particle number of each fracture face was determined from counting on at least ten photographs of different cells. The vertical line indicates SE. IFF, fracture face P; OFF, fracture face E; ⊙——⊙, total number of particles.

IV. DISCUSSION

    The role of auxins in the cold acclimation remains obscure although several attempts have been made to elucidate this by supplying auxins exogenously to intact plants (10)

and avena coleoptile (23). The responses of plants to exogenously supplied auxins are very diverse depending on the
physiological condition and type of tissue. Auxins at the
optimal concentration in the culture medium are required to
induce cell division and for a continuous growth of cultures
(8). In artichoke callus culture the optimal concentration
of 2,4-D is in the region of $10^{-6}M$ both for cell division and
fresh weight increase (26), while the concentration necessary
for enhancing maximum hardiness was observed to be approximately $5 \times 10^{-5}M$ (Figure 3). Apparently such a high concentration of 2,4-D is superoptimal for the callus growth at
26° C. However, such a concentration of 2,4-D induced appreciable callus growth even at the temperatures of 8 and 0° C
(Figure 2). Similar results were obtained in NAA (unpublished results).

Shannon *et al.* previously showed that higher concentration
of 2,4-D caused a marked decline of the optimal temperature
for growing of *Zea mays* mesocotyl (16). Cox and Levitt observed that the increase in hardiness of cabbage leaves was
proportional to growth rate during hardening period; e.g.,
younger leaves are hardened well (5). They considered that
hardening effect of cabbage leaves may be directly related to
leaf development, but not to growth *per se* (10).

In tissue culture, Bannier and Steponkus (1) and Ogolevets
(13) revealed that growth stage of culture is an important
factor for acclimating the callus fully. Actively growing
younger cells which are in early stages of culture develop
hardiness higher than those of old ones (advanced stage).
Such young cells were observed electron-microscopically to be
filled with a highly dense cytoplasm resulting from new protein synthesis, specific metabolic activities (21) and to have
a smaller cell size as compared with the cells in the advanced
stage.

Our previous study showed that the omission of 2,4-D
from the medium during hardening stimulated hardiness increase
of artichoke callus previously grown on the medium with $5 \times 10^{-5}M$ 2,4-D, while a high concentration of 2,4-D during hardening suppressed hardiness increase (22). However, the hardi-
ness of callus previously grown on the medium with $10^{-6}M$
2,4-D increased proportionally to growing time at 26° C on
the medium with $5 \times 10^{-5}M$ 2,4-D before hardening. In this
experiment, no appreciable increase in hardiness was detected
when 2,4-D was supplied to the callus only in the period of
hardening (unpublished results).

These results may suggest that 2,4-D at a high concentration in the culture medium plays an important role in the cold
acclimation process of artichoke callus, through the modification of callus growth and/or some metabolic shifts in a growing period just before being exposed to 0° C.

Most research attention to cold acclimation in plants has been focussed on the changes in cellular membranes (6, 9, 17, 18, 27, 28). Hardiness changes have been intimately associated with changes in cellular membrane components, especially phospholipids and proteins. Also, seasonal changes in cell membranes have been related to seasonal cycle of hardiness (14). In the present study, concentration of intramembrane particle on the fracture face E of plasma membrane was observed to decrease after hardening and was restored to the initial level after dehardening. These changes in particle concentration were intimately associated with hardiness change of artichoke callus. Such an ultrastructural change suggests that some alterations in hydrophobic region (3), especially in the half closest to the extracellular space of plasma membrane, occur during hardening and that temperature significantly controls particle concentration, probably due to lipid metabolism change and development of hardiness.

The decrease in particle concentration on freeze-fracturing face of cellular membrane during cold acclimation was first observed on the inner fracture face of chloroplast thylakoid membrane by Gaber and Steponkus (7, 19). In addition they observed a change of particle size distribution on this face, which might be based on the structural rearrangement of hydrophobic matrix of chloroplast thylakoid (19). In artichoke callus, however, little or no changes was observed in the size distribution of intramembrane particles both after hardening and dehardening (unpublished results). From these results it may also be considered that the cold acclimation involves the transition of plasma membrane, especially in the half closest to extracellular space from lipid depleted state to lipid enriched state as shown by Yoshida (28). However, we have little or no available data as to how lipid changes alter the membrane properties and how the altered membrane properties confer protection against freezing. Further studies are required to elucidate the causal relation between such a structural alteration of cell membrane and hardiness change.

REFERENCES

1. Bannier, L. J., and Steponkus, P., *J. Amer. Soc. Hort. Sci. 101*, 409-412 (1976).
2. Branton, D., and Park, R. B., *J. Ultrastructural Res. 19*, 283-303 (1967).
3. Branton, D., *Ann. Rev. Plant Physiol. 20*, 209-238 (1969).
4. Cocking, E. C., and Yemm, E. W., *Biochem. J. 58*, xii (1954).

5. Cox, W., and Levitt, J., *Plant Physiol. 44*, 923–928 (1969).
6. de la Roche, I. A., Pomeroy, M. K., Weinberger, P., and Kates, M., *Can. J. Bot. 50*, 2401–2409 (1972).
7. Gaber, M. P., and Steponkus, P., *Plant Physiol. 57*, 681–686 (1976).
8. Gauthalet, R. T., *Ann. Rev. Plant Physiol. 6*, 433–486 (1955).
9. Kuiper, P. J. C., *Plant Physiol. 45*, 684 (1970).
10. Levitt, J., "Responses of Plants to Environmental Stresses," 697 pages. Academic Press, New York (1972).
11. Linsmaier, E. M., and Skoog, F., *Physiol. Plant. 18*, 100–127 (1965).
12. Moor, H., and Muhlethaler, K., *J. Cell Biol. 17*, 609–628 (1963).
13. Ogolevets, I. V., *Fiziol. Rast. 23*, 139–145 (1976).
14. Pomeroy, M. K., and Siminovitch, D., *Can. J. Bot. 49*, 787–795 (1971).
15. Sakai, A., and Sugawara, Y., *Plant Cell Physiol. 14*, 1201–1204 (1973).
16. Shannon, J. C., Hanson, J. B., and Wilson, C. M., *Plant Physiol. 39*, 804–809 (1964).
17. Siminovitch, D., Rheaume, B., Pomeroy, K., and Lepage, M., *Cryobiol. 5*, 202–225 (1968).
18. Smith, D., *Cryobiol. 5*, 148–159 (1968).
19. Steponkus, P. L., Gaber, M. P., Myers, S. P., and Lineberger, R. D., *Cryobiol. 14*, 303–321 (1977).
20. Steponkus, P., and Lanphear, F. O., *Plant Physiol. 42*, 1423–1426 (1967).
21. Street, H. E., *In* "Biosynthesis and its Control in Plant" (V. B. Milborrow, Ed.), pp. 93–125. Academic Press, London (1973).
22. Sugawara, Y., and Sakai, A., *Low Temp. Sci., Ser. B 34*, 1–8 (in Japanese with English summary) (1976).
23. Tumanov, I. I., and Turunova, T. I., *Fiziol. Rast. 5*, 112–122 (1958).
24. Tumanov, I. I., Butenko, R. G., and Ogolevets, I. V., *Fiziol. Rast. 15*, 749–756 (1968).
25. Yasuda, T., Yajima, Y., and Yamada, Y., *Plant Cell Physiol. 15*, 321–329 (1974).
26. Yoemann, M. M., *In* "Plant Tissue and Cell Culture" (H. E. Street, ed.), pp. 31–58. Blackmell Publication, Oxford (1973).
27. Yoshida, S., and Sakai, A., *Plant Cell Physiol. 14*, 353–359 (1973).
28. Yoshida, S., *Plant Physiol. 57*, 710–715 (1976).
29. Weiser, C. J., *Science 169*, 1267–1278 (1970).

Part IV
Supercooling

THE OCCURRENCE OF DEEP UNDERCOOLING
IN THE GENERA PYRUS, PRUNUS AND ROSA:
A PRELIMINARY REPORT[1]

C. Rajashekar
M. J. Burke

Department of Horticulture
Colorado State University
Fort Collins, Colorado

I. INTRODUCTION

Plants that survive subfreezing temperatures do so by
several means. Two of the most important are tolerating cell
dehydration caused by extracellular ice formation or by merely
avoiding the freezing process in certain critical tissues.
The degree of tolerance to extracellular ice formation is
highly variable and depending on the species, stage of development and season may reach a temperature limit below liquid
nitrogen temperature ($-196°$ C) (7). Avoiding freezing is
accomplished by deep undercooling which only confers hardiness
to a limited extent and never below the homogeneous nucleation
temperature of the tissue solution, about $-41°$ C (5).
The deep undercooled water in plants exists as a meta-
stable condition which becomes unstable and freezes at the
homogeneous nucleation temperature or higher. This freezing
is rapid and leads to intracellular ice which is invariably
fatal. The tissues that deep undercool are the xylem ray
parenchyma and floral primordia in many hardwood trees and

[1]*These studies were supported in part by grants from the
National Science Foundation (BMS 74-23137), the Petroleum
Research Fund (PRF 9702-AC1, 6) the Research Corporation, the
Horticultural Research Institute and the Colorado Experiment
Station.*

FIGURE 1. Hardiness zone map of North America. The map is divided into three hardiness regions. Region A has average annual minimums below -40° C and extreme minimums much lower. Region B has finite probability of -40° C minimums. Region C has very little probability of -40° C. Deep undercooling trees only occur in the very southern parts of region A. However, they dominate the forests in regions B and C.

shrubs (4). Microscopic and DTA (differential thermal analysis) studies indicate that freezing of the deep undercooled water in these tissues is lethal and results in "black heart" injury in twigs and complete blackening of floral primordia.

Evergreen (conifer) and deciduous species of the northern Boreal Forest which covers Canada, Alaska and northern Eurasia do not deep undercool. Most deciduous species native to the Eastern Deciduous Forest covering the eastern half of the U.S. do deep undercool. George *et al.* (5) suggested that the undercooling characteristic is the causal factor determining the above differences in geographic distribution of native species. Quamme (9) and Strang and Stushnoff (14) suggested that establishment and cultivation of fruit species in northern regions were limited chiefly by the deep undercooling characteristic. The northern regions where -40° C or lower temperatures are probable are the regions where the deep undercooling plants do not occur (Figure 1) (5).

It is now clear that deep undercooling limits the cold hardiness of many important horticultural trees and shrubs to -40° C. If this temperature is exceeded, xylem ray parenchyma and floral primordia may be killed. Therefore, the deep undercooling characteristic is undesirable relative to other mechanisms of cold hardiness such as tolerance of extracellular freezing. Possibly the deep undercooling characteristic can be removed through breeding efforts. An important aspect of undercooling studies is finding if deep undercooling cultivated plants have relatives that do not have the deep undercooling characteristic with which they can be crossed. Is it possible that the deep undercooling characteristic can be removed through interspecific crosses between hardy non-undercoolers and cultivated undercoolers?

Here, results regarding the occurrence of deep undercooling are discussed relative to the species members of three horticulturally important genera--*Pyrus* (pear), *Prunus* (stone fruits) and *Rosa* (rose). Interest is in the absence of deep undercooling characteristic in closely related species which can potentially be crossed. Most of the cultivated taxa tested showed deep undercooling. In *Prunus* several very hardy native species do not deep undercool and these species might be important in breeding programs aimed at eliminating deep undercooling.

II. DEEP UNDERCOOLING IN THE GENUS *PYRUS*

Twigs of pear species native to various parts of the world, primarily Europe and Asia, have been studied using DTA techniques to detect the presence of the deep undercooling characteristic (11). Preliminary results of these studies are in Table I. All the *Pyrus* species tested had low temperature exotherms and therefore all the plants were deep undercoolers. Frequently, there were multiple exotherms reaching a

TABLE I.   *Results of DTA Studies of Pyrus Species Native to Various Parts of the World.*[a]

Pyrus species	Killing temperature of twig (^{O}C)	Twig exotherm temperature (^{O}C)	Leaf bud exotherm temperature (^{O}C)
P. caucasia	-33	-26, -36	
P. hondoensis	-32	-30	-30
P. dimorphophylla	-32	-24, -36	
P. pyraster	-30	-31, -38	
P. nivalis	-29	-29, -40	
P. communis	-29	-21, -30	-30
P. betulaefolia	-29	-22, -28 -36	-14, -22, -26
P. ussuriensis	-28	-27, -40	
P. pyrifolia	-28	-24, -39	
Kansu pear	-28	-27	
P. elaeagrifolia	-28	-25, -39	
P. heterophylla	-28	-29, -38	
P. longipes	-28	-18, -22 -30, -40	-23, -28 -31
P. salicifolia	-27	-25	
P. amygdaliformis	-27	-25	
P. fauriei	-27	-27, -34	
P. calleryana (24-38)	-27	-26, -38	-34
P. calleryana (11)	-25	-24, -40	
P. calleryana (7)	-25	-24, -29, -34	-28
P. cordata	-24	-23, -37	
P. syriaca	-22	-18, -28	-11 - - -27
P. koehnei	-18	-18	
P. pashia	-16	-32	

[a]*Twigs and leaf buds used in the study were obtained in January, 1977 from the collection of Pyrus species maintained at Oregon State University by Dr. M. Westwood. DTA's and viability tests of stem sections and leaf buds were made as described by George et al. (5). DTA patterns of twigs for old and new growth were similar. The exotherm temperature indicates the freezing temperature of deep undercooled twig or bud water (11).*

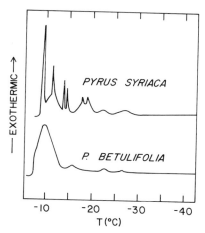

*FIGURE 2.   DTA profiles of leaf buds of two Pyrus species.
The peaks in the DTA profiles are called exotherms and they
result from the freezing of undercooled water.   The high temperature exotherms at about -10° C are not associated with injury to the bud.   The smaller low temperature exotherms are
associated with bud injury (see text).*

maximum number of four in *P. longipes*.   Quamme *et al.* (10)
have also observed multiple low temperature exotherms, as
many as four in Haralson apple.   They show that the number of
exotherms is highly variable, being primarily dependent on the
season.   Viability testing cold stressed tissues using visual
browning and 2,3,5-triphenyl-tetrazolium chloride methods (16)
indicate that xylem is the most susceptible tissue to cold
injury followed by pith.   The injury usually coincided with
the initiation of the low temperature exotherms.   This will be
discussed in more detail later.
     The results of DTA experiments on leaf buds of selected
Pyrus species are summarized in Table I.   Of the species
studied, *P. elaeagrifolia* and *P. nivalis* did not show any
undercooling in their leaf buds.   As in xylem, leaf buds of a
few species gave multiple exotherms.   However, *P. syriaca* was
the only species that had a DTA profile typical of buds where
the primordia deep undercool (e.g. compare Figures 2 and 4).
Low temperature exotherms in the buds of other *Pyrus* species
probably originate from deep undercooling in the xylem of
the bud axis (Figure 2).   There is good precedence for this.
Pierquet *et al.* (8) showed that the low temperature exotherm
in grape bud resulted from xylem associated with the bud and
a similar result was obtained in azalea (6).   In any case the
low temperature exotherm in the buds was associated with the
killing temperature of the bud as determined using tissue
browning as a viability test.

All species tested were killed at -33° C or above and all had the deep undercooling characteristic. Therefore, it seems unlikely that any of these species will be of value in breeding efforts to eliminate deep undercooling in *Pyrus*.   *P. ussuriensis*, possibly the hardiest *Pyrus* species, adapted to cold siberian winters and known to have breeding potential, does in fact deep undercool to -38° C.

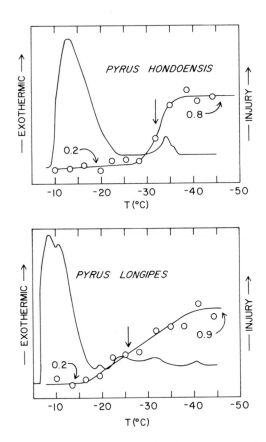

FIGURE 3.   DTA profiles and relative electrolyte leakage for two Pyrus species. The solid curve gives the DTA and the peaks indicate freezing points of undercooled water. The curve connecting circles shows fractional cell disruption as measured by electrolyte leakage. The numbers on this latter plot give fractional electrolyte leakage for the temperatures indicated. The arrow gives the visually estimated killing temperature using browning of stressed tissue to estimate tissue injury.

III.   COLD INJURY AND DEEP UNDERCOOLING IN *PYRUS*

It is now known that many native and cultivated woody
species including pear (9) have the low temperature exotherm
at the xylem killing temperature (5, 10). However, freezing
of undercooled cellular water in multiple events may not always result in a perfect concurrence of the lowest exotherm
with the death of xylem. Mainly this is because intracellular
freezing is spread over a broad range of temperatures and obviously it becomes difficult to find a well defined killing
point.
The loss of electrolytes from frozen tissues has long
been used in assessing the frost hardiness of plants (2) and
recent workers have developed conductivity methods to quanta-
tively measure the tissue injury as a viability test (3, 15).
These conductivity methods are particularly useful for woody
tissues which are complex. Since the low temperature exotherms correspond to the freezing of deep undercooled water
within the cells and perhaps to cell lysis, it is interesting
to relate these freezing events to the electrolyte loss from
the injured *Pyrus* twig tissues. In *P. hondoensis* (Figure 3)
the inflection in the electrolyte loss curve is sharp and
there is more than one clearly defined low temperature exotherm. When there is more than one low temperature exotherm
as in *P. longipes*, the inflection tends to broaden (Figure 3).
This perhaps suggests that lysing of cells occurs over a
broad range of temperatures in *P. longipes* and that each low
temperature exotherm contributes to injury of different sets
of xylem cells in the twig.

IV.   DEEP UNDERCOOLING IN THE GENUS *PRUNUS*

DTA studies of twigs of native and cultivated taxa of
Prunus were conducted by Burke and Stushnoff (1) and this is
summarized in Table II. The natural geographic distribution
of the species depends on the occurrence of the deep undercooling characteristic in the twigs. For example, *P. virgini-*
ana and *P. pensylvanica* had no low temperature exotherms or
deep undercooling and these species are distributed in the
Boreal Forest and other northern regions. The other native
species all had low temperature twig exotherms and deep undercooling and these species had Deciduous Forest or other more
southern distributions. Except for *P. padus* and *P. maacki* all
the cultivated *Prunus* taxa had low temperature exotherms and
deep undercooling. Burke and Stushnoff (1) suggest that

TABLE II.   Results of DTA Studies on Twigs and Flower
Buds of Prunus Taxa[a]

Prunus taxa	Twig exotherm temperature (^{O}C)	Flower bud exotherm temperatures (^{O}C)
NATIVE		
P. virginiana	absent	absent
P. pensylvanica	absent	-13 to -21
P. serotina	-38	absent
P. nigra	-45	-18 to -20
P. besseyi	-40	-12 to -23
P. americana	-43	-10 to -14
CULTIVATED		
P. padus	absent	absent
P. maacki	absent	absent
P. japonica	-41	-11 to -15
South Dakota plum (P. americana)	-42	-12 to -18
Redglow plum (P. salicina x P. munsoniana)	-41	-7 to -13
Sapalta cherry plum (P. besseyi x P. salicina)	-41	-11 to -17
Muckle plum (P. nigra x P. tenella)	-43	-17 to -18
Pipestone plum (P. salicina x {P. salicina x P. americana})	-38	-20
Meteor cherry (P. cerasus)	-40	-20 to -24
Mt. Royal plum (P. domestica)	-41	-22 to -27
Northstar cherry (P. cerasus)	-40	absent
Brilliant plum (P. salicina)	-41	-19 to -22
Superior plum (P. salicina x {P. americana x P. simoni})	-42	-12
Stanley plum (P. domestica)	-36	absent
Mesabi cherry (P. cerasus x P. avium)	-41	-19 to -20
Deep purple plum (P. besseyi x {P. salicina x})	-40	-14 to -17

[a]Samples were taken in January, 1974 from the Minnesota
Landscape Arboretum or the Horticultural Research Center of
the University of Minnesota (1).

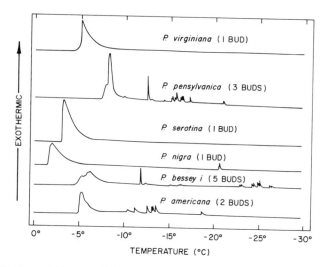

FIGURE 4.   DTA profiles of flower buds of native Prunus species.   See Table II for description of plants used.

P. *pensylvanica* may make excellent parental material for breeding efforts aimed at eliminating the deep undercooling characteristic in twigs.

Deep undercooling was found in the flower buds of many of the *Prunus* taxa (Table II, Figure 4). However, unlike the relation between geographic distribution and twig undercooling, there was no obvious relation between geographic distribution and flower bud undercooling (1).

For example, P. *virginiana* and P. *pensylvanica* are both distributed widely including the Boreal Forest; however, one has flower bud deep undercooling and the other does not. The other four native species have more southern geographic distributions, yet one of them, P. *serotina*, does not have flower bud deep undercooling. The occurrence or absence of flower bud deep undercooling is related to the flower bud morphology. Taxa with raceme flowers (P. *virginiana*, P. *padus*, P. *serotina*, P. *maacki*) do not have bud deep undercooling while all others do.

Burke and Stushnoff (1) report capability of very rapid cold acclimation in flower buds of some of the *Prunus* taxa. Typical results are in Figure 5. When flower buds are held at subfreezing temperatures (but not below the killing temperature) for short times the floral primordia killing exotherms move to lower temperatures and become progressively smaller. After more prolonged times at subfreezing temperatures the flower bud deep undercooling disappears and in some of the species buds treated in this way survive as low as $-80°$ C.

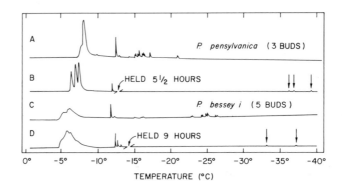

FIGURE 5. The effect of prefreezing on the DTA's of two Prunus species. The uppermost and third uppermost tracings are taken from Figure 4. The second and fourth tracings, P. pensylvanica and P. besseyi, respectively, are for similar samples treated in the same way except, in these two cases, the samples were held for the times and at the prefreezing temperatures indicated in the figure. As can be seen, prefreezing lowers the exotherm temperature (indicated by arrows in the figure) and reduces the exotherm size.

Burke and Stushnoff (1) have classified the Prunus species into five groups. The first group includes the very hardy species without deep undercooling in twigs or flower buds. The three species in this group, P. virginiana, P. padus and P. maacki, all have raceme flower buds. The second less hardy group includes one species, P. pensylvanica, which has deep undercooling only in the flower buds and the buds are capable of rapid cold acclimation. The remaining groups all have deep undercooling of twigs and a hardiness limitation of -40° C. The third of these groups includes only one species, P. serotina, which has no flower bud deep undercooling. The fourth group includes those species that have deep undercooling in the flower buds, but are also capable of rapid cold acclimation to low temperatures (P. nigra, P. besseyi, P. americana and the cultivated taxa in Table II between P. japonica and Pipestone plum). The fifth group probably includes most cultivated Prunus taxa of major importance in horticulture. Members of this group have deep undercooling in flower buds and are not capable of rapid cold acclimation.

TABLE III.   *Results of DTA Studies on Canes of Rosa Taxa[a]*

Species class and name[b]	Cane exotherm temperature (°C)
R. acicularis	-42
R. woodsii	-37
R. arkansana	-38
R. blanda	-25, -34
R. rugosa	-22, -38
R. canina	-19, -36, -40
R. rubrifolia	-19, -25, -34
R. palustris	-21, -36
R. foetida 'Austrian copper'	-26, -31
Martin Frobisher (shrub)	-20, -29
Prairie Dawn (shrub)	-23, -31
Prairie Wren (shrub)	-
Assiniboine (shrub)	-35
Salet Albert	-36
Lillian Gibson (shrub)	-32
Therese Bugnet (shrub)	-
Louis Gimard (moss)	-25, -31, -36
Grandiflora (Gallica)	-21, -29
Alain Blanchard (Gallica)	-28, -38, -43, -46

[a]Samples were taken in January 1977 from native plantings in Colorado (R. acicularis, R. woodsii and R. arkansana) and from the Minnesota Landscape Arboretum of the University of Minnesota (12).

[b]Classification according to Snyder (1975). Shrub and old-fashioned roses, Agricultural Extension Service, University of Minnesota Arboretum Review 22.

V.   DEEP UNDERCOOLING IN THE GENUS *ROSA*

The genus *Rosa* is not only interesting because of its considerable horticultural importance and the wide geographic distribution of its native and cultivated taxa, but also because it is easy to hybridize the species members of this genus. Due to this *Rosa* is an ideal genus for checking into variation of the deep undercooling characteristic and then using breeding techniques for altering the characteristic.

DTA studies of twigs of native and cultivated roses have been conducted by Rajashekar *et al.* (12) and results of their study are summarized in Table III. Unfortunately all the cold hardy taxa studied had the deep undercooling characteristic. This is surprising. The three native species (*R. acicularis*, *R. woodsii* and *R. arkansana*) have vastly different geographic distributions and we expected significant differences in hardiness level. *R. acicularis* extends northward to the Arctic Circle and southward to the Rocky Mountains of Colorado where its distribution runs to tree line (approximately 11,000 ft elevation in northern Colorado). *R. woodsii* is less widely distributed and is found primarily in the western part of the U.S., southwest Canada and in Colorado it can be found as high as 9,500 ft above sea level. *R. arkansana* is found in the prairie regions of the U.S. and southern Canada and in Colorado at lower elevations not exceeding 9,000 ft above sea level. It is herbaceous at these elevations. Deep undercooling is unexpected for *R. acicularis* which survives in regions subjected to much lower than -40° C. Possibly it survives under snow cover, thus avoiding low air temperature. *R. acicularis* native to Hokkaido, Japan was investigated by Sakai (13). His results are consistent with our observation that xylem in this species is injured when twigs are frozen below -40° C. The hardiest *Rosa* species studied by Sakai (13) showed xylem injury between -30° and -45° C.

Although this initial search for absence of the deep undercooling characteristic in *Rosa* is discouraging, considerably more study in this genus will be necessary.

REFERENCES

1. Burke, M. J., and Stushnoff, C., *in* "Stress Physiology of Crop Plants" (H. Mussell and R. C. Staples, eds.). Wiley - Interscience, New York (1978).
2. Dexter, S. T., Tottingham, W. E., and Graber, L. F., *Plant Physiol. 7*, 63 (1932).
3. Flint, H. L., Boyce, B. R., and Beattie, D. J., *Can. J. Plant Sci. 47*, 229 (1967).
4. George, M. F., and Burke, M. J., *Current Adv. Plant Sci. 8*, 349 (1976).
5. George, M. F., Burke, M. J., Pellett, H. M., and Johnson, A. G., *HortSci. 9*, 519 (1974).
6. George, M. F., Burke, M. J., and Weiser, C. J., *Plant Physiol. 54*, 29 (1974).
7. Levitt, J., "Responses of Plants to Environmental Stresses," 697 pages. Academic Press, New York (1972).
8. Pierquet, P., Stushnoff, C., and Burke, M. J., *J. Amer. Soc. Hort. Sci. 102*, 54 (1977).

9. Quamme, H. A., *Can. J. Plant Sci. 56*, 493 (1976).
10. Quamme, H. A., Stushnoff, C., and Weiser, C. J., *J. Amer. Soc. Hort. Sci. 97*, 608 (1972).
11. Rajashekar, C., Burke, M. J., and Westwood, M., in preparation.
12. Rajashekar, C., Burke, M. J., and Pellett, H., in preparation.
13. Sakai, A., *J. Hort. Assoc. Japan 28*, 70 (1959).
14. Strang, H. G., and Stushnoff, C., *Fruit Var. J. 29*, 78 (1975).
15. Sukumaran, N. P., and Weiser, C. J., *HortSci. 7*, 467 (1972).
16. Towill, L. E., and Mazur, P., *Can. J. Bot. 53*, 1096 (1975).

LOW TEMPERATURE EXOTHERMS IN XYLEMS
OF EVERGREEN AND DECIDUOUS
BROAD-LEAVED TREES IN JAPAN WITH REFERENCE
TO FREEZING RESISTANCE AND DISTRIBUTION RANGE

S. Kaku
M. Iwaya

Biological Laboratory
College of General Education
Kyushu University Ropponmatsu
Fukuoka

Deep supercooling of LT[1] exotherm found in overwintering
floral buds of some ornamental species and xylem ray parenchyma of deciduous hardwood trees in North America has recently attracted special interest as a mechanism of frost avoidance in plants (1, 2). To examine the possibility of widespread existence of LT exotherm in xylem, George *et al.* (3)
performed DTA experiments during the midwinter on the wood of
49 species native to North America. The wood of 25 species
exhibited LT exotherms between -41° and -47° C which correlated with their xylem injury. These 25 species having a LT
exotherm were all indigeneous to the Eastern Deciduous Forest
where environmental temperatures below -40° C are not commonly
observed. No LT exotherm in xylems were found in Angiosperm
species native to the northern range of the boreal forest.
Thus, they anticipated that supercooling may play a role in
limiting the northern distribution of many Angiosperm species.
Although it is probable that deep supercooling is seen to be a
common phenomenon as a mechanism of frost resistance in xylem
throughout deciduous tree species of North America, further
research is required on tree species showing xylem injury
above -40° C to understand the details of ecological implications of this supercooling. There has been little work done

[1]*Abbreviation: LT--low temperature; DTA--differential
thermal analysis; HT--high temperature.*

on tree species having LT exotherms ranging from -20° to -40° C and suffering injury to xylem tissues at about the same range of temperatures. Furthermore, no experiments have been carried out on the occurrence of LT exotherms in xylems of Asian tree species in spite of their widespread existence in North America. In the present paper, the existence of LT exotherms in xylems of evergreen and deciduous broad-leaved trees native to Japan was examined with regard to their different freezing resistances and the northern limit of their distribution in Japan. Also, the seasonal changes of LT exotherms in xylem of these trees during cold deacclimation from winter to spring were investigated.

Evergreen and deciduous broad-leaved trees indigeneous to Japan have been classified into four groups on the basis of their observed freezing resistance and their distribution range (7): (A) Evergreen. (B) Less hardy deciduous trees. In these two groups the northern limits of the natural ranges are located in the northern Kanto District of Honshu (Japanese mainland). They resisted freezing down to -10° C to -15° C. (C) Hardy deciduous trees. They ranged in northern Honshu and southern Hokkaido, and survived freezing to -20° to -30° C. (D) Very hardy deciduous trees. They were more widely distributed and extended to inland Hokkaido and survived freezing to -30° C or below. The 16 species used as the materials for the present work were selected to include four species from each of four tree groups.

I. MATERIALS AND METHODS

Stems of 16 woody plants as shown in Table I were collected as required from native forests in Fukuoka city and plantings at our university campus. Growth of two to three years was collected from an individual plant to complete this experiment. Pieces of twigs, 15 to 20 mm in length and about 300 mg in weight, were cut from the internodal region of the twigs and dissected longitudinally to two pieces, and the bark and pith were removed. The surfaces of wood pieces were well blotted by filter paper to prevent ice nucleation at the cut surface of the material.

The apparatus of DTA is a modification of that described by Quamme *et al.* (5). The thermocouples were made on 40-gauge copper-constantan wires. Each xylem piece was supported on cotton threads with the tip of the thermojunction touching the central part of the xylem piece surface. Contact between the sample and the thermocouple junction was maintained by the tension of the thermocouple leads. The thermocouple was placed on the surface of the fresh xylem pieces. Another

TABLE I. *Physical Properties of Xylems in Midwinter*

	Species	Northern limit of distribution	Temp. of xylem injury (°C)[b]	Exotherm (°C) HT	Size	Exotherm (°C) LT	Size	Water Content (%)	Wood Morphology[c]
Evergreen	Myrica rubura	Kanto	-8	-10	L	-17	S	47	d
	Cinnamomum japonicum[a]	Kanto	-10			-19	L	50	d
	Quercus glauca[a]	Tohoku	-15	-13	M	-20	L	45	r
	Camellia japonica	Tohoku	-18	-11	L	-19	L	47	d
Less hardy deciduous	Melia Azedarach	Shikoku	-10	-14	L			46	r
	Rhus succedanea	Kanto	-15	-11	M	-16	L	41	r
	Albizzia Julibrissin	Tohoku	-17	-11	M	-19	L	43	Semi-r
	Celtis sinensis	Tohoku	-17	-10	S	-20	L	40	r
Hardy deciduous	Quercus acutissima[a]	Tohoku	-25	-14	S	-25	S	37	r
	Zelkova serrata[a]	Tohoku	-27	-12	M	-23	S	36	r
	Quercus serrata	Southern Hokkaido	-23	-11	L	-25	S	40	r
	Castanea crenata	Southern Hokkaido	-27	-11	L	-24	S	45	r
Very hardy deciduous	Cornus controversa	Inland Hokkaido	-25	-9	L	-25	S	49	r
	Magnolia obovata	Inland Hokkaido	-25	-10	L	-34	S	50	d
	Sorbus commixta[a]	Inland Hokkaido	-35	-10	L	-30	M	47	d
	Alnus hirsuta	Inland Hokkdido	-40	-10	L	-31	M	50	d

(Japanese Mainland (Honshu))

[a] Planting.
[b] Sakai's determination (1972, 1975).
[c] Wood porosity: r=ring-porus, d=diffuse-porus, Semi-r=semiring-porus.

junction, connected in series with the first, was placed on
another dry piece of the same dimensions which had been previously dried in an oven. The dry piece was used as a reference to detect temperature changes occurring in the sample
during freezing. An additional thermocouple was also placed
in the reference pieces to record the reference temperature
during freezing. The differential temperature output ampli-
fied through a DC μV amplifier before being fed into the re-
cording unit. To slow down the rate of cooling to 1 to 2° C/
min (below 0° C) the assembly was enclosed within two concentric glass vials and was placed in an insulated foam rubber
box which acted as a temperature stabilizer during freezing.
This box was cooled in a freezer which could be regulated to
-90° C. The DTA profiles were recorded on 5 samples for each
run.

Water content was determined by drying the samples at
90° C for 24 hr and was expressed as percentage of fresh
weight.

II. RESULTS

A. *DTA Profiles of Xylem in Midwinter and the Relation
    Between the Initiation Temperatures of LT Exotherms and
    Xylem Injury Temperatures*

Typical profiles of DTA in xylem pieces in midwinter are
shown in Figure 1. Two exotherms were observed in 14 species of the 16 trees and the major exotherm occurred suddenly
immediately after the initiation of freezing at relatively
high subzero temperatures, about -10° C. The second exotherm
occurred gradually and was delayed somewhat. These two exotherms have been called "high temperature exotherm" and "low
temperature exotherm" (3). No HT exotherms were found in
Cinnamomum, while in *Melia* the LT exotherm did not occur.
Although the HT exotherm generally occurred between -9° and
-14° C regardless of tree group difference, the initiation
temperature of LT exotherms differed among tree groups in relation to their freezing resistance and their northern limits
of distribution. In the evergreen and the less hardy deciduous trees the LT exotherm occurred between -16° and -20° C,
and -23° to -25° C in the hardy deciduous trees, and -25° to
-34° C in the very hardy deciduous trees. The initiation of
LT exotherms in most of the evergreen and the less hardy deciduous trees was rather lower than their xylem injury temperatures as determined by Sakai (6, 7). On the other hand,
the LT exotherms in some of the hardy deciduous trees such as
Zelkova, *Castanea*, and *Sorbus* and *Alnus* in the very hardy

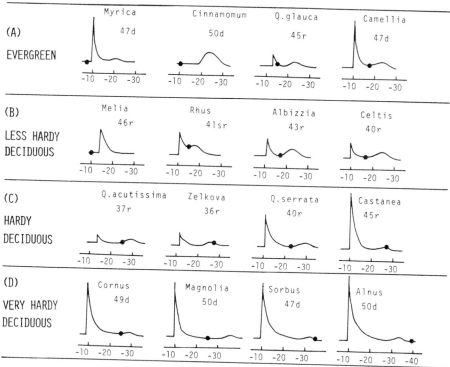

FIGURE 1.   Typical profiles of DTA in xylem pieces in mid-winter.   The solid circles in the DTA tracings indicate the xylem injury temperature as determined by Sakai (1972, 1975). In each profile the number and the letter show water content and morphology of xylems, respectively.   Abbreviations of the wood porosity are the same as in Table I.

deciduous trees occurred at temperatures higher than their xylem injury temperatures (Table I).   Therefore, it can be seen that the initiation temperature of an LT exotherm in each species did not necessarily show good agreement with the temperature of xylem injury, but the temperature ranges of LT exotherms in each tree group were well correlated with their freezing resistance and the northern limit of distribution (Table II).

The sizes of LT exotherms in the evergreen and the less hardy deciduous trees were relatively larger than those of the hardy and the very hardy deciduous trees (Table I).

TABLE II. Relationships between Freezing Resistance of Broad-leaved Tree Species, the Average Annual Minimum Temperatures at the Northern Limits of Natural Ranges and the Initiation Temperatures of LT Exotherms in Midwinter.

Relative hardiness classification	Northern limits of natural ranges	Average minimum temp. at the northern limits of growth (°C)a	Observed freezing resistance (°C)a	Initiation temp. of LT exotherm (°C)
Evergreen	Northern Kanto, Tohoku of Honshu (Japanese mainland)	$-10 \sim -15$	$-10 \sim -15$	$-17 \sim -20$
Less hardy deciduous	Northern Kanto, Tohoku of Honshu (Japanese mainland)	$-7 \sim -15$	$-7 \sim -15$	$-16 \sim -20$
Hardy deciduous	Northern Honshu, Southern Hokkaido	$-20 \sim -25$	$-20 \sim -30$	$-23 \sim -25$
Very hardy deciduous	Inland Hokkaido	below -30	below -30	$-25 \sim -34$

aBased on Sakai (1975).

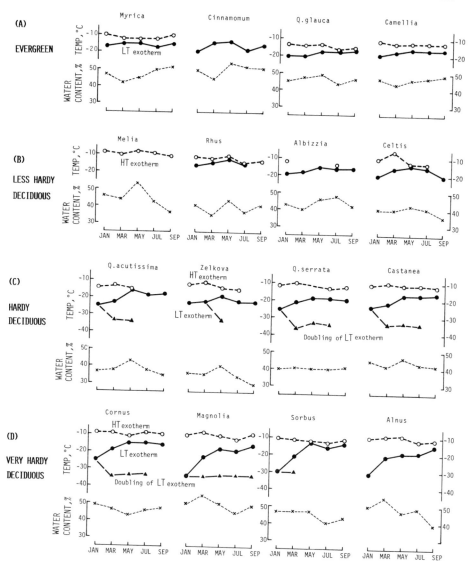

FIGURE 2. Seasonal variations of HT (-o-), LT (-●-), 2nd LT (-▲-) exotherms and water content (-x-) in xylem pieces during cold deacclimation.

B. Seasonal Variations of LT Exotherms during Cold-deacclimation

DTA was carried out every other month from midwinter to autumn. As shown in Figure 2, the HT exotherms remained between -10° and -15° C throughout the season in all of the tree

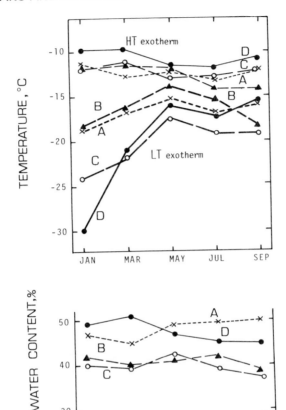

FIGURE 3. *The general trends of variations in LT and HT exotherms and water content from midwinter to autumn illustrated from the average value of each tree group. (A) Evergreen, (B) Less hardy deciduous, (C) Hardy deciduous, (D) Very hardy deciduous. Symbols are the same as in Figure 2.*

species used, and no marked seasonal changes were observed. On the other hand, the initiation temperature of LT exotherms rose as spring approached. The extent of this seasonal ascension varied among the tree groups. In the evergreen and the less hardy deciduous trees it was generally slight and the extent of temperature changes was within 5° C from winter to spring. In the hardy deciduous trees, however, the initiation temperature of LT exotherm shifted upwards by about 5° to 10° C and ranged between -15° and -25° C. Pronounced ascension of LT exotherm was observed in the very hardy deciduous trees, about 10° to 15° C in each species, in the same temperature range as the hardy deciduous trees. Accordingly, the marked difference of the initiation temperature of LT exotherm among tree groups during midwinter gradually diminished from spring to summer. The tendency of variations in HT and LT exotherms and water content from winter to autumn was summarized from the average value of each tree group and illustrated in Figure 3.

As seen in Figure 3, the water content of xylems in each species did not change seasonally, indicating that the freezing properties of xylems of the hardy and the very hardy deciduous trees are significantly altered without much alteration in the overall water content of xylems.

C. Doubling of LT Exotherm in Xylems of the Hardy Deciduous Broad-leaved Trees during Deacclimation from Cold

In midwinter, the LT exotherm in all hardy deciduous trees including the very hardy deciduous trees occurred as a single rebound, but appeared to change into two exotherms in the spring (except for *Alnus*). As shown in Figure 4, xylems of *Quercus serrata* and *Cornus* in March showed a large HT exotherm at about -10° C and the first LT exotherm (LT_1) occurred at about -20° C and the second one (LT_2) occurred at -33° C. The occurrence of these double LT exotherms in spring xylems was not observed in the evergreen and the less hardy deciduous trees. In these double LT exotherms, the major exotherm was always the first and the second appeared as a step on the shoulder of the first. The initiation temperature of this second LT exotherm was almost the same temperature, approximately -33° C, in each species and did not change throughout spring to summer. In *Sorbus*, the doubling of the exotherm was observed only in March and thereafter disappeared. In *Zelkova* it was not observed in March but was temporarily observed in May. Although *Alnus* is one of the very hardy deciduous trees, the LT exotherm remained single throughout the season (Figure 2).

The detail of the occurrence of a second LT exotherm in spring to summer was distinct in two of the groups (Figure 2). Of the four species in the hardy deciduous trees and *Cornus* in the very hardy deciduous trees, the second LT exotherm occurred at somewhat lower temperatures than the initiation temperature of LT exotherm in winter, whereas those of *Magnolia* and *Sorbus* in the very hardy deciduous group occurred at almost the same temperatures as in winter. In other words, the second LT exotherm in spring xylems of the former appeared to represent newly manifested ability to supercool and those of the latter represent residual ability following winter.

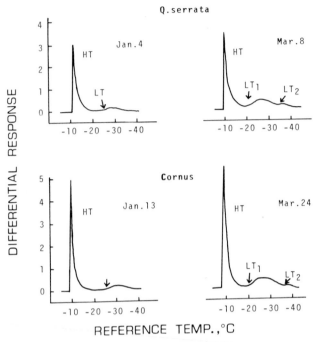

FIGURE 4.   *Typical profiles of DTA in xylem pieces of Quercus serrata and Cornus controversa in January and March. In spring Xylems of both trees, LT_1 and LT_2 show doubling of LT exotherm accompanying deacclimation from cold.*

III. DISCUSSION

George *et al.* (3) have observed the widespread existence of LT exotherms in xylems of 25 native woody species of North America. The initiation temperature of LT exotherms of these plants ranged between $-41°$ and $-47°$ C and they were killed at the same temperatures. In the present work, it was ascertained that the LT exotherm exists widely in xylems of the evergreen and the deciduous broad-leaved trees in Japan and the initiation temperature of the LT exotherms in midwinter varied within tree groups, correlating with their freezing resistance and northern limit of distribution; $-16°$ to $-20°$ C in the evergreen and the less hardy deciduous trees, $-23°$ to $-25°$ C in the hardy deciduous trees, and $-25°$ to $-34°$ C in the very hardy deciduous trees. In general, the LT exotherms in the hardy tree species occurred at lower temperatures than the least hardy species studied.

Although HT exotherms occurred at almost the same temperatures, $-10°$ to $-15°$ C, in all tree species throughout the season, the initiation temperature of LT exotherm in the hardy and the very hardy deciduous trees shifted upwards from winter to spring, whereas no marked seasonal changes were observed in the evergreen and the less hardy deciduous trees. These seasonal changes that occur upon deacclimating from cold are in agreement with similar DTA analysis in apple twigs by Quamme *et al.* (5) and that of shagbark hickory by George *et al.* (3) and George and Burke (4).

In previous studies, it was demonstrated that the HT exotherm appeared to represent the freezing of bulk water in the vascular system of xylems, and the LT exotherm was derived from the freezing of a fraction of water localized in xylem ram parenchyma which remain unfrozen well below the temperature at which the bulk of the xylem water freezes. Therefore, it can be considered that the seasonal rise of the LT exotherm from winter to summer that can be observed in the hardy and the very hardy deciduous trees is probably caused by the increase of the ease of freezing in xylem ray parenchyma in deacclimating woods, and that lack of seasonal variation in the HT exotherm of all tree species is related to the status of water as bulk in the vascular system of xylems.

It should be noted that seasonal rise in the initiation temperature of LT exotherms accompanying deacclimating from cold is more pronounced in hardy tree species than less hardy ones, and also that the occurrence of two LT exotherms can be observed in the hardy and the very hardy deciduous trees. The DTA experiment on shagbark hickory woods deacclimating from cold was carried out by Burke *et al.* (1) and George and Burke (4). The midpoint of LT exotherm in xylems of this tree was about $-48°$ C in February and shifted gradually to a higher temperature, $-22°$ C, in May as the plants became less hardy. The doubling of LT exotherms has been shown in the DTA profile of this tree in May. Although Burke *et al.* (1) did not pay special attention to the existence of this second LT exotherm which occurred at about $-35°$ C, it appeared to correspond to the doubling of the LT exotherm in spring xylems just described above. Therefore, it is probable that the double low exotherm is a widespread phenomenon which is observed in hardy tree species during deacclimation from cold. The rise in initiation temperatures of LT exotherms may be related to the decrease of supercooling ability in most of the xylem ray parenchyma, and it can be assumed that the high supercooling ability in very restricted small areas in xylem ray parenchyma results in the double LT exotherm in deacclimating xylems. As stated above, there are two types of second LT exotherm: first, the second LT exotherm occurred at lower temperatures than the initiation temperature of LT exotherm in midwinter; second, it occurred at almost the same temperature as that of midwinter. The former can be generally observed in the hardy deciduous trees, and the latter is in the very hardy ones. Accordingly, there seem to be two different de-acclimation processes from cold in these hardy deciduous trees. The mechanism is unknown at present; however, the doubling of the LT exotherm during deacclimation from cold may suggest further resolution is needed on the phsiological changes that occur upon acclimation and deacclimation of living woods.

IV. CONCLUSION

The existence of LT exotherms in xylems was widely observed in the evergreen and deciduous broad-leaved trees native to Japan. In midwinter, the initiation temperature of LT exotherms in xylems of these trees was different from four groups classified according to differences of freezing resistance and the northern limit of distribution: it occurred at -16° to -20° C in the evergreen and the less hardy deciduous trees, at -23° to -25° C in the hardy deciduous trees, and at -25° to -34° C in the very hardy deciduous trees. The initiation temperatures of LT exotherms gradually rose as spring progressed. The extent of this rise was more pronounced in hardier tree groups than less hardy ones. No marked differences occurred in LT exotherms of any tree group in summer. During deacclimation from cold, two LT exotherms were observed in the hardy and the very hardy deciduous trees and this was characterized as a deacclimation process in xylems of these trees because the same phenomena did not occur in the evergreen and the less hardy deciduous trees. Thus, the widespread existence of LT exotherms in xylems of hardwood trees indigenous to Japan suggest that deep supercooling may indeed play a more extensive role in the frost avoidance of these plants than previously thought.

ACKNOWLEDGMENT

This work was supported in part by a grant to S. Kaku from the Ministry of Education of Japan (No. 254227).

REFERENCES

1. Burke, M. J., George, M. F., and Bryant, R. G., *in* "Water Relation of Foods" (R. B. Duckword, ed.). Academic Press, New York (1975).
2. Burke, M. J., Gusta, L. V., Quamme, H. A., Weiser, C. J., and Li, P. H., *Ann. Rev. Plant Physiol.* *27*, 507 (1976).
3. George, M. F., Burke, M. J., Pellett, H. M., and Johnson, A. G., *HortScience* *9*, 519 (1974).
4. George, M. F., and Burke, M. J., *Plant Physiol.* *59*, 319 (1977).
5. Quamme, H., Stushnoff, C., and Weiser, C. J., *J. Amer. Soc. Hort. Sci.* *97*, 608 (1977).
6. Sakai, A., *J. Jap. Forestry Soc.* *54*, 333 (1972).
7. Sakai, A., *Jap. J. Ecol.* *25*, 101 (1975).

RESISTANCE TO LOW TEMPERATURE INJURY
IN HYDRATED LETTUCE SEED BY SUPERCOOLING

C. Stushnoff

Department of Horticultural Science
and Landscape Architecture
University of Minnesota
St. Paul, Minnesota

O. Junttila

Institute of Biology and Geology
University of Tromso
Tromso, Norway

Dry seeds of several species have been shown to survive
exposure to extremely low temperatures, including immersion
in liquid air for up to 60 days (4, 6, 7). On the other
hand, germinating seedlings are killed by only a few degrees
of frost. Fully hardened twigs of several northern tree species tolerate freezing in liquid nitrogen, -196° C (10),
and deep supercooling seems to be an important mechanism for
the avoidance of freezing in both xylem ray parenchyma and
flower buds of many other hardy but less cold resistant,
deciduous woody species (1, 2).
    Resistance to freezing stress in seeds has been suggested
to be related to water content (4, 5, 8). More recently,
Junttila and Stushnoff (3) studied the relationship between
stage of germination and degree of freezing resistance and
the nature of freezing resistance in lettuce seed. Hydrated
lettuce seeds (achenes) avoid injury by supercooling and the
level of resistance can be determined precisely by differential thermal analysis (DTA). The degree of resistance to
low temperature is related to the degree of hydration in the
20 to 40% moisture range and the killing point is detectable
by DTA. Furthermore, supercooling as an avoidance mechanism
is dependent on the intact structure of the endosperm (3).

241

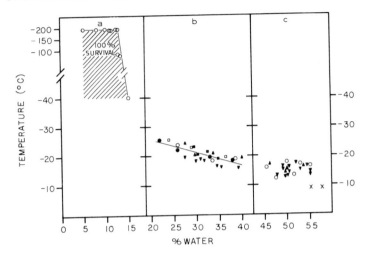

FIGURE 1.   Relationship between seed water content and hardiness in Grand Rapids lettuce seeds.   a: Limits for 100% survival of seeds stored at 18° C or at 4° C in controlled humidities for 10-14 days.   Hardiness at -196° C was tested by keeping the seeds in liquid nitrogen for 30 min.   b: The mean temperature of the secondary exotherms in seeds imbibed at 30° C in various concentrations of PEG 6000 (0.06-0.12 molal). Open symbols: 5 min red (R) light, 4.2 μW cm⁻² at 660 nm, given after 2 h imbibition in darkness; closed symbols: darkness ▢, ▪, 12 h imbibition; ○, ● 24 h; ▼, 72 h; ▲, 96 h. Regression equation: y = -0.43x + 33.9, r = -0.84, P < 0.001. c: Mean temperature of secondary exotherms in seeds imbibed at 30° C in water for 4-72 h.   O, R; ▲, R plus 5 min far-red (FR), 24 μW cm⁻² at 730 nm; ▼, darkness; X, germinated seeds.

Experiments were conducted with seed of Grand Rapids lettuce. Moisture content was controlled by leaving seeds to imbibe water for specific intervals; by leaving them to imbibe in different concentrations of polyethylene glycol (molecular weight 6,000, PEG 6000) for specific intervals, and by storing them at controlled relative humidities in closed vials on top of various concentrations of H_2SO_4. Moist seeds were dried rapidly between filter papers and samples were prepared for freezing and for determination of water content. Seeds that had imbibed in polyethylene glycol were washed three times with distilled water before preparation. Dry weight was determined after drying at 105° C for 24 hr. Moisture content was calculated as percentage water of fresh weight. Supercooling nucleation points were determined by DTA with an automatically controlled, liquid-N_2-cooled Cryosan

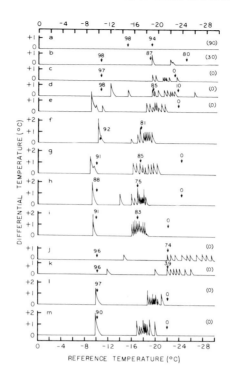

FIGURE 2.   Differential thermal analysis of lettuce seeds
with moisture contents at various levels.   Ten seeds in each
sample.   Figures and arrows show the percentage or survival
(germination) in samples removed at temperatures indicated,
those in parenthesis show survival of samples frozen down to
-35° C.   Freezing rate 20° C per h in all cases.   To obtain
the various moisture levels seeds were imbibed at 30° C in
darkness either in water for various times (a-i) or in various
molal concentrations of PEG 6000 (j-m) for 24 h.   a, 0 or 5
min imbibition, 5.6-16% water content; b, 10 min, 20%; c,
20 min, 26%; d, 30 min, 32%; e, 60 min, 41%; f, 4 h, 49.7%;
g, 8 h, 49.5%; h, 12 h, 48.8%; i, 24 h, 51.2%; j, 0.12 molal,
21.9% water; k, 0.10 molal, 26.0%; l, 0.08 molal, 33.3%; m,
0.06 molal, 38.2%.

freezing system, with a constant freezing rate of 20° C per
hr.   Exotherms were recorded on potentiometric recorders, at
0.5 mV, connected to the samples with 0.5 mm copper-constant
thermocouples.   Between 1 and 10 seeds were used with each re-
cording channel.   Samples (10 to about 100 seeds) for survival
tests were removed from the freezer at predetermined

temperatures and kept at 2° C for 2 to 4 hr before ability to germinate was tested in light at 24° C. Seeds were counted after 3 days and the percentage of germinated and growing seeds was used as a criterion for viability.

Lettuce seeds with 5 to 13% water content were not injured by freezing to -196° C, but with 13 to 16% moisture, the threshold of freezing tolerance declined to -196° C (Figure 1A). No exotherms were detected by DTA in either case but two types of exotherms were found in intact seeds containing more than 20% water (Figure 2). The first appeared as a single peak when the temperature of the sample was -10 ± 2° C regardless of the number of seeds in the sample, indicating that the nucleation point for this peak is remarkably uniform. This exotherm is not related to the injury of intact seed and we suggest that it represents freezing of extracellular or bulk water in the outer layers, similar to extracellular freezing of other plant tissue (1). Each seed frozen also showed a secondary supercooling exotherm. These secondary exotherms represent the killing point of individual seeds. In seeds imbibed in PEG 6000 the mean exotherm temperature was linearly correlated ($r = -0.84$, $P < 0.001$) with the moisture content within the limits of 20 to 40% (Figure 1B). A plateau was reached at 40 to 50% moisture where the mean exotherm temperature of intact seeds remained at -16 ± 2° C ($r = 0.004$), irrespective of imbibition time (4 to 72 hr). There was no effect of red (5 min, 4.2 μW cm^{-2} nm^{-1} at 660 nm) or far red (5 min 24μW cm^{-2} nm^{-1} at 730 nm) irradiation on the position of the secondary exotherms up to 8 hr after exposure of seeds.

The freezing pattern changed considerably, however, after disruption of the endosperm-seed coat envelope. As soon as the radicle emerged the secondary exotherms disappeared and the first exotherm represented the killing point of the seed (Figure 3). Similarly, embryos excised from the seeds after imbibition for 2 to 12 hr produced only 1 exotherm and did not survive below this nucleation temperature (Figure 3). The occurrence of secondary exotherms depends on the intact structure of the endosperm. If fruit coats were excised carefully without injury to the endosperm envelope, embryos did supercool as detected by the presence of secondary peaks at the nucleation points, but if the endosperm was damaged, for example with a needle, deep supercooling ceased and death was coincident with the first peak (Figure 3).

Therefore the integrity of the endosperm envelope seems to facilitate deep supercooling which concurrently imparts freezing avoidance to -12° to -18° C in intact, imbibed lettuce seeds. Deep supercooling of lettuce seeds is not, however, an exclusive property of live seeds: it also occurred with intact, imbibed seeds which had been killed by low

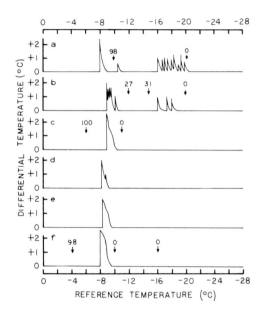

FIGURE 3. *Differential thermal analysis of lettuce seeds.*
a: Intact seeds; b: seed coats removed; c: excised embryos;
d: endosperms, e: intact seeds stuck four times with a needle;
f: germinated seeds. Figures and arrows are as in Figure 2.
Seed coats and endosperm were removed with a scalpel under
a stereoscope in light after an imbibition at 30° C for 16
h. Seeds stuck with a needle were imbibed in water for 4 h
at 30° C. There were ten seeds or embryos in each sample.

or high temperatures. Work with apple stems has shown that
intact dead twigs yield exotherms if rehydrated to 20% water
but not if ground to a fine powder (9).

The mechanism by which the intact endosperm imparts
freezing avoidance to the embryo is not clear. Perhaps disruption of the endosperm facilitates intimate contact of
relatively pure water in the coats with the water in the
embryo, thus providing ice crystal formation in all parts at
the first exotherm. Germinated seeds contain slightly more
water than imbibed non-germinated seeds, and perhaps endosperm injury increases embryo water content. But even embryos
excised from seeds imbibed in 0.10 molal PEG 6000 and kept in
0.12 molal PEG 6000 during freezing did not yield secondary
exotherms.

Extremely rapid freezing rates of 80°, 110° and 240° C/hr
shifted the free water peak slightly and no appreciable change

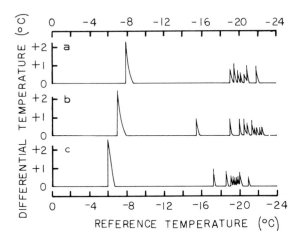

FIGURE 4.   Differential thermal analysis of lettuce seeds.   a: 80° C/hr; b: 100° C/h; c: 240° C/h.

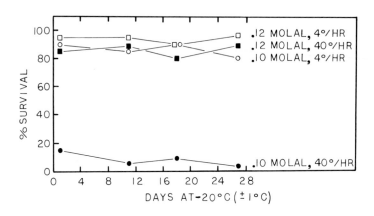

FIGURE 5.   Survival of hydrated lettuce seeds during long term freezing stress at -20° C.   Seeds were imbibed in water, 0.10 and 0.12 molal PEG 6000 at 30° C for 20 h prior to placement at -20° C storage.   Water imbibed seeds (47.3% H₂O) were all dead after 1 day at -20° C.   0.10 molal and 0.12 molal PEG imbibed seeds contained 28.3% and 21.6% H₂O respectively, determined just prior to freezing.

occurred in the secondary peaks (Figure 4). Supercooling in lettuce seed embryos thus appears to be non-sensitive to freezing rate as reported for ray parenchyma of apple and other hardwoods (1, 9).

A long term freezing tolerance stress was imparted by storing seeds up to 28 days at -20° C following hydration to controlled moisture levels. Seeds with 28.3% moisture or less survived a freezing rate of 4° C/hr but not 40° C/hr prior to long term storage. However, at 21.8% moisture, freezing rate was not a factor and high survival also occurred at 40° C/hr. Water imbibed seeds 47.3% moisture could not tolerate even one day at -20° C (Figure 5).

Lettuce seeds tolerate extreme cold at low water content (5 to 13%). The mechanism of resistance to freezing apparently includes avoidance of ice formation by supercooling at 20 to 50% water content, which is lost if the endosperm envelope is damaged or radicle emergence begins. The ease of manipulating hydration-germination response and freezing resistance appears to offer interesting possibilities for further study of cryo-stress in plant tissues.

REFERENCES

1. Burke, M. J., Gusta, L. V., Quamme, H. A., Weiser, C. J., and Li, P. H., *Ann. Rev. Plant Physiol. 27*, 507 (1976).
2. George, M. F., and Burke, M. J., *Curr. Adv. Plant Sci. 8*, 349 (1976).
3. Junttila, O., and Stushnoff, C., *Nature 269*, 325 (1977).
4. Kiesselbach, T. A., and Ratcliff, J. A., *Nebr. Agric. Exp. Sta. Bull. 163*, 1 (1918).
5. Levitt, J., "Responses of Plants to Environmental Stresses." Academic Press, New York (1972).
6. Lipman, C. B., and Lewis, G. N., *Plant Physiol. 9*, 392 (1934).
7. Lipman, C. B., and Lewis, G. N., *Plant Physiol. 11*, 201 (1936).
8. Lockett, M. C., and Luyet, B. J., *Biodynamica 7*, 67 (1951).
9. Quamme, H., Weiser, C. J., and Stushnoff, C., *Plant Physiol. 51*, 273 (1973).
10. Sakai, A., *Nature 185*, 392 (1960).

THE ROLE OF BACTERIAL ICE NUCLEI
IN FROST INJURY TO SENSITIVE PLANTS[1]

S. E. Lindow
D. C. Arny
C. D. Upper

Department of Plant Pathology
University of Wisconsin
Madison, Wisconsin

W. R. Barchet

Department of Meteorology
University of Wisconsin
Madison, Wisconsin

ABSTRACT

No intrinsic ice nuclei active above about $-10°$ C were
found associated with the leaves of several plant species.
Strains of two species of epiphytic bacteria, *Pseudomonas
syringae* and *Erwinia herbicola*, are efficient ice nuclei between -2 and $-5°$ C. Leaves of most plants collected from
several geographically different areas and during different
seasons of the year had substantial numbers of these ice
nucleation active (INA) bacteria. The ice nucleus content of

[1]*Research cooperative with the College of Agricultural
and Life Sciences, University of Wisconsin, Madison, and the
Agricultural Research Service, United States Department of
Agriculture.*
*Mention of a trademark, proprietary product, or vendor
does not constitute a guarantee or warranty of the product or
vendor by the U. S. Department of Agriculture and does not
imply its approval to the exclusion of others that may also
be suitable.*

field-grown leaves responded to various treatments as if the
ice nuclei active at -5° C on those leaves were INA bacteria.
Frost injury to field-grown corn leaves at -5° C was directly
proportional to the logarithm of the INA bacterial populations on these leaves. On the basis of these observations, we
conclude that epiphytic INA bacteria are incitants of frost
injury to tender plants.

For ice to form at temperatures only a few degrees below
0° C, heterogeneous ice nuclei must be present to initiate
crystallization of supercooled water. Frost injures plants
only if ice is formed (2, 13). Thus, ice nucleation is a
requisite for frost injury to tender plants. Frost injury
to tender plants under field conditions usually occurs within
the temperature range of ca. -2 to -5° C (14). Thus, since
injury does occur, supercooling to temperatures below -2 to
-5° C in tender plants does not occur, and ice nuclei active
in this temperature range must be associated with these
plants.
The role of ice nucleation in frost injury has been
recognized for some time (3). Ice nuclei have been associated with the surfaces of plant parts (10). Kaku (4-6) has
shown that ice nuclei are not uniformly distributed on a given
leaf, and that they vary in quantity with maturity of leaves,
and vary among plant species. Although atmospheric ice nuclei had been assumed to be sources of ice nuclei on plants,
Marcellos and Single (12) have presented evidence that
external ice nucleation on leaves by air borne particles is
unlikely and suggested that an alternative source of ice
nuclei on leaf surfaces be sought. Maki and his co-workers
(11) found that living cells of the bacterium *Pseudomonas
syringae* van Hall are very efficient ice nuclei. Vali and
Schnell and their co-workers have discussed the importance of
ice nuclei that arise from (or on) leaves (15, 17). We have
found that certain epiphytic bacteria are abundant, efficient
ice nuclei on plant leaves (1, 7). These ice nucleation
active (INA) bacteria are *P. syringae* and *Erwinia herbicola*
(Löhnis) Dye.
This paper will review our arguments in support of the
following hypothesis: Bacterial ice nuclei, present on
leaves of tender plants, prevent supercooling below -2 to -5°
C by initiating ice formation and thus incite frost injury in
this temperature range.
For this hypothesis to be correct, plants lacking ice
nuclei should not freeze in the temperature range -2 to -5° C;
plants on which INA bacteria have been established should
freeze under these same conditions. Plants grown in our
growth chamber do not have detectable populations of INA

FIGURE 1. Frost damage at -5° C to bean plants with and without leaf populations of ice nucleation active bacteria. The plants on the left were sprayed with a suspension of approximately 10⁶ cells/ml of Pseudomonas syringae isolate #31 in phosphate buffer and the plants on the right with buffer alone. Both groups of plants were incubated in a mist chamber for 24 hours before they were cooled to -5° C. This picture was taken 24 hours after the plants were exposed to -5° C.

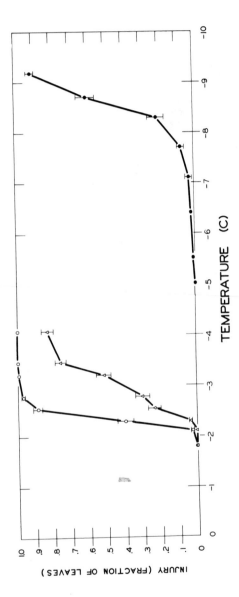

FIGURE 2. Frost damage as a function of temperature to corn seedlings with and without leaf populations of Pseudomonas syringae. Plants were sprayed with P. syringae suspensions (~ 0.5 ml/plant) of 3 x 10^8 cells/ml (○) or 3 x 10^5 cells/ml (△) and incubated in a mist chamber for 24 hours. Other plants remained untreated (●). Then the plants were placed in a growth chamber at 0° C and cooled at 0.050 C/minute. Groups of plants were removed from the chamber at the temperatures shown on the abscissa. Data are represented as means ± standard error (vertical bars).

bacteria established on their leaves. These plants survive esposure to -5° C (1). If populations of INA bacteria are established on the leaves of these plants before they are cooled to -5° C, the plants freeze (1, 7), and if they are tender plants, they are killed. This phenomenon is illustrated in Figure 1, which shows snap beans (*Phaseolus vulgaris* L.) with and without leaf surface populations of *P. syringae*, after exposure to -5° C. Plants were typically maintained at temperatures below 0° C for two or more hours in experiments of this type (7). More than 10 species of "tender" plants have been tested in this manner, and the results have always been the same: frost injury was limited to those plants that harbored populations of INA bacteria. The extent of injury of corn (*Zea mays* L.) plants as a function of temperature is shown in Figure 2. All leaves on plants harboring high epiphytic populations of *P. syringae* were killed between -2 and -3° C; those with lower populations of *P. syringae* were severely injured between -2 and -4° C, whereas those lacking populations of INA bacteria were not injured or killed until the temperature had reached -8 to -9° C.

The ice nucleus content of axenic leaves was compared with that of similar leaves on which *P. syringae* was established. In the experiment illustrated in Figure 3, 3-mm corn leaf discs were immersed in droplets of water and cooled. The temperature at which each droplet froze was recorded. The cumulative ice nucleus content per gram of corn leaf tissue was calculated according to the equation of Vali (16) except that the mass of the leaf disc was substituted for the volume of the droplet in the calculations. Ice nuclei were not detected in axenic corn leaves at temperatures above -9 to -10° C. On the other hand, ice nuclei active between -2 and -3° C were detected in discs harboring populations of *P. syringae*. Similar experiments with several other plant species yielded essentially the same results. Thus, leaves do not contain intrinsic ice nuclei active above about -10° C, but ice nuclei active at higher temperatures are associated with leaves harboring INA bacteria.

The ice nucleation in corn leaf discs harboring *P. syringae* occurs at the same temperatures as in suspensions of *P. syringae* (Figure 4). Ice nucleus content per ml of suspension at a given temperature was determined by the drop-freezing method of Vali (16). Ice nucleation activity in this suspension was detectable slightly below -2° C. The probability that any given cell will function as an ice nucleus increases by a factor or more than 10^5 between -2° C and -5° C. At -5° C only about one cell in 100 was active as an ice nucleus; below -5° C little additional increase in nucleation activity was observed with additional cooling. Most of the

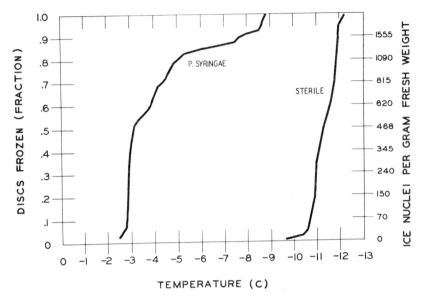

FIGURE 3. *Comparison of ice nucleation activity of leaf discs from corn plants with leaf surface Pseudomonas syringae populations with those grown on sterile Hoagland's nutrient solution.*

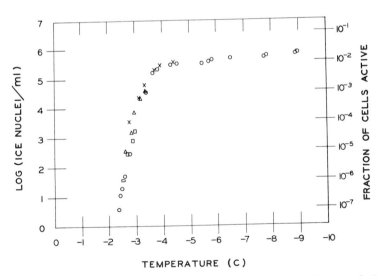

FIGURE 4. *Cumulative ice nucleus concentration and fraction of cells in a Pseudomonas syringae cell suspension that were active in ice nucleation as a function of temperature. The different symbols represent determinations from different dilutions.*

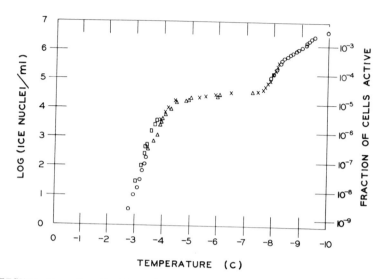

FIGURE 5. Cumulative ice nucleus concentration and fraction of cells in Erwinia herbicola cell suspensions that were active in ice nucleation as a function of temperature. The different symbols represent determinations from different dilutions.

nuclei in this suspension were active between −2 and −5° C—the temperature range in which frost injury to tender plants occurs. *Erwinia herbicola* is somewhat less efficient than *P. syringae* as an ice nucleus (7). However, as shown in Figure 5, suspensions of *E. herbicola* also contain substantial numbers of ice nuclei active between −2 and −5° C.

Thus far, we have established that exogenous ice nuclei are necessary for frost injury to tender plants, that INA bacteria are active in the appropriate temperature range, and that INA bacteria can serve as the ice nuclei for frost injury to plants in growth-chamber experiments. The question remains: Are epiphytic INA bacteria the ice nuclei that incite frost injury to tender plants in the field?

If INA bacteria are responsible for frost injury to plants in the field, then

(1) They must be broadly distributed, with respect to both plant species and geographical region.

(2) Ice nuclei on plant leaves must be INA bacteria (or derived from INA bacteria).

(3) Frost injury to leaves must be predictable on the basis of numbers of epiphytic INA bacteria on the leaves.

TABLE I. Plants Harboring INA Bacteria

	Number of plant species	
Source of sample	With INA bacteria	Without INA bacteria
California[a]	9	1
Florida[b]	16	2
Wisconsin[c]	45	10
Total[d]	55	12

[a]*Sampled 6-8 April 1976. All INA bacteria were Pseudomonas syringae.*
[b]*Sampled 29 December 1976 to 13 March 1977. Both Erwinia herbicola and P. syringae were found.*
[c]*Sampled during May and early October, 1976. Both P. syringae and E. herbicola were found.*
[d]*Corrected for species sampled in more than one location.*

We have sampled plants for the presence of INA bacteria at several locations in the United States and at different times of the year (9). A summary of the results from three locations is shown in Table I. Leaves were washed in buffer and the washings dilution-plated on suitable media. Numbers of INA bacteria were determined by a replica freezing technique. Replicas of colonies on agar plates were transferred to paraffin-coated aluminum foil by means of a velvet pad, and the foil was cooled to -5° C and sprayed with a fine mist of water. Patches of ice represented colonies of INA bacteria. Of the 67 species of plants sampled in Florida, California, and Wisconsin, 55 harbored INA bacteria. Half of the species that did not were conifers. Active bacteria were detected from only one of the seven species of conifers sampled.

Pseudomonas syringae was isolated from nearly all samples tested. *Erwinia herbicola* was present on the plants from Florida and on the plants sampled in Wisconsin in the fall, but not on those sampled in the spring in either Wisconsin or California.

Populations of INA bacteria vary during the growing season as illustrated in Figure 6 for corn near Arlington, Wisconsin during 1976. Shortly after the plants emerged in early summer, both the total bacterial population and the population of INA bacteria were quite low; both increased slowly to

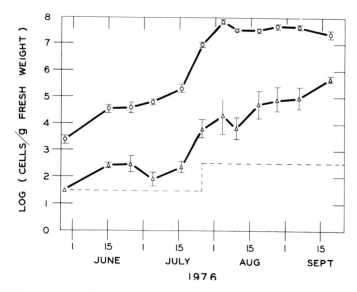

FIGURE 6. Total (O) and ice nucleation active (INA) (Δ) bacterial populations on corn leaves during the 1976 growing season from samples of entire, healthy, nontreated corn leaves from plots near Arlington, Wisconsin. The dashed line represents the estimated limit of detection of INA bacteria. Vertical bars represent the standard error of the mean of log population.

mid-July. The abrupt increase in late July coincided with pollination of the corn. The total population of bacteria stabilized near 10^8 cells per gram in early August, but the population of INA bacteria continued to increase and was nearly 10^6 cells per gram at the time the plants were killed by frost in late September.

The number of INA bacteria per gram of leaf can be influenced by treatment with bactericides. Plants treated with streptomycin (Figure 7) had about the same total bacterial population as the untreated controls (Figure 6), but had substantially lower numbers of INA bacteria. Samples from which no INA colonies were isolated are represented as closed circles below the limit of sensitivity (dashed line). Treatment with cupric hydroxide (not shown) decreased the numbers of INA bacteria as effectively as did treatment with streptomycin and also caused about a 10-fold decrease in the total population of bacteria compared to the untreated controls.

The number of INA bacteria per gram of corn leaf also can be influenced by competing bacteria. A non-INA strain of *E. herbicola* (M232A) competes successfully with INA strains of

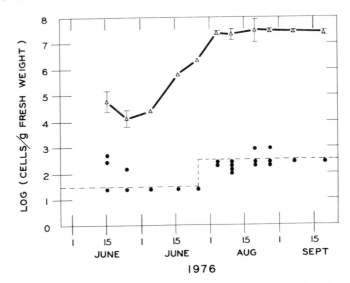

FIGURE 7.   *Total (Δ) and ice nucleation active (INA) (●)
bacterial populations on corn leaves sprayed with streptomycin
throughout the 1976 growing season.   Each individual determination of INA bacteria is represented (●).   The dashed line
represents the estimated limit of detection of INA bacteria.
Values plotted below the detection limit represent samples
from which INA bacteria were not detected.   Vertical bars
represent the standard error of the mean of log population.*

E. herbicola and *P. syringae* (8).   Populations of M232A were
established on corn by spraying plants with suspensions of
M232A several times during the growing season.   Numbers of
total and INA bacteria on leaves of these plants are shown
in Figure 8.   Treatments with M232A suspensions elevated the
total bacterial population relative to untreated controls
during the early part of the growing season.   However, by
late July, the total population was essentially the same on
treated as on untreated leaves.   The number of INA bacteria
was substantially lower in the presence of M232A (Figure 8)
than on untreated controls (Figure 6) throughout the growing
season.   Indeed, no INA bacteria were found on leaves of
plants treated with M232A until late July.
     The number of ice nuclei on leaves of field-grown corn
plants is affected by treatments that alter populations of
INA bacteria.   Leaves of plants sprayed with either streptomycin or cupric hydroxide several times during the growing
season had significantly fewer ice nuclei per gram than leaves
of untreated controls, or leaves that had been treated with a
bactericide (RH6401) that is not effective in decreasing INA

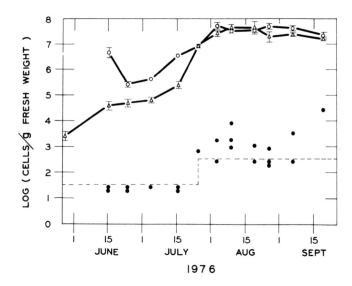

*FIGURE 8. Total (O) and ice nucleation active (INA) (●)
bacterial populations on corn leaves sprayed with Erwinia herbicola strain M232A throughout the 1976 growing season and
total bacteria on unsprayed leaves (△). Each individual determination of INA bacteria is represented (●). The dashed
line represents the estimated limits of detection of INA
bacteria. Values plotted below the detection limit represent
samples for which INA bacteria were not detected. Vertical
bars represent the standard error of the mean of log
population.*

bacterial populations (Table II). In addition, treatments
that inactivate bacterial ice nuclei (11), such as HCN,
chloroform, crystal violet and cetylpyridinium bromide, also
decreased the number of ice nuclei per gram of leaf tissue.

Changes in frost sensitivity to field-grown leaves
throughout the growing season were measured by subjecting
detached leaves to -5° C in a growth chamber. The number of
untreated corn leaves injured at -5° C (Figure 9) was relatively low early in the season when INA bacterial populations
were low, and increased throughout the season. Frost injury
to leaves treated with a bactericide (RH6401) that had decreased neither the numbers of INA bacteria, nor the numbers
of ice nuclei per gram, was not different from that to control
leaves. However, frost injury to leaves treated with the
competing bacterium M232A, or with the bactericides that did
decrease numbers of INA bacteria was substantially less than
that to control leaves (Figure 9).

TABLE II.   Effects of Various Treatments on Ice Nucleus
Concentration of Corn Leaf Discs at -5.2° C

Treatment	Ice nucleus concentration (ice nuclei/gram)
Plants treated in the field[a]	
Cupric hydroxide (1.25 g/liter Kocide 101)	1.5
Streptomycin (500 ppm)	5.8
RH6401[b] (0.5 g/liter)	46.9
Untreated control	69.2
Leaves treated in the laboratory[c]	
HCN (gas)	None detected
Chloroform (gas)	3.05 + 1.53
Crystal violet (0.3 mg/ml)	5.03 + 0.65
Cetylpyridinium bromide (0.01 M)	5.19 + 0.01
Washed with water	45.6 + 10.2
Untreated control	49.4 + 6.4

[a]Treatment materials were sprayed on plants 14 times
(Cupric hydroxide and streptomycin) or 12 times (RH6401) during the growing season. LSD 5% = 28.7.
[b]Rohm and Haas experimental bactericide 6401.
[c]Leaves were from untreated, field-grown plants. Leaves
were treated with gaseous reagents before discs were removed.
Discs were soaked in liquid reagents for 30 min. Mean + SE.

The relationship between population of epiphytic INA
bacteria and frost injury is shown in Figure 10. The regression line is drawn for data taken from control (untreated)
leaves (0) at various times throughout the season. Data from
the various treatments (bactericides, competing bacteria) are
also shown (□) and fit the line quite well. Thus, injury
to detached corn leaves at -5° C is directly proportional
to the logarithm of the population of INA bacteria on the
leaves.

For three years we have examined various treatments for
their effect on radiation frost injury to tender plants in
field plots. We have found significant differences between
the amount of frost injury to control plants and to plants
given various treatments that effectively decreased the number of active bacterial ice nuclei on their leaves.

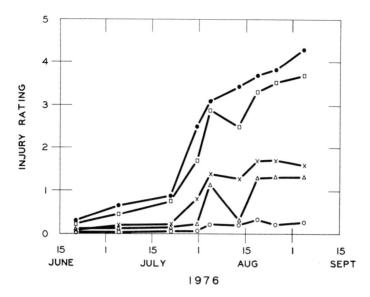

FIGURE 9.   Frost damage at -4.5 to -5.0° C to detached
leaves of corn plants that had been given various treatments
during the 1976 growing season. Leaves were detached from
(□) unsprayed control corn plants, and from plants receiving
applications of (●) Rohm and Haas experimental bactericide
6401; (x) competing bacteria; (Δ) streptomycin; or (0) cupric
hydroxide, during the 1976 growing season. Injury is reported
as the mean number of leaves per 5-leaf bundle which froze
at -4.5 to -5.0° C.

In summary, plants do not have intrinsic ice nuclei
active above about $-10°$ C associated with their leaves. Iso-
lates of two species of epiphytic bacteria are efficient ice
nuclei between -2 and $-5°$ C. Leaves of most plants from
several geographic areas and collected during different seasons of the year had substantial numbers of INA bacteria.
The ice nucleus content of field-grown leaves responds to
various treatments as if the ice nuclei active at $-5°$ C
on those leaves were INA bacteria. The extent of injury under
field conditions is modified by treatments that modify
numbers of INA bacteria on leaves and, finally, frost injury
to field-grown leaves at $-5°$ C is directly proportional to
the logarithm of the INA bacterial population on those
leaves. On the basis of these observations, we conclude that
epiphytic INA bacteria are incitants of frost injury to
tender plants.

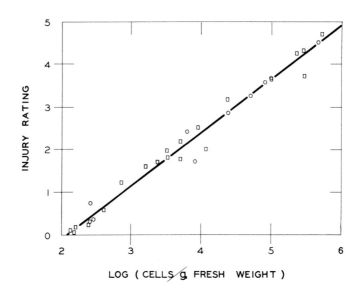

FIGURE 10.   Relationship between log ice nucleation active bacterial population and frost injury at -4.5 to -5.0° C. The mean number of detached corn leaves per 5-leaf bundle which froze at -4.5 to -5.0° C from unsprayed control plots (0) or from plants receiving bactericides or antagonistic bacteria treatments (□). The line drawn represents the linear regression Y = 1.24 x -2.54; r = 0.983, P < 0.01, for data from untreated leaves.

ACKNOWLEDGMENT

We thank Steven Vicen for his aid in preparation of the figures for this manuscript.

REFERENCES

1.   Arny, D. C., Lindow, S. E., and Upper, C. D., *Nature 262*, 282 (1976).
2.   Burke, M. J., Gusta, L. V., Quamme, H. A., Weiser, C. J., and Li, P. H., *Ann. Rev. Plant Physiol. 27*, 507 (1976).
3.   Chandler, W. H., *Proc. Amer. Soc. Hort. Sci. 64*, 552 (1954).
4.   Kaku, S., *Plant Cell Physiol. 12*, 147 (1971).
5.   Kaku, S., *Plant Cell Physiol. 14*, 1035 (1973).
6.   Kaku, S., *Cryobiology 12*, 154 (1975).

7. Lindow, S. E., Arny, D. C., and Upper, C. D., *Phyto-pathology 67*, in press (1977).
8. Lindow, S. E., Arny, D. C., and Upper, C. D., *Proc. Amer. Phytopath. Soc. 4*, Abstract, in press (1977).
9. Lindow, S. E., Arny, D. C., and Upper, C. D., *Proc. Amer. Phytopath. Soc. 4*, Abstract, in press (1977).
10. Lucas, J. W., *Plant Physiol. 29*, 245 (1954).
11. Maki, L. R., Galyon, E. L., Chang-Chien, M., and Cald-well, D. R., *Appl. Microbiol. 28*, 456 (1974).
12. Marcellos, H., and Single, W. V., *Agr. Meteorol. 16*, 125 (1976).
13. Mazur, P., *Ann. Rev. Plant Physiol. 20*, 419 (1969).
14. Modlibowska, I., *Roy. Hort. Soc. London 5*, 180 (1962).
15. Schnell, R. C., and Vali, G., *J. Atmos. Sci. 33*, 1554 (1976).
16. Vali, G., *J. Atmos. Sci. 28*, 402 (1971).
17. Vali, G., Christensen, M., Fresh, R. W., Galyon, E. L., Maki, L. R., and Schnell, R. C., *J. Atmos. Sci. 33*, 1565 (1976).

Part V
Survival and Breeding

ADAPTING COLD HARDINESS CONCEPTS
TO DECIDUOUS FRUIT CULTURE

E. L. Proebsting

Irrigated Agriculture Research
and Extension Center
Washington State University
Prosser, Washington

I. INTRODUCTION

Description of the relationships among environment, cultural practices and plant physiology that determine the responses of deciduous fruit plants to low temperatures
represents a marriage of ideas born in the plant physiology
laboratories with concepts that have grown from countless observations in the field. It is at this level that information
from research succeeds or fails in improving human
enterprises.

Deciduous fruit culture has evolved through the centuries
practices that minimize low temperature injury in most years.
Genetic constitution is the principal factor in determining
plant responses to low temperatures. The geographical distribution of deciduous fruit culture has developed through
human experience. Through physiological research, much of it
on woody plants other than the fruit plants of commerce, concepts have been developed that form a theoretical base to explain why certain orchard practices are successful, to describe and predict the seasonal progression of changes in
resistance to cold injury, and to provide information necessary to improve protective measures.

This paper will summarize significant research develop-
ments at this practical level of cold hardiness research.

267

II. METHODOLOGY

Systematic study of response to low temperature in deciduous fruit plants required development of methods that
would permit frequent assessment of resistance rather than
sporadic observations when damaging temperatures occurred in
the field. By using refrigeration and developing programmed
systems which provide repeatable time-temperature regimes it
became possible to expose plant tissues to damaging temperatures at will. This simple step has been a most important
development in recent low temperature research on deciduous
fruits.
     After exposure to damaging low temperatures the viability
of the treated tissues must be evaluated. The more familiar
tests include 1) observation of browning or regrowth, 2)
specific conductivity (7), 3) triphenyl tetrazolium chloride
(44), and 4) multiple freezing points (26). Comparing these
methods, Stergios and Howell (43) concluded that browning or
regrowth was the most reliable and that the validity and prac-
ticality of other methods should be established within the
limits of each individual study.
     After conducting the freezing test, its significance must
be evaluated in terms of the probably effect on cropping or on
survival of the tree. A continuing difficulty in cold hardiness studies of deciduous tree fruits is the fact that a tree
comprises many tissues and organs that vary in resistance, acclimation rates, heat exchange properties, exposure and mode
of resistance. The methods of sampling and evaluating the
resulting injury must depend upon the objectives of the particular research. In general, 1) the sample should be large
enough to minimize natural variability, 2) the material should
be amenable to rapid quantitative evaluation of injury and 3)
it should be possible to extrapolate the results to whole tree
scale. Fruit buds in *Prunus* are well adapted to systematic
study because 1) significant numbers can be accommodated in a
freezing chamber, 2) injury is easily and rapidly evaluated
in quantitative terms by visual inspection, and 3) results
are quite reliably extrapolated to field conditions if sampling is done correctly. Woody tissues are anatomically more
complex than fruit buds and present more problems of evaluation. Dormant one-year-old shoots provide material that meets
requirement 1 satisfactorily, meets requirement 2 if evaluated
by conductivity, and meets requirement 3 if applied where
treatments or varieties are compared. Correlation with older
woody tissues is good where treatments are compared (33) but
may be deficient in representing seasonal changes when extrapolated to the whole tree.

Because of limitations in laboratory testing cold hardiness concepts for deciduous fruit trees are based mostly on data from 1) "test winters" in the field, 2) basic concepts developed on woody species from other plant families, and 3) direct, systematic evaluations that are limited mostly to fruit buds and one-year-old shoots.

III. BASIC CONCEPTS

A. *Rest and Dormancy*

Before resistance to very low temperatures can develop, active growth must cease (49). In fruit trees the cessation of growth by different plant parts is not synchronized. Fruit buds are formed during the summer and develop slowly but are not in active growth. Terminal growth ceases early in some shoots, but vigorous shoots may continue to grow until late summer and early fall. Cambial activity is believed to cease first in the periphery of the tree, last in the trunk and crotches. The age of the tree, soil fertility, soil moisture, growth-regulating chemicals, and fall temperatures can modify the time of cessation of growth. Early winter freezes tend to injure the tissues that are the last to become dormant (28).

Pomologists have found that the concept of rest is very useful in explaining late winter injury to trees, especially peaches. Much of our early pomological literature on cold injury deals with increased susceptibility of tissues when early growth activity follows the end of rest (2, 28). This is proving to be very important as a factor in the short tree life problem in the peach areas of southeastern U.S. (28). Trunks and crotches are injured by low temperatures in the -10° to -15° C range, mild by northern standards, during late winter. Any practice that advances the end of rest or stimulates growth activity after rest is completed makes the trees more susceptible to low temperature injury and subsequent development of other problems.

In colder climates, as in central Washington, the end of rest does not seem to be an important direct factor in low temperature resistance (32, 36). While in rest, hardiness may fluctuate repeatedly (30). Usually rest is completed while the weather is cold enough to maintain the trees in a quiescent state. Cold resistance is not lost in these circumstances until enough heat has accumulated to start tissue development. This normally occurs after the probability of damaging temperatures is very low in central Washington.

The end of rest is a key element in using overtree sprinkling to delay bloom (41). Water is wasted if it is applied before the buds are able to respond to warm weather. If the start is delayed the total possible increment of bloom delay is reduced.

B. *Three-Stage Acclimation*

The concept of acclimation developing in three stages has been most useful in describing the behavior of deciduous fruit tree hardiness. Stage 1 is initiated by shortening days; in Stage 2 frost triggers major metabolic changes; and in Stage 3 further reversible increases occur while tissues are in the frozen state. The basic data were developed with *Cornus* (49) and *Hedera* (20), but the ideas have proved to be compatible with field observations of *Malus*, *Pyrus* and *Prunus* species. Apple seems to follow a 3-stage cycle similar to that described for *Cornus stolonifera* (18, 19). With *Prunus avium* it is not unequivocally clear that Stage 2 is initiated by low temperatures. In any event, both Stages 1 and 2 are initiated before damaging temperatures are likely to occur. The most significant factors in determining survival of most freezes in the Pacific Northwest are: 1) how far Stage 2 has progressed at the time of the freeze, and 2) how much Stage 3 acclimation has been superimposed on the normal cycle before the damaging temperatures occur (50).

Exposure to subfreezing temperatures can greatly increase the cold resistance of dormant, acclimated woody plants (43, 47). This phenomenon is very important to deciduous fruit culture.

The temperature required to kill half the buds (T_{50}) of peach and sweet cherry is normally $-20°$ to $-23°$ C by mid-November (3, 30, 32, 36). This is soon after normal defoliation. This level of resistance doesn't change appreciably during the dormant period unless the buds are exposed to prolonged subfreezing temperatures. After rest has ended bud development begins in response to warm temperatures. As development progresses the T_{50} rises slowly. It rarely rises above $-15°$ C before the buds begin the rapid swelling that leads to anthesis. We have referred to this relatively constant value of T_{50} as the minimum hardiness level (MHL).

At any time after completing Stage 2 buds will develop additional resistance when exposed continuously to temperatures below about $-3°$ C. Hardiness increases at a rate of about $1°$ to $2°$ per day, more rapidly with sweet cherry than with peach. When the buds thaw this added increment of hardiness is lost almost immediately. Sweet cherries were observed to lose $6°$ of resistance in 4 hours when exposed to

24° C. This loss of hardiness is limited by the MHL. Buds may go through repeated cycles of Stage 3 hardening and de-hardening but each time the hardiness is lost only to the MHL and no further. Prolonged periods of sub-freezing tempera-tures lasting several days result in continued added small increments of resistance. T_{50}'s of -34° C for cherries and -27° C for peaches have been observed. There appears to be a limit beyond which no further hardening will occur. This limit is reached after several days. The number of days and the limit reached both may be variables. In peach the capa-bility to respond to Stage 3 hardening seems to diminish after rest is completed and some development begins.

Regular monitoring of fruit bud hardiness levels in peach and cherry orchards has shown that there are differ-ences in cold hardiness between orchards, perhaps associated with soil type. The use of laboratory techniques has per-mitted synoptic evaluation of cold hardiness in samples from several orchards collected on the same day and injured under identical conditions. The technique permits one to avoid the questions often faced in field conditions of whether differences between orchards reflected temperature differences or cultural differences.

Having established a systematic pattern for the frequent changes in cold resistance of *Prunus* fruit buds it is at-tractive to extend this concept to other tissues. For the time being it is useful to assume that other tissues and other species do approximate this behavior. This assumption was helpful to workers in the western United States in interpret-ing the variable effects of a damaging region-wide freeze in 1972 (50).

Several new concepts have been presented that can be used to explain differences in hardiness.
1. The minimum hardiness level (MHL).
2. The duration of constant MHL.
3. The temperature below which Stage 3 hardening occurs.
4. The rate of Stage 3 hardening.
5. The rate at which State 3-hardened tissue loses resistance.
6. The shape of the temperature survival curve.

C. Deep Supercooling

In 1971 Graham and Mullin (13) reported that dormant azalea flower buds showed small exotherms at temperatures as low as -43° C, one for each floret. This important observa-tion introduced a new mechanism for explaining low tempera-ture survival and appeared to be particularly applicable to the *Prunus* fruit bud phenomena (39). It explains why we never observe partially-injured dormant buds in *Prunus*. It explains why one individual floret can be killed while others in the same bud are uninjured.

In peach fruit buds ice nucleation in the flower primordium appears to be delayed by barriers at the surface of the primordium and a dry region at its base which is formed by withdrawal of water into the bud scales during the first stage of freezing (40). Stage 3 acclimation must be assumed to increase either the ability of the primordia to resist spontaneous nucleation or the effectivness of the barriers. Continued transport of water out of the dry region during extended periods of continual freezing seems reasonable. We need quantitative data to show that it would produce the nearly linear increase in resistance of the buds over a period of several days that is frequently observed.

IV.   CULTURAL MEASURES TO REDUCE INJURY

Horticulturists deal with growers. Mere explanation of plant responses is not sufficient to solve cultural problems. Systems must be devised to improve chances for survival of the crop and the trees.

The basis for cold resistance is in the genetic constitution of the crop. Breeding for improved resistance is the preferred approach to improving survival. However, commercial horticulture resists change because of strong orientation within the marketing system toward uniformity of product. Thus, an apple cultivar would not be readily adopted in Washington State unless the fruit was identical to 'Delicious' or 'Golden Delicious'.

Because of this resistance to change in cultivars and because all the main deciduous cultivars are grown extensively in areas where losses to low temperatures occur, it is critical to the success of an orchard enterprise that growing practices minimize losses to low temperature.

Practices have evolved that reduce the chances of injury. They generally fall into two categories, environmental and physiological. A third approach, chemical, may still emerge as a practical means to reduce cold injury.

A.   *Environmental*

Environments favoring successful fruit culture can be selected, e.g. by planting on sites with good air drainage; or they can be created, e.g. by heating the air or sprinkling. Modification of the environment for protection against losses

from low temperatures goes beyond the scope of this review.
The horticulturist plays an important role in the application
of such protective measures by providing estimates of the
temperatures at which injury is likely to occur.

Such estimates are called "critical temperatures". They
evolved during the early years of orchard heating. Today's
values can be traced directly to USDA Farmers Bulletin 1096,
published in 1920, written by Floyd Young of the U.S. Weather
Bureau (52). These values were based on field observations.
Although modified frequently over the years they are still our
standard. More recently, laboratory freezing procedures have
made it possible to describe the distribution of susceptibility within a population of fruit buds and relate this to
development of the floral structures (31, 34).

The old critical temperatures were defined as the
temperatures that the buds could endure without injury for 30
minutes or less. New information gives the average temperatures that kill 10% and 90% of the buds. This makes it possible to modify orchard heating decisions taking into consideration the element of risk (1). It is now known that in
the later stages of floral bud development 1° to 1.5^{o} separates slight from severe injury. This provides much less
margin for error than does the 6° to 8° range that is found
in the earlier visually observable stages of bud development.
One can take advantage of this information by taking risks
early in the season in order to save fuel that may be used
to reduce risk-taking later.

Laboratory evaluations have also shown that there are
variations in bud resistance with time and between orchards
that need to be explained. During the very early visually
observable stages of bud development laboratory evaluation of
cold resistance produced very erratic data. Part of this
variability is due to the method of expressing the data. The
temperature-response curve, steep during dormancy and near
full bloom, is quite flat during the early developmental
stages (31). A minor shift in susceptibility of the population of buds can cause a shift of several degrees in the T_{50}.
Some of the variability is due to sampling error. Sampling
is much more critical at a stage when buds are killed over a
wide range of temperatures than when all are killed within a
very narrow range of temperatures. Also, temperature during
the day preceeding the low temperature probably has an effect,
but the relationships are not clear. The Stage 3 effect that
is so striking during dormancy doesn't exist at full bloom.
Freezing temperatures will kill the buds. Differences
between orchards exist at this time also. Injury from spring
frosts in the field is primarily determined by the minimum
temperature reached and by the stage of development. Both of
these factors vary with location. Additional differences

associated with cultural practices, and perhaps soil type, also exist.

Elevated tissue water content following overtree sprinkling has a major effect on resistance. Wet tissues may be several degrees more tender than dry tissues at a comparable stage of development. This factor has caused several instances of unexpectedly severe injury.

Critical temperature information is available in tabular form. More accurate information could be provided if an orchard were tested just prior to a critical night, but this is usually not feasible. It is feasible and necessary for growers who desire to activate protection systems on critical nights during the dormant period. Laboratory measurements of bud resistance during dormancy are sufficiently precise that reasonably accurate forecasts can be made on the basis of previous fixes on hardiness plus forecasts of temperatures for the period in question.

B. *Physiological*

Most of the acclimation of deciduous fruit trees is preordained by the genetic constitution of the plant and by the normal environmental events to which the plant responds. The fruit grower must manipulate those environmental factors that are under his control to obtain whatever added advantage possible. Usually it is a very small increment compared with the natural progression of acclimation. However, because of the steep response-temperature curve (34), a small increment can appear as a major advantage if the right temperature should occur. Too low, everything is killed; too high, no damage is observed.

There is strong evidence that reduced water content of plant tissues increases cold resistance and that such changes occur during natural cold acclimation (24, 25).

A great deal of emphasis has been placed on orchard practices that will minimize growth during late fall and early winter. It is important that the shortening day-length signal not be overridden by excessive nitrogen or water, particularly on young trees of species that may be less day length sensitive.

Pruning prior to the occurrence of low temperatures tends to increase injury (9). In many areas pruning is deferred until after the threat of damaging low temperatures is past, particularly with peaches (29).

Acclimation is an active metabolic process that requires the products of photosynthesis (2). Foliage should be protected from pests. Early thinning of fruit increases cold

resistance (8). Early maturing cultivars often survive low
fall temperatures better than cultivars with fruit on the tree
or with the crop recently harvested (28).

In regions of low rainfall where irrigation is practiced
soil moisture can be controlled. This may permit some additional control of cold resistance. Present practice is to
reduce applications in late summer and early fall to help the
tree stop growth. Then, near the time of defoliation the
trees are irrigated so that the roots are moist through the
winter.

This final irrigation serves to slow the rate of fall of soil
temperature below freezing and thus help prevent cold injury
to the roots. It also guards against those winters in which
natural precipitation is insufficient to assure normal growth
and development in the spring.

On the other hand, there is some evidence that fruit buds
are more hardy if soil moisture is low during the winter.
Therefore, one is faced with a value judgment as to which is
the greater risk. Common practice is to protect against root
injury and a dry winter at the cost of reduced bud hardiness.
At some locations bud losses may occur five times more frequently than root injury. We have insufficient information
to make such judgments properly.

C. Chemical

Chemical control of insects, diseases and weeds has
been an important part of the post-war agricultural successes.
Growth-regulating chemicals have found significant roles in
deciduous fruit production. The hope that comparable success
might be realized in cold protection has not materialized.

There are several points in the cycle of acclimation and
deacclimation at which chemical control might be exerted.

The earliest attempts were through control of dormancy
and rest. There were early observations indicating that NAA
delayed bloom (16). Maleic hydrazide prolonged dormancy of
potatoes and onions but did not prove to be as effective on
fruit plants (51). Identification of the "inhibitor β" complex in the buds of woody species (48) led to identification
of naringenin (14) and other phenolic compounds and abscisic
acid (ABA) (42) which were endogenous growth inhibitors. This
led to further expectations that cold hardiness in deciduous
woody plants could be improved through chemical control of
dormancy. As of today this has not happened. Holubowicz and
Boe (17) have reported some increased cold resistance in apple
from ABA. ABA may still prove to be effective but can't be
useful unless it is available in commercial quantities.

The existence of hardiness-promoting and inhibiting chemicals has been postulated, and evidence for their existence has been presented (11, 21). This line of research will be watched carefully by pomologists. Isolation, identification and synthesis fo such chemicals could lead to direct methods for controlling hardiness, but there is no reason for immediate optimism.

Interest has developed in the cell membrane as the site of injury and in changes in physical properties of membranes with changes in the proportion of saturation of fatty acids. This led to work designed to use these concepts to improve resistance. Kuiper (23) applied decenylsuccinic acid at 10^{-3} M to apple and pear blossoms and increased cold resistance by an estimated 2° C. Subsequent extensive horticultural trials failed to confirm his results (15).

Cryoprotectants, acquired from research into freeze preservation of living tissues, provide another possible avenue for field protection of plant tissues. Success has been reported by Ketchie (22) with vegetative tissues of apple and pear trees. Adoption of the treatment by commercial growers is still in the future, with many practical questions still unanswered, but development is continuing.

Plant growth regulators, applied topically, profoundly influence the behavior of the treated plant. It was not unreasonable to screen these compounds for plant responses that would reduce cold injury. Some success has been achieved on an empirical basis (12). GA_3 applied in late summer increased resistance of peach fruit buds slightly and delayed bloom slightly (3). This has been confirmed from at least two locations (4, 10). GA_3 is too costly for routine commercial use, and undesirable effects on hardiness of vegetative tissues would limit its usefulness even if cost were not a factor. Ethephon, applied in late summer or early fall, increases resistance of sweet cherry fruit buds and is in the early phase of commercial development in central Washington (35, 38). However, it has not been successful in tests at other locations (6). ABA has been mentioned already. No reports on effects of cytokinins on cold resistance have yet appeared in the pomological literature. Auxins were first examined with respect to dormancy. Subsequently, an auxin-like compound, 2,4,5-T, was shown to reduce frost injury with apricots when used as a pre-frost spray and also to reduce the drop of fruit with dead embryos but uninjured pericarp when used as a post-frost treatment (5). This approach, reported 25 years ago, still seems valid but has not been developed commercially. Gibberellic acid and auxins have been used to retain on the tree fruits of certain cultivars of pear and apple after frost injury (27, 46). The general concept of treatment to reduce fruit drop after injury has occurred has strong roots in horticultural practice but has not been subjected to much research.

Much of fruit growing consists of coordinating measures to induce the trees to perform in the most advantageous and profitable manner. The avoidance of low temperature injury is a critical factor in the success of many North American fruit-growing districts. Fruit growing is successful where satisfactory relationships among environment, cultural practices and plant physiology have been developed, largely empirically. Plant physiologists have a continuing opportunity to profide new understanding of low temperature injury and its physiological control. The pomologist must stand ready to adapt the basic concepts derived from research to the realities of the fruit-growing world.

REFERENCES

1. Ballard, J. K., and Proebsting, E. L., *Wash. State Univ. Ext. Bul. 634* (1972).
2. Chandler, W. H., *Proc. Amer. Soc. Hort. Sci. 64*, 552 (1954).
3. Chaplin, C. E., *Proc. Amer. Soc. Hort. Sci. 52*, 121 (1948).
4. Corgan, J. N., and Widmoyer, F. B., *J. Amer. Soc. Hort. Sci. 96*, 54 (1971).
5. Crane, J. C., *Proc. Amer. Soc. Hort. Sci. 64*, 125 (1954).
6. Dennis, F. G., Jr., *Amer. Sci. Hort. Sci. 101*, 241 (1976).
7. Dexter, S. E., Tottingham, W., and Garber, L. R., *Plant Physiol. 1*, 63 (1932).
8. Edgerton, L. J., *Proc. Amer. Soc. Hort. Sci. 52*, 112 (1948).
9. Edgerton, L. J., and Shaulis, N. J., *Proc. Amer. Soc. Hort. Sci. 62*, 209 (1953).
10. Edgerton, L. J., *Proc. Amer. Soc. Hort. Sci. 88*, 197 (1966).
11. Fuchigami, L. M., Evert, D. R., and Weiser, C. J., *Plant Physiol. 47*, 164 (1971).
12. Gambrell, C. E., Jr., and Rhodes, W. H., *S. Car. Agric. Expt. Sta. Tech. Bul. 1030* (1969).
13. Graham, P. R., and Mullin, R., *J. Amer. Soc. Hort. Sci. 101*, 7 (1976).
14. Hendershott, C. H., Jr., and Walker, D. R., *Science 130*, 798 (1959).
15. Hilborn, M. T., *Research in the Life Sciences, Univ. of Maine 21*, 2 (1973).

16. Hitchcock, A. E., and Zimmerman, P. W., *Proc. Amer. Soc. Hort. Sci. 42*, 141 (1943).

17. Holubowicz, T., and Boe, A. A., *J. Amer. Soc. Hort. Sci. 94*, 661 (1970).

18. Howell, G. S., and Weiser, C. J., *J. Amer. Soc. Hort. Sci. 95*, 190 (1970).

19. Howell, G. S., and Weiser, C. J., *Plant Physiol. 45*, 390 (1970).

20. Irving, R. M., and Lanphear, F. O., *Plant Physiol. 42*, 1191 (1967).

21. Irving, R. M., and Lanphear, F. O., *Plant Physiol. 42*, 1384 (1967).

22. Ketchie, D. O., and Murren, C., *J. Amer. Soc. Hort. Sci. 101*, 57 (1976).

23. Kuiper, P. J. C., *Science 146*, 544 (1964).

24. Li, P. H., and Weiser, C. J., *Cryobiology 8*, 108 (1971).

25. McKenzie, J. S., Weiser, C. J., and Li, P. H., *J. Amer. Soc. Hort. Sci. 99*, 223 (1974).

26. McLeester, R. C., Weiser, C. J., and Hall, T. C., *Plant Physiol. 44*, 37 (1968).

27. Modlibowska, I., *J. Hort. Sci. 50*, 21 (1975).

28. Potter, G. F., *Proc. Amer. Soc. Hort. Sci. 36*, 184 (1939).

29. Prince, V. E., *Proc. Amer. Soc. Hort. Sci. 88*, 190 (1966).

30. Proebsting, E. L., Jr., *Proc. Amer. Soc. Hort. Sci. 74*, 144 (1959).

31. Proebsting, E. L., Jr., and Mills, H. H., *Proc. Amer. Soc. Hort. Sci. 78*, 104 (1961).

32. Proebsting, E. L., Jr., *Proc. Amer. Soc. Hort. Sci. 83*, 259 (1963).

33. Proebsting, E. L., Jr., and Mills, H. H., *Proc. Amer. Soc. Hort. Sci. 85*, 134 (1964).

34. Proebsting, E. L., Jr., Mills, H. H., and Russell, T. S., *Proc. Amer. Soc. Hort. Sci. 89*, 85 (1966).

35. Proebsting, E. L., Jr., and Mills, H. H., *HortScience 4*, 254 (1969).

36. Proebsting, E. L., Jr., and Mills, H. H., *J. Amer. Soc. Hort. Sci. 97*, 802 (1972).

37. Proebsting, E. L., Jr., and Mills, H. H., *J. Amer. Soc. Hort. Sci. 99*, 464 (1974).

38. Proebsting, E. L., Jr., and Mills, H. H., *J. Amer. Soc. Hort. Sci. 101*, 31 (1976).

39. Quamme, H. A., *J. Amer. Soc. Hort. Sci. 99*, 315 (1974).

40. Quamme, H. A., *J. Amer. Soc. Hort. Sci.* (in press).

41. Richardson, E. A., Seeley, S. D., Walker, D. R., Anderson, J. L., and Ashcroft, G. L., *HortScience 10*, 236 (1975).

42. Robinson, P. M., Wareing, P. F., and Thomas, T. H., *Nature 199*, 875 (1963).
43. Sakai, A., *Plant Physiol. 41*, 353 (1966).
44. Steponkus, P. L., and Lanphear, F. O., *Plant Physiol. 42*, 1423 (1967).
45. Stergios, B. G., and Howell, G. S., Jr., *J. Amer. Soc. Hort. Sci. 98*, 325 (1973).
46. Swarbrick, T., *Nature 156*, 691 (1945).
47. Tumanov, I. I., and Krasavtsev, O. A., *Soviet Plant Physiol. 6*, 663 (1959).
48. Wareing, P. F., and Saunders, P. F., *Ann. Rev. Pl. Physiol. 22*, 261 (1971).
49. Weiser, C. J., *Science 169*, 1269 (1970).
50. West. Regional Coop. Proj. 130., *Wash. Agric. Res. Ctr. Bul. 813* (1975).
51. White, D. G., and Kennard, W. C., *Proc. Amer. Soc. Hort. Sci. 55*, 147 (1950).
52. Young, F. D., *USDA Farmer's Bul. 1096* (1920).

EFFECT OF SOLAR RADIATION
ON FROST DAMAGE TO YOUNG CRYPTOMERIAS[1]

T. Horiuchi

Ibaraki Prefectural
Forest Experimental Station
Naka, Ibaraki

A. Sakai

The Institute
of Low Temperature Science
Hokkaido University
Sapporo

ABSTRACT

The basal stem of young cryoptomerias, especially cambium sustains frost damage in early winter when the basal stem is exposed to large daily fluctuations in temperature. The hardiness of the basal stem tissue is invariably lower than that of the higher part and that of the cambium is the least hardy tissue of the stem. Hardiness does not increase in basal tissue that is exposed to day temperatures higher than about 15° C even when subsequently exposed to 0° C at night during a 12 day period of day-night alternating temperature regimes. This suggests that the increased day temperature of the basal stem due to solar radiation may be related to prevention of increase in hardiness in early winter. This consideration is supported by the fact that frost damage of young cryptomerias is avoided by either shading or shortening the length of time during which the basal stem tissue is exposed to solar radiation.

[1]*The Institute of Low Temperature Science, Hokkaido University, Sapporo.*

I.   INTRODUCTION

Basal stems of many young trees including cryotomeria,
Japanese cypress, tea plant, chestnut, apple tree and other
deciduous fruit trees often sustain frost damage in frost
basins, flat lands, plateaus and sunny slopes. Frost damage
of this type greatly reduces growth of individual plants and
plantation yield, not only in the northern climates but also
in the milder climates of southern Japan. Since protection
from such basal damage is one of the practical aims in sil-
viculture and horticulture, this type of frost damage has
been studied in several countries (1, 3, 14, 19). Among the
environmental factors causing the frost damage, it is well
known that little or no basal damage occurs in cryptomerias
planted on shady slopes, while serious damage is observed in
cryptomerias planted on sunny slopes. Similarly, frost damage
of mature deciduous trees often occurs only on the sunny side
of the stems (9, 12, 17). These observations suggest that
sunshine may be involved in hardiness changes of wintering
trees. This possibility is examined in the present study of
the effects of solar radiation on the level of hardiness in
young cryptomeria plants.

II.   MATERIALS AND METHODS

Field experiments were carried out at Koibuchi nursery, a
suburb of Mito city (36° 23') located about 150 km northeast
from Tokyo. Laboratory experiments were conducted in the
Institute of Low Temperature Science, Sapporo. In all experiments, young trees of cryptomeria (*Cryptomeria japonica*), 1.0
to 2.0 cm basal diameter near the ground surface, 30 to 50 cm
height, were used as experimental materials. For hardiness
evaluation, stem pieces 2 or 3 cm long, were cut from each
stem, which was enclosed in polyethylene bags and frozen in a
cold room at −5° C. Thereafter, frozen stem pieces were
cooled down to different temperatures at 2 or 2.5° C increments at hourly intervals. The frozen stem pieces were held
at the various test temperatures for 4 hrs before being rewarmed slowly in air at 0° C. To evaluate viability after
freezing, the thawed stem pieces were placed in polyethylene
bags saturated with water vapour at room temperature for 20
days when freezing damage was evaluated visually or microscopically. Browning was used as the criterion for rating
damage. The lowest survival temperature at which no browning
occurred was taken as the hardiness of the tissues.

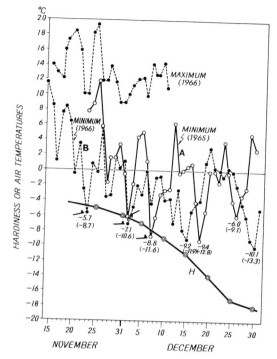

FIGURE 1. Time at which basal stem of cryptomerias sustained frost damage in Koibuchi. A. B.: Daily minimum air
temperatures in 1965 (A) and 1966 (B) in Koibuchi. Temperatures in brackets: Air temperatures at 30 cm above the ground
surface. Arrows indicate that frost damage on the basal stem
was observed.

To investigate the effect of daily temperature fluctuation
on the frost hardiness of stems, various temperature regimes
were given to a segment of the basal stem (5 cm long) of potted cryptomerias using the thermoelectric apparatus (15) or
infrared radiation. Stem temperatures were measured with 0.2
mm copper-constantan thermocouples and recorded.

III. RESULTS

Typical observations at the Koibuchi nursery, where young
cryptomerias growing on flat land have suffered from basal
frost damage on the cambium, are related to hardiness in
Figure 1. For the two years of observations illustrated in
Figure 1, the occurrence of minimum air temperatures below the

TABLE I. *Frost Hardiness in Different Heights on a Stem of Young Cryptomeria[a]*

	Hardiness (^{o}C)[b]	
Stem Part	November 20 cambium	December 11 cambium
Upper (45 to 50 cm)	-10	-15
Middle (25 to 30 cm)	-8	-13
Basal (3 to 8 cm)	-5	-10

[a]*Material: Young cryptomeria (60 cm height)*
[b]*Hardiness was represented as the minimum survival temperatures on the south side of basal stem tissues.*

basal stem hardiness temperature between mid-November and early December provides a satisfactory explanation for basal stem damage during this period. However, since the hardiness increases abruptly from mid-December to late January, reaching a high level of -25 to -27° C which was maintained until early or mid-February, the winter frosts which seldom fall below -15° C in this district do not cause basal stem damage. Furthermore, the basal stem was still hardy to -10° C in late March, when severe frosts below -10° C rarely occurred. Thus, early winter is a critical period for wintering of young cryptomerias in this district.

Frost hardiness of bark tissues at different heights on a stem of young cryptomeria was studied using 4-year cryptomerias grown in Koibuchi. The frost hardiness of the basal stem, south side, was invariably much lower than that of the higher parts except of unmatured tip parts (Table I). It was also observed that the hardiness of bark on the south side in a stem was much lower than that of the bark on the north side especially at the basal stem. Microscopic observation of tissue sections taken from the basal stem revealed that the cambium and the adjacent cells were the least hardy tissues of the stem of cryptomeria (14).

Daily fluctuations in bark temperatures at various heights of the stem (1 cm diameter) as determined by thermocouples, vary considerably at different heights above the snow surface (Figure 2). On a clear fine day, the bark temperature fell with the increase in height above the snow surface, while at night, it rose considerably with the increase in height. On a

FIGURE 2. Shelter belt. The Shelter belt was composed
of cryptomerias and Japanese cypress, 15 m height, 24 m wide,
and 94 m long from east to west. S--Shelter belt; C--3-year
cryptomerias. Cryptomerias were planted at different distances to the north of the shelter belt.

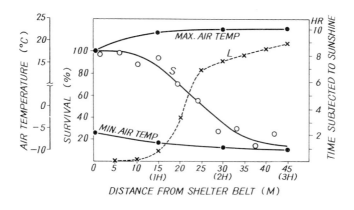

DISTANCE FROM SHELTER BELT (M)

FIGURE 3. Effect of length of time subjected to solar
radiation against frost damage of young cryptomeria at the
winter solstice. S--Percentage of Survival; L--Length of time
subjected to solar radiation at the winter solstice. Air
temperatures were determined at 30 cm in the shade above the
ground surface. H--Height of shelter belt (15 m). 1H, 2H,
3H--Distance from shelter belt expressed as one, two and three
times of tree height, respectively.

FIGURE 4. Protection of cryptomerias from frost damage by sheltering nets. Black color, polyethylene net: (A) 50% screen, 2/2x6 inch mesh; (B) 75% screen, 2/2x10 inch mesh. Sheltering nets were set at 25 cm south from cryptomerias.

clear night, the area near the snow surface was cooled greatly by radiation. Nearly the same results were obtained using the young cryptomerias at Koibuchi without snow cover where the trends in day and night temperature of the bark were affected by height above the ground (7). These results imply that the increased day temperature of tissues of the basal stem is responsible for preventing increase in the hardiness of cryptomeria in early winter.

To test this supposition, young cryptomerias were planted at different distances to the north of the shleter belt (Figure 3) so as to impose differences in the length of the time during which the basal stem tissues of cryptomerias were subjected to solar radiation. As shown in Figure 4, most of the young trees planted more than 30 m north from the shelter belt (15 m height, 94 m long east to west, 24 m wide) sustained frost damage, while only slight damage was observed in the trees planted within 15 m north from the shelter. Thus, the frost damage increased markedly in the area from about 15 m to about 30 m from the shelter belt. Despite these differences in frost damage, the temperature data in Table II show that air temperature 30 cm above the ground at different distances north away from the shelter belt varied only

TABLE II.   Air Temperatures in Mid-winter at Different
Distances North from the Shelter Belt

Air temperature (^oC)[a] (30 cm above the ground)	Distance south from the tree belt (m)			
	OH[b]	1H	2H	3H
Maximum	17.2	21.9	22.6	22.9
Minimum	-6.2	-8.3	-9.3	-10.2
Mean	5.5	6.8	6.7	6.4
Daily temperature difference	23.5	30.2	31.9	33.1

[a]Air temperatures were determined at 30 cm in the shade
above the ground surface.
[b]Shelter belt: 15 m height, 24 m wide, and 94 m long
from east to west (Refer to Figure 3).

slightly. It is especially important to note that the shelter
belt had little effect on maximum and minimum air temperatures
at the places between 15 m (1H) and 30 m (2H) north from the
shelter belt where marked differences were observed in frost
damage.   Thus, air temperature does not appear to have a major
role in this shelter belt effect. Rather, the increase in
the length of time during which the basal stems of 2H and 3H
trees were exposed to solar radiation, which would tend to
increase bark temperature during the day, seems likely to be
involved in this markedly increased frost damage.

To check this possibility, basal stem tissues of potted
3-year cryptomerias were exposed to 23° C while the rest of
the plant was at 10° C. The higher temperature which is the
daily maximum temperature prevailing in late November in
Koibuchi was applied for different lengths of time up to 12
days, using a narrow beam lamp or a thermoelectric apparatus.
From the results which are summarized in Table III, it can be
seen that decreases in hardiness of basal stem tissues were
noted only in trees exposed to 23° C over 8 hrs per day for 12
days.   The same result was obtained with cryptomerias held at
23° C for 8 hrs per day with either radiation from an infrared
lamp or a thermoelectric apparatus (Table III). This suggests
that the rise in temperature of the basal stem due to solar
radiation, may be responsible for preventing hardiness increases in early winter and in early spring.

TABLE III. *Effect of Length of Time Subjected to 23° C at Day Time on the Frost Hardiness of Basal Stem of Cryptomeria in Early Winter*

Daily temperature treatment (12 days) day--night	Frost hardiness (° C)			
	-5	-7.5	-10	-15
10° (control)	0	0	0	3
23° (4 h, IFR)--10° (20 h)	0	0	0	3
23° (6 h, IFR)--10° (18 h)	0	0	0	3
23° (8 h, IFR)--10° (16 h)	0	0	3	3
23° (8 h, TE)--10° (16 h)	0	0	3	3
23° (16 h, IFR)--10° (8 h)	0	3	3	3

*Material: Potted 3-year cryptomerias placed at 10° C.
Basal stem (10 cm) was locally held at 23° C by infrared lamp (IFR) or thermoelectric apparatus (TE).
Frost damage was represented by numerals: 0 (normal), 3 (serious or killed).*

TABLE IV. *Effect of Daily Temperature Fluctuation upon Frost Hardiness of Young Cryptomeria in Early Winter*

Day-night temperature regimes day (14 h)--night (10 h)	Frost hardiness (° C)					
	-5	-7.5	-10	-12.5	-15	-17
Control[a]	0	0	1	3	3	3
23--0	0	3	3			
19--0	0	1	3	3		
15--0	0	0	3	3		
10--0	0	0	0	2	3	
0	0	0	0	0	3	

[a]*Frost damage: 0 (normal), 1 (slight), 2 (medial), 3 (serious or killed).*
Basal stem (10 cm) of potted cryptomerias were locally subjected to day-night alternating temperature regimes for 10 days by thermoelectric apparatus and infrared lamp.

TABLE V.   Effect of Daily Temperature Fluctuations on
Frost Hardiness in Mid-winter

Temperature treatments (12 days)			Frost damage (°C)					
			Before treatment			After treatment		
Day (8 h)	0° (2 h)	Night (14 h)	-23	-20	-23	-20	-17	-15
1. 0°	-	-5° (5°)a	3b	0	3	0		
2. 5°	-	-5° (10°)	3	0	3	0		
3. 10°	-	-5° (15°)	3	0	3	0		
4. 10°	-	-10° (20°)	3	0	3	0		
5. 10°	-	-15° (25°)	3	0	3	0		
6. 13°	-	-5° (18°)	3	0	3	1		0
7. 20°	-	-5° (25°)	3	0	3	3		0
8. 20°	-	-10° (30°)	3	0	3	3		0
9.	0°		3	0	3	0		
10.	10°		3	0	3	1	0	
11.	13°		3	0	3	3	1	0
12.	20°		3	0	3	4	4	3

Material: Potted 4-year cryptomerias.
aDaily temperature fluctuation.
bFrost damage: 0 (normal), 1 (slight), 2 (medial), 3
(serious or killed).
Basal stem (10 cm) was locally subjected to day-night
alternative temperature regimes by thermoelectric apparatus.

Partially hardened cryptomerias were subjected to day-
night alternating temperature regimes to check the former
possibility. The results obtained are summarized in Table
IV. Daily fluctuations of temperature between 23 and 0° C
or 19 and 0° C caused a considerable decrease in hardiness,
while little or no decrease was noted in the cryptomerias
subjected to daily fluctuation between 15 and 0° C. Hardiness
increase was noted only in the cryptomerias subjected to the
daily alternating temperature regime between 10 and 0° C.
These results suggest that an increase in hardiness was prevented by day temperatures higher than 15° C even when subjected to subsequent low temperatures, 0° C at night. In the
basal stem of cryptomerias, the effect of sequential daily

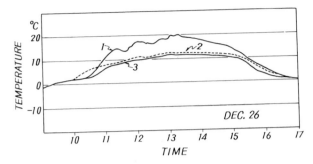

FIGURE 5. *Daily temperature fluctuations in the basal stem of young cryptomerias sheltered in different degrees. Tree height: 60 cm; diameter at basal stem: 1 cm.*
1. *Bark temperature of unsheltered control tree.*
2. *Bark temperature of a tree partially sheltered by board (45 cm²) set 25 cm south away from the stem.*
3. *Air temperature at 15 cm above the ground.*
Bark temperature was determined by thermocouples inserted in the cambium at 15 cm height and recorded. Air temperature at 15 cm height above the ground level was determined in special aluminum vessels set in the shade at 15 cm height.

temperature fluctuations on hardiness changes was further examined by varying the basal stem temperature or winter hardy, potted cryptomerias which had been hardened to the maximum level by exposure to -3° C for 10 days as noted in other tree species (4, 12, 13, 18). As shown in Table V, daily fluctuations of temperature between 10° C and -5° C or -15° C (15° C or 20° C) for 12 days caused no decrease in hardiness of basal stem tissue. However, exposure to daily fluctuations of temperature between 20° C and -5° C or -10° C (25° C or 30° C) caused a considerable decrease in hardiness. Continuous exposure to constant temperature caused little or no decrease in hardiness, if below 10° C for 12 days, but if higher than about 13° there was a marked decrease in hardiness. Thus, exposure to day temperature higher than about 13°, rather than any particular temperature fluctuation, appears responsible for the decrease in hardiness of cryptomerias in winter.

The presumed rise in the bark temperature of the basal stem of the cryptomerias when exposed to solar radiation was prevented by positioning a plywood board (45 cm²) or a net (60 cm² size, 2/2x10 inch mesh) (Figure 5) at 25 cm south from cryptomerias. Frost damage to these shaded cryptomerias was completely prevented (Table VI). The bark temperature for the shaded plants was generally 7 to 10° C cooler than unshaded plants (Figure 6). A similar lack of damage also

TABLE VI. *Protection of Basal Stem of Young Cryptomeria from Frost Damage by Sheltering Direct Sunshine*

Treatment	Normal tree (%)	Trees seriously damaged or killed (%)
Untreated control	59.5	40.5
Sheltered by a net[a]	100	0
Sheltered by a board[b]	100	0

[a]*Sheltered by a dense net (60x60 cm², 2/2x10") set 25 cm south from cryptomeria (60 cm height) (Refer to Figure 4).*
[b]*Sheltered by a plywood board (45x45 cm²) set 25 cm south from cryptomeria.*

FIGURE 6. *Protection of young cryptomerias from frost damage by planting under pine canopies. Pine stand: 18-year Japanese red pine (8 m height, 30% stand density, 2,000 trees/ha). Grasses under pine trees were cut at 3 week intervals; during winter grasses withered.*

TABLE VII.   Protection of Young Cryptomerias from Frost
Damage by Planting under Pine Canopy

Treatment	Normal (%)	Damaged (%)	Killed (%)
Open (number of trees planted) 64	21.9	29.7	38.4
Under pine canopy (number of trees planted) 88	100	0	0

Pine forest: 8-year forest (8 m height, tree density
2,000/ha).

occurred in cryptomerias planted under pine canopies (Table
VII, Figure 7). The seasonal changes in hardiness and the
air temperature at 30 cm above the ground were compared in
young cryptomerias planted under pine-canopies and at
100 m out of the pine forest beginning in October and
extending to the mid-winter of the following year. The
pine-canopies moderated the minimum temperatures at night
and decreased the rise in temperature in the daytime. It
was also observed that the formation of winter buds and the
development of frost hardiness were two weeks earlier than
for those out of the canopies. It is interesting, however,
that the winter maximal level in hardiness of the cryptomerias planted under pine-canopies was 5° C lower than that out
of the pine forest.

IV. DISCUSSION

Proebsting (10, 11) clearly demonstrated that the winter
hardiness of peach flower buds varied with changes in environmental temperatures with rather short periods. He also
noted the interesting fact that until the end of the rest
period there was a minimum hardiness level (-16 to -18° C)
above which peach bud hardiness did not decrease despite
warm weather. Hewett (5) also reported that mean maximum
daily temperature of the preceding 7 days was more closely
related to hardiness of apricot buds in winter. However,
Sakai (18) reported that in extremely hardy twigs, any appreciable change in winter hardiness in buds and twigs was
not observed for 3 days at least, depending on the prevailing

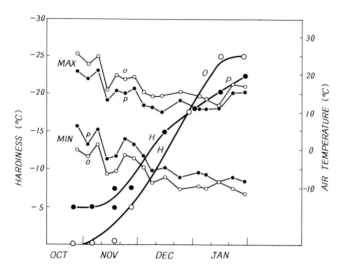

FIGURE 7. Hardiness increase of basal stem of cryptomerias planted under pine canopies and open area 100 m away from
the forest along with air temperatures. P.O.--Hardiness of
cryptomerias planted under pine canopies (P), and open areas
(O). p.o.--Maximal and minimal air temperatures at 30 cm
above the ground surface under pine canopies (p), and open
areas (o). Pine stand--Refer to Figure 6 (18-year Japanese
red pine, 8 m height, 30% stand density, 2,000 trees/ha).

environmental air temperatures. It seems likely that for
very hardy twigs the minimum hardiness level proposed by
Proebsting remains much lower and that changes in hardiness
during winter may differ greatly depending on the species and
climates in which plants are wintering.

In Saghalien fir, a conifer native to Hokkaido, the far
northern island of Japan, the basal stem was similar to cryptomeria in having the most susceptible tissue to frost
stress (14). In the seedlings of Saghalien fir which sustained damage by late frost, the cambial initials remained
alive and continued differentiating new phloem and xylem
mother cells while differentiating tracheids were killed,
leaving a permanent band, frost ring (14). This contrast with
cryptomeria suggests that the maximum cambium hardiness in
twigs is characteristic of tree species native to cold
climates and is unlike species growing in mild climates such
as chestnut, tea plant, citrus plant, Japanese cypress,
cryptomerias, etc.

Doi et al. (2, 7) clearly demonstrated that when lower
parts of young cryptomerias (50 to 70 cm height) were sheltered with other cryptomerias by the concentric circular

plantings, the frost damage of basal stem was nearly completely avoided. In their experiments, 24 young cryptomerias
were planted on each concentric circle with 8 different diameters, which increased by 0.35 times from 1.6 m to 20 m.
Nearly all of the cryptomerias on the innermost and second
circles survive, but almost all of the cryptomerias on the
outer circles sustained frost damage. They also found that
survival percentage could be estimated by the shading rate,
which was determined by net radiation near stems at night,
regardless of size of cryptomerias and planting distance,
and that 90% of cryptomerias survived when shading rate was
more than 40%, but in the case of less than 40%, the survival
decreased drastically. It was observed that cryptomerias
growing on the sunny slopes became frost hardy much later than
those on the shady slopes due to a delay in growth cessation
in late autumn. In contrast, in late winter especially in
mild climates, cryptomerias planted on the sunny slopes became less hardy earlier than those planted on the shady
slopes due to the dehardening effects of a marked increase in
basal stem temperature during the daytime. For this reason,
cryptomerias on the sunny slopes are more susceptible to frost
damage than those on the shady slopes.

   A relation between the shading treatment and frost hardiness was investigated in detail in citrus plants by Konakahara
(8). In the plot where the shading rate was kept at 83% by
covering with straw mat, the hardiness level of the leaves,
-7.0° C, was retained until early March, while the maximum
value of hardiness, -6.5° C, of non-shading plot gradually
decreased from mid-February, reaching to the level of -5.0° C
in early March. In the shading plants, the leaf temperature
in the daytime was kept 8 to 10° C lower than that in the
non-shading plot in winter, which enabled shading plants to
retain the maximum hardiness for a long time.

   In Hokkaido frost damage to basal stem and buds was often
observed in reforestations after clear cuttings in the period
from mid-May to early July. Planting conifers under canopies
of deciduous trees such as birch and oak may be the most
effective and practical method for protecting young conifers
from frost damage and winter desiccation (13), especially in
severe cold climates. Although practical methods for preventing frost damage differ considerably for fruit trees
and forest trees, shading plants from direct sunshine or
shortening the length of time during which plants are exposed
to direct sunshine seems to be a common requirement to prevent
hardiness loss during winter and to enhance hardiness sooner
in early winter.

REFERENCES

1. Day, W. R., and Pease, T. R., *Oxford Memoirs 16*, 4
    (1934).
2. Doi, K., Horiuchi, K., and Okanoue, M., *J. Jap. Forest
    Sci. 52*, 120 (1970).
3. Eiche, V., *Studia Forestalia Suecica 36*, 1 (1966).
4. Eiga, S., and Sakai, A., *J. Jap. Forest Sci. 54*, 412
    (1972).
5. Hewett, E. W., *New Zealand J. Agr. Res. 19*, 353 (1976).
6. Horiuchi, T., and Sakai, A., *J. Jap. Forest Sci. 55*,
    46 (1973).
7. Horiuchi, T., *Ibaraki Prefec. Forest Exp. Stat. Bull. 10*,
    1 (1976).
8. Konakahara, M., *Shizuoka Prefect. Citrus Exp. Stat.
    Special Bull. 3*, 1 (1975).
9. Levitt, J., "The Hardiness of Plants," 278 pages.
    Academic Press, New York (1956).
10. Proebsting, E. L., *Proc. Amer. Soc. Hort. Sci. 74*, 144
    (1959).
11. Proebsting, E. L., *Proc. Amer. Soc. Hort. Sci. 83*, 259
    (1963).
12. Sakai, A., *Physiol. Plant. 19*, 105 (1966).
13. Sakai, A., *Plant Physiol. 41*, 353 (1966).
14. Sakai, A., *Conf. Inst. Low Temp. Sci., 15*, 1 (1968).
15. Sakai, A., *J. Jap. Forest Sci. 50*, 79 (1968).
16. Sakai, A., *Ecology 51*, 657 (1970).
17. Sakai, A., and Horiuchi, T., *J. Jap. Forest Sci. 54*,
    379 (1972).
18. Sakai, A., *Plant and Cell Physiol. 14*, 1 (1973).
19. Toya, T., Kayumi, S., Katsuo, K., and Matsushita, S.,
    Bull. Nat. Res. Inst. of Tea 9, 1 (1974).

FREEZE SURVIVAL
OF CITRUS TREES IN FLORIDA

G. Yelenosky

Agricultural Research Service
United States Department of Agriculture
Orlando, Florida

Freeze injury is probably the most important factor
limiting citrus production in Florida, an area having some of
the most damaging freezes in world citrus production.
Florida is the leading citrus state in the United States and
produces more than two-thirds of the national output in citrus
crops valued at close to $1 billion. Freezes not only cause
economic hardships, but also inhibit agricultural development
of freeze-stressed areas.

Severe freezes in Florida cause millions of dollars of
damage to citrus fruit and trees. Production is affected
for several years and total recovery is slow. In the 1800's,
severe freezes threatened the very existence of the citrus
industry in Florida (65). Since 1958, fruit losses have
averaged about 7% annually. Losses greater than 30% occurred
during both the 1962 and 1977 freezes. The 1962 freeze cost
Florida one-fourth of its 52 million citrus trees, and estimated total damage approached $500 million.

I. GENERAL FREEZE PATTERNS

The duration of the freeze season in Florida is about
4 months, mid-November to mid-March. Critical periods extend
from late November to early February. A severe freeze for
citrus is expected in Florida when air temperatures approach
-6° C. In the citrus belt, this occurs on the average at
least once every 10 years. Lesser freezes, -3° C and warmer,
occur almost annually in some areas, and sometimes more than
once per freeze season. Recurring freezes become increasingly

297

damaging to citrus trees previously injured during the same
season. Total damage from multiple light-to-moderate freezes
usually is less than the damage from one severe freeze.

Severe freezes in Florida generally start as massive out-
breaks of very cold air in Canada. The cold air moves toward
Florida, usually under the influence of the jet stream or
high-altitude winds. The extent to which the jet stream dips
southward and the strength of a trough, created by a high-
pressure ridge in the mid-western United States and a low-
pressure region along the East Coast, will largely determine
how far the cold air will penetrate Florida (25). The cold
air moves into the peninsula with north-to-northwest wind
speeds generally greater than 30 mph, and marks the beginning
of an advective freeze. The cold winds rapidly remove heat
from citrus groves, and air temperatures cool 3^0 C to 4^0 C
per hour. These rates are moderated to less than 1^0 C per
hour as temperatures progressively get colder than 0^0 C.
Tree temperatures lag 1^0 C or more behind air temperatures.
Largely because of less mass and reserve heat, citrus leaves
are cooled more rapidly than either the fruit or the wood.
Snow and freezing rain accompanied the January 1977 freeze
into Florida, but this is a rare event.

Advective freeze conditions gradually phase into cold but
calm and clear weather as the front continues southward into
the peninsula. The air is dry, with 30% or less relative
humidity. These are favorable conditions for development of
nocturnal radiation freezes. Net radiation losses begin
shortly before sunset and gradually develop during the night
(Figure 1). Tree surfaces exposed to the sky are 0.5^0 C
and colder than ambient air. Low ground sites are as much as
3^0 C colder than high ground sites and rapidly develop into
"freeze pockets." Radiation-type freezes cause minimum
temperatures as low as -8.9^0 C in Florida, and tend to occur
on two or more consecutive nights immediately after an advective freeze.

The combination of advective and radiation-type freezes
constitutes a severe freeze condition in Florida. The duration of a severe freeze generally does not exceed three consecutive nights, although the sequence of freeze types may
vary. Severe freezes usually end with the radiation type,
with coldest temperatures shortly before sunrise. Shortly
after sunrise, thawing rates exceed 7^0 C per hours. Day
temperatures of 30^0 C and warmer generally follow severe
freezes in Florida. These temperatures cause freeze damage to
become readily evident within a few days. Citrus groves show
excessive fruit drop and spoilage, completely killed trees
with nonabscised dead leaves, and severely damaged trees with
pronounced wood dieback, longitudinal splits in the bark, and

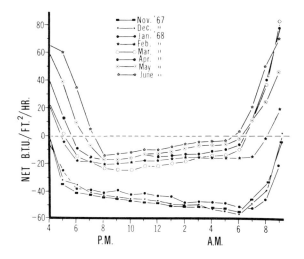

Figure 1. The average trend of net radiation loss in a citrus planting during different months in Florida. Redrawn from Yelenosky, 1972 (73).

extensive freeze cankers on the tree trunk and stems. Recovery of freeze-injured citrus groves is a long and costly venture. Freezes are less damaging when they are followed by 2 to 4 weeks of cool weather, which provides time for fruit salvage operations. This occurred after the January 1977 freeze in Florida, when 28 consecutive postfreeze days averaged less than 16.7° C maximum temperature.

Most citrus trees are cold-tender evergreens, and some freeze damage is expected annually in Florida. Commercial varieties are injured at -2.2° C, although some citrus trees survive -10° C (76). Most citrus trees do not have a pronounced and stable period of dormancy, but growth is reduced during the winter season. This period of reduced growth is the time when citrus trees are most cold tolerant (84). The growers must decide each year whether or not to protect their citrus, when, how, and to what extent.

II. FREEZE SURVIVAL

A. *Artificial Protection*

The science and technology of freeze protection has developed throughout the history of citriculture in the United States (62). Costly heating systems, wind machines, water,

and tree wraps provide limited protection. This is also the
case with artificial fogs (48) and protective foams (13).
More direct methods have not been developed to maintain or
increase the temperature of citrus tissues during critical
freezes. Attempts to generate heat in citrus trees through
electrical current have not progressed beyond the initial
stages (23). Also, in some of our tests, direct application
of heat to citrus wood or to the soil resulted in poor distribution of heat throughout the tree. Citrus trees sometimes have adequate heat reserves to offset heat losses during
freezes (63).

Efficient, protective action requires early freeze warnings and up-to-date information on freeze development. Space
technology has brought satellite systems which map land areas
according to minimum temperature-distribution patterns accurate to 1° C (61). This thermal imagery improves freeze
warnings through monitored development of damaging freezes and
utilizes the full advantes of various communication systems
in one integrated program of freeze protection. Satellite
systems also provide valuable site information on bodies of
water, net radiation losses, and suitability for customized
protection systems.

Freeze protection with applied substances is an attractive, but yet unsuccessful, approach with citrus trees. Many
of the cryoprotectants listed in freeze protection of plants
(36, 45) are largely ineffective when applied in field trials
with citrus. Maleic hydrazide (22) and others (9, 83) have
not found widespread use in commercial citriculture. Recent
findings by other workers, however, favor continued research
for cryoprotectants to increase freeze protection in citrus
trees. Ethephon, (2-chloroethyl)phosphonic acid, increases
the cold hardiness of cherry fruit buds and improves fruit
yield after spring frosts (50). Ethephon applications followed by naphthalene-acetic acid increase the cold hardiness
of apple wood.[1]

A combination of 15% polyvinyl pyrrolidone, 25% glycerol,
and 0.5% dimethylsulfoxide decreases the 50% freeze killing
temperature of apple shoots from -9° C to -12° C (30). Yet
untried on citrus are the fluorinated compounds, such as
1,5-difluoro-2,4-dinitrobenzene, which protect young bean
plants at -3° C for 8 hours (33). Herbicides are used extensively in citriculture and need freeze-protection evaluations, since some herbicides lower the cold resistance of

[1]*Personal communication with J. T. Raese, Plant Physiolo-
gist, U.S. Department of Agriculture, Agricultural Research
Service, Wenatchee, Washington.*

other plants (58). Untested are the bromacil and terbacil
herbicides which increase fruit yields of citrus trees
(11).

B. *Inherent Tree Factors*

The use of different scion varieties and rootstocks which
vary in cold hardiness is additional protection against
devastating freeze losses. This protection increases with
tree age and tissue maturity, if trees are healthy. Lemon
types are the most cold sensitive, and mandarin or tangerine
types are the most cold hardy in commercial citriculture in
Florida. Grapefruit and orange types are intermediate in
this regard. However, there are situations when the more
cold-sensitive citrus types survive better than the cold-
hardy types (79).
Developing varieties through screening procedures (88),
colchicine-induced polyploidy (3), cell and tissue culture
(31), X-ray mutations (17), and broadening the genetic base
with new germplasm through field collections, all may aid in
developing cold-hardy types. Several properties of mitochondria are apparently related to general plant cold hardiness. Incorporation of cold hardiness into new genotypes
through cytoplasmic inheritance supposedly will depend
partially on the proper selection of the female parent (28,
68). In all these attempts, basic guidelines are formulated
according to the capacity of different citrus selections to
a) avoid freezing (supercooling), b) resist freeze damage
(ice tolerance), and c) recover after freeze injury (regenerative capacity of the tissues).
Citrus tissues supercool, but considerably less than the
-40° C associated with the winter survival of northern trees
(7). Some of the coldest supercooling temperatures observed
in citrus are -6.1° C in grapefruit and sweet orange leaves
during the January 1977 freeze in Florida, -10.2° C in the
wood of a sweet orange seedling during controlled freeze
tests (80), and -12° C in isolated vesicles of lemon fruit
(38). Whole fruit, mature sweet oranges, supercooled 30
minutes at -4.4° C and 8 hr at -3.3° C during controlled
freeze tests (21). The practical benefits of supercooling
in citrus are not known, but are probably limited to freezes
warmer than -4.4° C in Florida. Little is known about supercooling in citrus tissues during natural freezes. Consequently, the fact that citrus tissues supercool cannot yet be
fully utilized in freeze-protection practices. Osmotic
pressure is only one important factor that influences supercooling in cold-tender plants (10). There are also anatomical

and cytological-biochemical factors (27, 44, 53, 60). In some of our controlled tests, 1-week-old sweet orange leaves supercooled for 5 hr at -3.9° C, whereas 4-week-old leaves froze within 3 hr. The development of the vascular system might be a factor. Also, leaves occasionally supercooled more than the wood, and three consecutive 2-minute freezes (time lapsed after initial ice formation) decreased supercooling in 1-year-old sweet orange trees on lemon rootstock from -6.7 to -4.9° C. Supercooling protects if maintained in citrus. Otherwise, supercooling probably causes very destructive, nonequilibrium freezing (46), especially in unhardened citrus tissues. Cold-hardening citrus at low temperatures increases both the time and depth of supercooling (91). This relationship may not be evident in some northern plant types (42).

It is not known whether there are effective nucleating agents such as bacteria (2) in citrus groves. Bacteria, other disease-related agents and airborne contaminants merit evaluation, as well as various chemical sprays used in citriculture. Some mineral dusts are efficient ice nucleators, along with diazines (19) and some amino acids (49) in physical systems. Using similar procedures, we found phenazine to be 2.5 times more efficient than tryptophan, which in turn was twice as efficient as sucrose.

Unhardened citrus trees show no apparent chill injury during our standard cold hardening, 2 weeks of 21.1° C (12 hours of light) and 10° C nights followed by 2 weeks of 15.6° C days and 4.4° C nights. However, unreported data indicate that chill-type injury can be induced in citrus leaves within 5 continuous weeks at about 1.7° C. Light is essential. Bleached leaves, increased leakage of amino acids from leaves, and decreased O_2 uptake in leaf disks, all indicate damage to membranes. Similar disorders occur in the cells of blue-green algae, which are directly dependent on light for metabolic activities other than photosynthesis (5). Chilling injury is characteristic of tropical and subtropical plants (39), but chilling injury is not a factor in Florida citriculture.

Ice injures citrus trees during natural freezes. Water-soaking (90), which is a field marker for the start of ice formation in citrus leaves, may or may not be a lethal sign, depending on tissue condition and severity of the freeze. In controlled freeze tests, the start of freezing is readily followed with exotherms and, if needed, electrical impulses (Figure 2). Resistance to freeze damage, or ice tolerance, is largely a function of cold hardening, a phenomenon not adequately understood in citrus trees. Because most citrus are cold-sensitive evergreens growing in mild climates, only parts of plant cold-hardiness schemes in reviews and similar

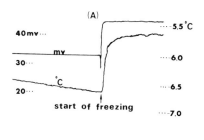

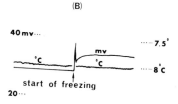

FIGURE 2. (A) represents simultaneous measurements of freezing in the wood of unhardened citrus seedlings. °C is the thermocouple technique (heat of fusion) and mv is the millivolt signal at the same site of measurement. Temperature was decreased at a rate of 4° hr⁻¹. The chart speed was 2.54 cm min⁻¹. (B) represents the failure of the thermocouple technique (°C) in contrast to the millivolt signal to indicate freezing in the wood of hardened-off citrus seedlings at temperature decreases of 1° hr⁻¹ or less. Redrawn from Yelenosky et al., 1970 (81).

reports are applicable to citrus (8, 11, 35, 40, 66). Low-temperature-induced cold hardening separates citrus types and is most efficient with light (75). Withholding water through controlled irrigation to maintain nongrowing conditions and reduce freeze damage is risky. This is largely due to natural rains and warm temperatures readily "triggering" cold-tender, new growth in water-stressed citrus trees. Also, "fruit splitting" may be increased and normal development of fruit inhibited. Experience teaches that excessive watering is not advisable during the winter season, but neither is excessive water stress. Water stress has reduced freeze injury in controlled studies (Table I), but too much stress decreases the low-temperature-induced cold-hardening potential in citrus (71).

Efficient cold protection practices are dependent on prevailing temperature patterns immediately before natural freezes. Cold hardening starts in citrus at about 13° C and apparently continues at temperatures as low as -3.3° C (84). Citrus leaves contribute to the cold hardening of the woody

TABLE I. *Effect of Dehydration on Freeze Damage to Unhardened 'Pineapple' Orange Seedlings*

Wilt symptoms	Ω^a ($\cdot 100$ K)	O.P.^{b} (atm)	$H_2O/O.D.^{wt^c}$ (g)	Leaf kill (%)	Stem kill (%)
	H_2O stress			-6.7^o C for 4 hr	
None	1.4	6	2.1	100	100
Slight	1.7	16	1.5	100	99
Moderate	2.3	26	1.3	100	82
Severe	3.8	34	1.1	76	30

^{a}Conductivity method - stems.
^{b}Dye method - leaves.
^{c}Weight measurements - leaves.

stems (71). Prevailing cool temperatures, 10° C and colder, for 10 to 97 hr per week for 11 weeks cold-hardened citrus trees and helped minimize tree injury during the January 1977 freeze in Florida. There are yet no satisfactory substitutes for cold-hardening temperatures to increase ice tolerance in citrus trees. When cold-hardening temperatures prevail under natural conditions, cold-hardy citrus types survive -8.9 and -10° C. Otherwise, the trees can be killed at -6.7^o C (Table II). Freezing point determinations (69), conductivity measurements (29), bark slippage (91), and colorimetric determinations (72) are several approaches to the problem of identifying degrees of cold hardiness in citrus.

In addition to visible kill of tree parts, ice tolerance is measured according to cell function and organization after freezes. Both are destroyed in severely frozen citrus tissues. Photosynthesis (CO_2 uptake) virtually stops in grapefruit leaves at -6.7^o C for 4 hr (87). Temperature and duration also cause severe cellular disorganization and membrane destruction in sour orange leaves (89). Three hours of -6.7^o C will stop O_2 uptake in sweet orange leaves (70). Within the first few minutes after ice forms in young, succulent citrus tissues, irreversible damage occurs. Solute leakage tests show that membranes are injured. In older and cold-hardened tissues, irreversible damage develops over a time span of 1 hr after ice forms (70). The quickness of lethal injury, membrane leakage, and extent of cellular disorganization suggest that freezing is both extra- and

TABLE II. *Average 30-day Air Temperatures Immediately Before Freezes and Survival of Satsuma Orange Trees[a] on Trifoliate Orange Rootstock at Byron, Georgia, Latitude 31°15'N*

	Prefreeze temp		Min Freeze temp	Trees killed
	Max	*Min*		
	(°C)		*(°C)*	*(%)*
1967-68	*15.6*	*2.2*	*-10.0*	*0*
1968-69	*11.74*	*1.7*	*- 8.9*	*0*
1969-70	*19.4*	*7.8*	*- 6.7*	*100.0*

[a]*50 trees budded in April 1967.*

intracellular in unconditioned citrus trees, especially in young, fully hydrated tissues. However, the delay in lethal injury suggests that extracellular freezing is initially dominant in cold-hardened citrus trees. Intracellular freezing is assumed to be lethal, whereas the effects of extracellular freezing stresses have not yet been determined (47). Double- or multiple-type exotherms are not characteristic of lethal freezing in citrus and, thus, have limited use in separating extracellular from intracellular freezing or locating ice in specific tissues and cells (16, 18).

Cold hardening in citrus is associated with multiple events. Seemingly, solute accumulation coupled with decreased water absorption leads to more closely knit, liquid-solute interfaces which are reinforced with increases and new additions to structure and function of membranes. All of these events are considered both temperature- and time-dependent and, presumably, result in less ice formed per unit time and lesser rates of dehydration and ice penetration. In plant cold hardiness, the individual and combined roles of carbohydrates, proteins, and lipids are many as they relate to osmotic pressure (52), hyperosmolality (43), "water-binding" (24, 67), new glycoproteins (6), membrane function (57, 82), replication (56), and denaturation (37). In citrus, more is known about carbohydrates than either proteins or lipids during cold hardening.

TABLE III. Estimated Killing Temperatures According to Assumed Lethal Dehydration Relationships between Sucrose Molalities in Plant Cells and Equivalent Molalities Attributed to Freezing Temperatures

°C and Eq Molality[a]	-1 0.54	-2 1.08	-3 1.61	-4 2.15	-5 2.69	-6 3.23	-7 3.76	-8 4.30	-9 4.84	Killing temp[b]
Molality in tissues				Dehydration (%)						
0.2	59	79	87	89	92	94	--	--	--	-1.8
0.4	13	59	75	79	84	88	89	90	91	-3.0
0.6	0	40	59	70	76	79	82	85	88	-4.8
0.8	0	18	50	59	67	75	77	79	81	-6.0
1.0	0	4	30	52	59	65	73	76	78	-7.7

[a] Based on 22.7 atm (ideally a 1 molal nonionized solution) and -12.2 atm stress for every 1° C colder than 0° C.
[b] Based on an assumed 75% lethal dehydration.

Carbohydrate accumulation is virtually universal in cold hardiness of plants. Sugars, especially sucrose, are accumulated in citrus, but sugar increase alone is not necessarily a good criterion for determining cold hardiness in citrus (64, 85). However, increases in sugar are favorable on several accounts. High sugar concentrations are not injurious in citrus tissues, although their role in operator-repressor mechanisms on gene activity is unknown. Sugars readily accumulate in citrus during favorable cold-hardening temperatures. Monosaccharides are probably most important in increasing osmotic pressures during the initial stages of cold hardening with disaccharides later functioning in a water-binding role. Both are factors in delaying dehydration stresses. Estimated dehydration to a preset level, and as a result of sucrose concentrations at different temperatures, is within the cold hardiness tolerances of citrus (Table III). Sugars also protect cell membranes (54). Cell membranes help to block extracellular ice from causing intracellular freezing at -10° C (41). Concentration and molecular weight of sugars influence the rate of ice formation in physical systems (14, 74); changing solute influences water potential in citrus (12). In cold-hardened citrus stems, ice traveled about one-third as fast as that measured in unhardened stems (71). Slower rates of ice crystal formation presumably lessen the chances of ice penetrating membranes. High sugar concentrations are also favorable in protecting against high concentrations of electrolytes as a result of ice-dehydration in plants (55). Sorbitol, which increases with sugar in dormant apple shoots (51), is yet untested in citrus.

Starch hydrolysis is greater in the more cold-hardy citrus types, and increases in citrus cold hardiness are likely through greater net accumulation of sugars at low temperatures (1, 59, 78). In some of our unreported controlled environment tests, 1-year-old sweet orange trees cold hardened sufficiently to withstand -6.7° C without injury. These trees showed sugar:starch ratios of 20:1 in the leaves and 17:1 in the wood. Unconditioned trees, 100% killed at -6.7° C, showed sugar:starch ratios of 1.8:1 in the leaves and 1:1 in the wood. During these tests, $^{14}CO_2$ assimilation resulted in more pronounced changes in quantity than quality of the sugar fraction. Translocation and distribution of the label were slowed at the lower cold-hardening temperatures, and cold hardiness of the test trees increased as the label became more evenly distributed in the trees. Field observations indicate that excessive fruiting depletes carbohydrates in citrus trees, thus making them more susceptible to freeze damage (32, 64).

Less is known about proteins than sugars in cold hardening of citrus trees. Water-soluble proteins did not increase
in grapefruit leaves during cold-hardening tests (86), and
increases in sweet orange trees were found to be poorly correlated with freezing points (15). In some of our tests, only
3 of 15 amino acids determined increased during cold hardening
of sweet orange trees at 10° C. These three amino acids were
proline, glutamic acid, and valine. Free proline, low molecular weight protein fractions, and glycoproteins are associated
with plant cold hardiness (4, 26). This is less evident with
changes in hydroxyproline in citrus leaves (77). Little is
known about the role of proteins, lipids (34), and enzyme
activities (20) in cold-hardening citrus trees.

The viable citrus industry in Florida attests to the potential of citrus to survive in high-risk freeze areas. This
potential has to be developed and translated into management
practices for maximum benefits to commercial citriculture.
Freeze avoidance, either through artificial protection or
supercooling, is less attractive than developing increased
cold hardiness through greater tolerance to ice. It is still
unclear how best to synchronize cold hardening processes with
natural winter conditions. In lieu of long-range breeding for
cold-hardy citrus types, applied cryoprotective substances
are probably most compatible with citrus grove management
practices. However, freeze survival of citrus trees results
from multiple events. Further advances are considered evident
in more detailed examinations of ice formation within citrus
tissues and resulting dehydration stresses on chemical bonding
and membrane integrity.

REFERENCES

1. Anderson, G., *in* "Gleerupska Universitetsbokhandeln,"
    p. 1. Lund (1977).
2. Arny, D. C., Lindow, S. E., and Upper, C. D., *Nature 262*,
    282 (1976).
3. Barrett, H. C., *Bot. Gaz. 135*, 29 (1974).
4. Benko, B., *Biologia Plant. (Praha) 11*, 334 (1968).
5. Brand, J. J., *Plant Physiol. 59*, 970 (1977).
6. Brown, G. N., and Bixby, J. A., *Physiol. Plant. 34*, 187
    (1975).
7. Burke, M. J., George, M. F., and Bryant, R. G., *in*
    "Proc. Int. Symp. on Water Relations of Foods" (R. B.
    Duckworth, ed.), p. 111 (1975).
8. Burke, M. J., Gusta, L. V., Quamme, H. A., Weiser, C. J.,
    and Li, P. H., *Ann. Rev. Plant Physiol. 27*, 507 (1976).

9. Burns, R. M. *Congreso Mundial Citricultura 3*, 167 (1977).
10. Cary, J. W., and Mayland, H. F., *Agron. J. 62*, 715 (1970).
11. Constantin, R. J., and Brown, R. T., *J. Amer. Soc. Hort. Sci. 101*, 26 (1976).
12. Darbyshire, B., and Steer, B. T., *Aust. J. Biol. Sci. 26*, 591 (1973).
13. Desjardins, R. L., and Siminovitch, D., *Agr. Meteorol. 5*, 291 (1968).
14. Doebbler, G. F., *Cryobiology 3*, 2 (1966).
15. Ghazaleh, M. Z. S., and Hendershott, C. H., *J. Amer. Soc. Hort. Sci. 90*, 93 (1967).
16. Graham, P. R., and Mullin, R., *J. Amer. Soc. Hort. Sci. 101*, 3 (1976).
17. Gregory, W. C., and Gregory, M. P., *Soil and Crop Sci. of Fla. 25*, 390 (1965).
18. Hatakeyama, I., and Kato, J., *Planta (Berl.) 65*, 259 (1965).
19. Head, R. B., *Nature 196*, 736 (1962).
20. Heber, U., *Cryobiology 5*, 188 (1968).
21. Hendershott, C. H., *Proc. Amer. Soc. Hort. Sci. 80*, 247 (1962).
22. Hendershott, C. H., *Proc. Amer. Soc. Hort. Sci. 80*, 241 (1962).
23. Horanic, G. E., and Yelenosky, G., *Proc. First. Int. Citrus Symp. 2*, 539 (1969).
24. Johansson, N. -O., *Nat. Swed. Inst. Plant Protect. Contr. 14*, 364 (1970).
25. Johnson, W. O., *Proc. Fla. State Hort. Soc. 75*, 89 (1962).
26. Kacperska-Palacz, A., Jasinska, M., Sobczyk, E. A., and Wcislinska, B., *Biologia Plant. (Praha) 19*, 18 (1977).
27. Kaku, S., *Plant and Cell Physiol. 12*, 147 (1971).
28. Kenefick, D. G., *Agric. Sci. Rev. 2*, 21 (1964).
29. Ketchie, D. O., *Proc. First Int. Citrus Symp. 2*, 559 (1969).
30. Ketchie, D. O., and Murren, C., *J. Amer. Soc. Hort. Sci. 101*, 57 (1976).
31. Kochba, J., and Spiegel-Roy, P., *HortScience 12*, 110 (1977).
32. Konakahara, M., *Shizuoka Prefectural Citrus Expt. Sta. Bull. 3*, 156. Komagoe, Shimizu-City, Japan (1975).
33. Kuiper, P. J. C., *Meded. Landbouwhogeschool Wageningen 67*, 1 (1967).
34. Kuiper, P. J. C., *Proc. First Int. Citrus Symp. 2* (1969).

35. Larcher, W., Heber, U., and Santarius, K. A., *in*
    "Temperature and Life" (H. Precht, J. Christopherson, H.
    Hensel, W. Larcher, eds.), p. 195. Springer-Verlag,
    New York (1973).
36. Levitt, J., *in* "Physiological Ecology" (T. T. Kozlowski,
    ed.), p. 168. Academic Press, New York (1972).
37. Levitt, J., and Dear, J., *in* "Ciba Foundation Symp. on
    the Frozen Cell" (G. E. W. Wolstenholme, M. O'Connor,
    eds.), p. 149. J. and A. Churchill, London (1970).
38. Lucas, J. W., *Plant Physiol. 29*, 245 (1954).
39. Lyons, J. M., *Ann. Rev. Plant Physiol. 24*, 445 (1973).
40. Mayland, H. F., and Cary, J. W., *Adv. Agron. 22*, 203
    (1970).
41. Mazur, P., *Cryobiology 14*, 251 (1977).
42. McLeester, R. C., Weiser, C. J., and Hall, T. C., *Plant
    and Cell Physiol. 9*, 807 (1968).
43. Meryman, H. T., Williams, R. J., and Douglas, M. St. J.,
    Cryobiology 14, 287 (1977).
44. Miftakhutinova, F. G., and Anisimov, A. V., *Soviet Plant
    Physiol. 23*, 671 (1977).
45. Moblidowska, I., *Cryobiology 5*, 175 (1968).
46. Olien, C. R., *Ann. Rev. Plant Physiol. 18*, 387 (1967).
47. Olien, C. R., *Crop Sci. 1*, 26 (1961).
48. Palmer, T. Y., *Proc. First Int. Citrus Symp. 2*, 579
    (1969).
49. Power, B. A., and Power, R. F., *Nature 194*, 1170 (1962).
50. Proebsting, E. L., Jr., and Mills, H. H., *J. Amer. Soc.
    Hort. Sci. 101*, 31 (1976).
51. Raese, J. T., Williams, M. W., and Billingsley, H. D.,
    Cryobiology 14, 373 (1977).
52. Sakai, A., and Yoshida, S., *Cryobiology 5*, 160 (1968).
53. Salazar, C. G., *J. Agric. Univ. PR 50*, 316 (1966).
54. Santarius, K. A., *Planta (Berl.) 113*, 105 (1973).
55. Santarius, K. A., *Planta (Berl.) 89*, 23 (1969).
56. Siminovitch, D., Rheaume, B., Pomeroy, K., and Lepage,
    M., *Cryobiology 5*, 202 (1968).
57. Steponkus, P. L., Garber, M. P., Myers, S. P., and
    Lineberger, R. D., *Cryobiology 14*, 303 (1977).
58. St. John, J. B., and Christiansen, M. N., *Plant Physiol.
    57*, 257 (1976).
59. Surkova, L. I., *Fiziol. Rast. 9*, 607 (1962).
60. Taniguchi, T., *Shizuoka Prefectural Citrus Expt. Sta.
    Bull. 5*, 15 (1965).
61. Taylor, S. E., Davis, D. R., and Jensen, R. E., *Proc.
    Fla. State Hort. Soc. 88*, 44 (1975).
62. Turrell, F. M., *in* "Citrus Industry" 3 (W. Reuther, ed.),
    p. 338. Univ. Calif. (1973).

63. Turrell, R. M., and Austin, S. W., *Ecology 46*, 25 (1965).
64. Vasil'yev, I. M., *in* "Wintering of Plants" (J. Levitt, ed. of English translation), p. 1. Amer. Inst. Biol. Sci. (1961).
65. Webber, H. J., *in* "Yearbook of U. S. Dept. of Agric.," p. 159. Washington, D. C. (1895).
66. Weiser, C. J., *Science 169*, 1269 (1970).
67. Williams, J. M., and Williams, R. J., *Plant Physiol. 58*, 243 (1976).
68. Wilner, J., *Can. J. Plant Sci. 45*, 67 (1965).
69. Wiltbank, W. J., *Congreso Mundial De Citricultura 3*, 137 (1977).
70. Yelenosky, G., *Cryobiology 13*, 243 (1976).
71. Yelenosky, G., *Plant Physiol. 56*, 540 (1975a).
72. Yelenosky, G., *HortScience 10*, 384 (1975b).
73. Yelenosky, G., *Proc. Fla. State Hort. Soc. 85*, 71 (1972).
74. Yelenosky, G., *Biodynamica 11*, 95 (1971a).
75. Yelenosky, G., *HortScience 6*, 234 (1971b).
76. Yelenosky, G., Brown, R. T., and Hearn, C. J., *Proc. Fla. State Hort. Soc. 86*, 99 (1973).
77. Yelenosky, G., and Gilbert, W., *HortScience 9*, 375 (1974).
78. Yelenosky, G., and Guy C. L., *Bot. Gaz. 138*, 13 (1977).
79. Yelenosky, G., and Hearn, C. J., *Proc. Fla. State Hort. Soc. 80*, 53 (1968).
80. Yelenosky, G., and Horanic, G., *Cryobiology 5*, 281 (1969).
81. Yelenosky, G., Horanic, G., and Galena, F., *HortScience 5*, 270 (1970).
82. Yoshida, S., *Plant Physiol. 57*, 710 (1976).
83. Young, R., *J. Amer. Soc. Hort. Sci. 96*, 708 (1971).
84. Young, R., *HortScience 5*, 411 (1970).
85. Young, R., *J. Amer. Soc. Hort. Sci. 94*, 612 (1969a).
86. Young, R., *J. Amer. Soc. Hort. Sci. 94*, 252 (1969b).
87. Young, R., *Proc. First Int. Citrus Symp. 2*, 553 (1969c).
88. Young, R., and Hearn, C. J., *HortScience 7*, 14 (1972).
89. Young, R., and Mann, M., *J. Amer. Soc. Hort. Sci. 99*, 403 (1974).
90. Young, R., and Peynado, A., *Proc. Amer. Soc. Hort. Sci. 91*, 157 (1967).
91. Young, R., and Peynado, A., *Proc. Amer. Soc. Hort. Sci. 86*, 244 (1965).

BREEDING AND SELECTING
TEMPERATE FRUIT CROPS
FOR COLD HARDINESS

H. A. Quamme

Agriculture Canada Research Station
Harrow, Ontario

Most temperate fruit crops are subject to periodic winter cold and spring frosts which cause severe economic losses. The risk of cold injury ranges from almost negligible in some regions to a degree severe enough to prevent fruit production in others. In many northern regions of North America, Europe and Asia, there are adequate soils and rainfall, but the risk of cold injury prevents fruit production.

Domestication of many important fruit crops first occurred within species adapted to some of the warmest regions of the temperate zone, i.e. *Malus pumila* Mill. (apple), *Pyrus communis* L. (pear), *Prunus domestica* L. (European plum), *Prunus avium* L. (sweet cherry) and *P. cerasus* L. (sour cherry), *Vitis vinifera* L. (grape) near the Caspian and Black Seas; *P. armeniaca* L. (apricot), *P. persica* (L.) Batch. (peach) and *P. salicina* Lindl. (Japanese plum) in Central China. Cultivation of these crops spread to Europe and warm regions of North America where varietal improvement, especially in fruit quality, size and appearance reached a high level of perfection. The level of cold hardiness in these crops has been frequently inadequate for further expansion of production into colder regions, i.e. the Northern Great Plains region of the United States, Canada and the Russian steppes. Much progress has been made in breeding for improved cold hardiness, but much remains to be accomplished. The improvement of cold hardiness is a major objective in many fruit breeding programs (23).

In improving cold hardiness through breeding, it is important to have adequate methods to measure cold hardiness. There are a number of methods of cold hardiness measurement

available, but not all are satisfactory for breeding work.
There are several criteria for determining the usefulness of
a method for measuring cold hardiness in breeding programs:
1) the method must allow precise differentiation of hardy
cultivars from tender ones, 2) it must be adaptable to measuring large populations and 3) it must be carried out with
reasonable expenditure of time and money. One of the objectives of this paper is to review the methods used for measuring cold hardiness with emphasis on their usefulness in
temperate fruit breeding programs. Another objective is to
review the field of crop hardiness breeding with emphasis on
the potential germplasm for cold hardiness improvement in
temperate fruit crops.

I.   MEASUREMENT AND EVALUATION OF COLD HARDINESS

     Cold hardiness is an all inclusive term which covers a
wide array of types of injury caused by low temperature. Injury may occur to the roots, shoots or reproductive organs.
The type of injury may vary with the season and type of
freezing stress. Apple serves best to illustrate some of the
common types of cold injury and freezing stress. Immature
apical buds of apple may be susceptible to fall frosts (55).
Severe freezes in late fall may also injure the trunk and
scaffold parts of the tree (32). During this period the bark
and xylem tissues of apple appear to be similar in hardiness,
but in the mid-winter the bark may survive to lower temperatures. If the cambium survives it can produce new wood. The
darkening of xylem following winter freezes is called 'black
heart.' In the advanced state, occlusions fill the vessels
and they become blocked. The tree may not be killed outright,
but wood rotting organisms often invade the injured sites.
The combination of factors reduce productivity and longevity
(7). 'Sunscald' is a type of bark injury that is found on
the southwest side of apple trees and is thought to be produced by radiant heating and cooling during periods of high
sunlight in the winter (31). Apple trees may be completely
killed by low soil temperatures which injure the roots. This
form of injury is common in years in which snow cover is
lacking (9). The flower buds of apple are rarely injured
during the winter and become susceptible to spring frost
during expansion in the spring. Knowledge of the type of
winter injury and the stress producing it is necessary for
designing tests which reproduce naturally induced injury and
thus accurately measure cold hardiness.

Cold hardiness is not a static entity and is under control of a number of variables. Plant hardiness changes considerably with season. During active growth, plants usually have little cold hardiness, but in autumn growth ceases and the plant hardens. The development of cold hardiness is termed acclimation and the loss of hardiness is termed deacclimation. Acclimation and deacclimation are active developmental processes and under control of certain environmental stimuli. Research on factors controlling acclimation and deacclimation has been adequately reviewed by Levitt (31) and Weiser (69). In short, the plant determines the approach of winter by sensing the shortening days and low temperatures. This induces dormancy and the onset of acclimation. Fall frosts further induce the maximal degree of hardiness. Fully acclimated plants can deacclimate on exposure to warm temperatures, but deacclimation may be reversible by subsequent exposure to cold. Deacclimation becomes irreversible when growth resumes in the spring. Considerable variation occurs in the rate of acclimation and depth among different organs and tissues of the plant. Furthermore, the rate of acclimation and depth depends on the age, vigor and health of the plant. In the design of sampling methods for cold hardiness tests, it is important to take the wide fluctuations of cold hardiness into account. Varietal comparison should be made with tests conducted in the same season and on the same date. Preconditioning the material at a constant temperature can be used to overcome variations in hardiness that arise from variations in collection time (52).

In spite of wide fluctuations in cold hardiness that can occur, the rank in hardiness of varieties may remain relatively constant (6, 20, 47, 48). A constant relationship in hardiness among varieties alleviates the need for extensive sampling. Breeders often select the time to conduct freezing tests when the greatest separation in varieties is possible. This appers to vary with fruit species, i.e. in hardiness tests conducted on apple twigs the greatest varietal separation was obtained in early winter (27), whereas with peach flower buds varietal separation in the mid-winter was best (68). Although it may be possible to select a single date to conduct hardiness tests, hardiness tests conducted on a single date may give misleading results, i.e. 'Siberian C' peach may rank the most hardy in the mid-winter, but in the spring it is less hardy than other varieties because it deacclimates earlier (29). When there is a possibility that varietal rank changes with season it seems advisable to test varietal hardiness at more than one time.

II.   METHODS FOR MEASURING PLANT HARDINESS

A. *Determination of Cold Hardiness Following Natural Freezing*

Useful information can be obtained by assessing survival after natural freezes. Natural freezes not only allow an opportunity for varietal evaluation, but allow an opportunity to determine the nature of injury and its frequency. Such information is useful for designing cold hardiness test procedures. In comparing varietal performance plants should be of a similar age and be subjected to similar propagation and cultural practices. The degree of injury can be scaled using a numerical rating system (31).

Special test sites may be chosen where the cold stress is more extreme or frequent. Care should be taken in interpreting information from special test sites, because relative hardiness among varieties can change from location to location (65). Natural stress may be enhanced by removing protective snow cover. Sweeping the snow off test plots has been used to select cold hardy apple rootstocks (37).

B. *Determination of Cold Hardiness by Freezing Plants in situ with a Portable Freezing Chamber*

The disadvantage of cold hardiness determination after natural freezes is that test winters may be infrequent and variable. In an attempt to overcome these disadvantages, researchers have built portable freezing chambers to simulate test winter stress conditions of plants *in situ* in the field (16, 36, 51, 54). In practice, most freezing chambers are limited in size because of the refrigeration capacity required to cool them and ease of portability. As a result, there are limitations to the number of plants that can be frozen at one time, and sampling large numbers of plants is restricted because of time and space. Most chambers will not accommodate large trees and are limited to freezing trees of nursery size. Juvenile trees may not respond in the same way as mature ones (26).

The portable freezing chamber has an advantage over freezing plants in the laboratory in that the plants are left *in situ* and damage or change in hardiness incurred in moving them to the laboratory is avoided. Moving rooted plants in the winter when there is difficulty in getting them out of the frozen ground can also be avoided. Furthermore, the ability

of the plant to recover freezing injury is often better determined in plants *in situ*, than in plants or plant parts
taken into the laboratory.

C. Determination of Cold Hardiness by Freezing Plants or Plant Parts in Laboratory Freezing Chambers

The most common method of determining cold hardiness is
to freeze plants or plant parts in laboratory test chambers
and evaluate damage by recovery, tissue browning or one of
several objective methods of measuring cell vitality. Laboratory tests have long been used by plant physiologists and
are well adapted to determining cold hardiness in plant breeding programs. Laboratory techniques can be adapted to sampling large numbers of plants in a short period of time. More
precise temperature control can be usually achieved in the
laboratory than in the field. Furthermore, laboratory tests
are more adaptable to determining lethal dosage temperatures
which can be used to express absolute cold hardiness.

There are many designs for freezing chambers for freezing
plant material. Mechanical refrigeration is usually used for
cooling chambers (54), but systems using dry ice or liquid
nitrogen (67) have also been developed. Automatic control
systems are available to precisely carry out any selected
program (49).

The expression of cold hardiness is affected by the
freezing and thawing rates used in the test. Freezing and
thawing rates are usually chosen to maintain the equilibrium
between liquid water and ice in the tissue close to zero
(about 2° C/hr). Equilibrium freezing and thawing is used to
allow full expression of resistance to freezing stress (31).
In peach (67) and strawberry (18), slow cooling rates (5°
C/hr or less) gave better varietal separation than more rapid
ones greater than 5° C/hr.

When plant tissue is cooled below 0° C, freezing doesn't
usually occur immediately, but the tissue remains supercooled.
In conducting freezing tests it is advisable to seed the plant
to prevent excessive supercooling. Detached parts tend to
supercool more than whole ones because nucleation is a chance
event and depends on the number of nucleation sites. Nucleation can spread throughout the plant from a single site.
Whole plants have more potential sites for nucleation than detached parts and thus have less potential for supercooling
(5). The explosive growth of ice which accompanies excessive
supercooling may be injurious to certain tissues (40). Seeding is achieved by attaching the plant to moistened cloth
(33) inducing ice formation in tissue cooled below 0° C with

fine droplets of ice water or ice crystals. A pipe cleaner dipped in liquid nitrogen provides a good source of ice crystals for seeding the tissue with ice.

Several methods are used for determining injury in plant material after it has been subjected to plant stress. The most direct method of measuring freezing injury is to observe if the plant recovers. Rooted plants in pots can be forced in the greenhouse or detached plants without roots can be forced in water to determine recovery. Some plant materials, however, do not lend themselves to the recovery test. Plants in deep rest cannot be forced and detached parts of some plants are not readily forced without roots.

An alternative to recovery is to use the degree of tissue browning after incubation as an estimate of freezing injury. Determination of injury by browning is relatively easy if the whole tissue is affected and if the injury occurs over a narrow range of temperatures. For example, injury to peach flower buds is localized in the flower bud primordia and occurs below some critical temperature. The flower bud primordia can be scored as alive or dead after exposure to low temperature and the number of surviving flower primordia counted. In other parts of the plant, such as the stem, several tissues may be affected and the degree of injury may be progressive over a wide range of temperatures. When injury is progressive with temperature and varies in the tissue, the degree of browning can be expressed using a numerical scoring system. These scores can be used accurately to differentiate among varieties which are close in degree of cold hardiness (31).

Objective methods have been developed to avoid subjective scoring. Dexter (10) were among the first to use conductivity of plant extracts as a measure of injury. Freezing stress results in membrane damage and leakage of cellular electrolytes. The cell leakage and thus cellular damage can be determined by measuring the conductivity of water extracts of the tissue. The conductivity method has been used with variable success to determine varietal hardiness of fruit species. In varietal comparison of strawberries, conductivity measurements were found to be a better method of separating cultivars of different cold hardiness than recovery method or the ninhydrin determination of amino acids in plant extracts (18). In apple, varietal separation in cold hardiness has been achieved by the conductivity method, but in comparison with scoring browning or recovery of twigs forced in water the conductivity method was not as sensitive (27). Conductivity was not found to give a good indication of cultivar hardiness in peach (53).

The measurement of amino acids released into water extracts of plant material by the ninhydrin method has been reported to be more sensitive than the conductivity method for measuring cellular damage (57). In tests on strawberry, however, the ninhydrin test was found to be less sensitive than conductivity measurements for separting cultivars of different hardiness levels and involved considerably more labor (18).

The reduction of colorless trephenyl tetrazolium chloride (TTC) to the red form on cut surfaces of plant material has been used as an aid in determining cold hardiness in apple and pear (28). Living tissue possesses reducing capacity, but dead tissue does not. The reduced form of TTC can be estimated visually or can be extracted and determined with a colorimetric method (59). The colorimetric method has been used to measure cold hardiness of woody ornamentals, but not fruit species.

Freezing tests conducted using a single test temperature are useful for separating relative hardiness and are adaptable to freezing large numbers of selections. However, tests in which the plant material is frozen to a series of temperatures can be used to determine a survival curve and these curves can be interpolated to express hardiness in absolute terms. Indices such as an incipient injury (Figure 1) or a lethal dose temperature can be derived from the survival curve and used to express absolute hardiness. The temperature at which 50 percent injury occurs (T_{50}) is commonly used to express absolute hardiness (31). Probit analysis (45) and Spearman Kärber method (2) are two statisticl procedures which can be used to determine the T_{50} index. More complex equipment and procedures are required to determine absolute hardiness than relative hardiness, but the increased sensitivity may outweigh this disadvantage (43). Furthermore, it may be possible to relate absolute hardiness indices to environmental survival (46).

D. Determination of Cold Hardiness by Exotherm Analysis

In plant tissues which deep supercool, injury is associated with a sudden freezing of a supercooled fraction of water. Sudden freezing and release of heat of fusion can be detected as a deflection on the time-temperature profile (exotherm) during constant cooling. Differential thermal analysis is a more sensitive method for detecting the exotherm in which the temperature of the sample is compared to a dried reference during freezing (5). Exotherm analysis by observing time-temperature profiles has been developed as a practical technique for determining the flower bud hardiness of *Prunus* (50). Differential thermal analysis has been used

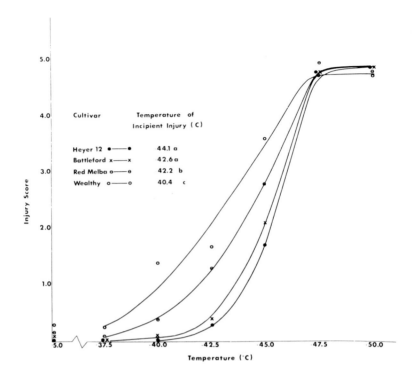

FIGURE 1.   The survival curve for the xylem of twigs from four apple cultivars. The twigs were frozen according to Quamme (46), and injury was scaled from 0 to 5. An incipient injury temperature was determined by interpolating the curve at a score of 1.0, a trace of injury. The mean temperature of incipient injury determined from four replications is represented. The unlike letters indicate mean separation by Duncan's multiple range test, 5% level.

to determine varietal differences in wood hardiness of apple and pear cultivars (46), but appears to be too time consuming for practical use in determining the resistance of large numbers of cultivars.

E.   Determination of Cold Hardiness by Indirect Measurements

Many researchers have searched for methods of determining cold hardiness without freezing the plant. Electrical impedance has been found to increase with hardiness and impedance measurements have been used in an attempt to measure varietal differences in hardiness (13, 62, 68, 71). Changes in impedance seem to reflect increases in membrane permeability associated with acclimation (13, 17). Impedance measurements are made using a metal probe which is inserted into the tissue. The impedance method has the advantage that the measurements are rapid and sacrifice of the plants is avoided. However, the ability to discriminate among varieties has been low and highly dependent on stage of acclimation (8, 17, 62, 68). The use of impedance after freezing to assess injury appears a better use than its direct use for predicting it (14).

The ability to withstand plasmolysis has been also used for indirect measurements of plant hardiness (56). Apparently, it is not useful for separating cultivars which are close in cold hardiness (31).

The plant undergoes many morphological and biochemical changes which are correlated with acclimation, but only in a few cases have the correlations been strong enough to discriminate among cultivars (31).

III. STUDIES CONDUCTED TO DETERMINE THE INHERITANCE OF COLD HARDINESS

Studies conducted to determine the inheritance of cold hardiness are few in number. Most observations have been made on progenies in breeding programs where cultivar development is the prime objective. As a result, these progenies are derived from a limited genetic base. Studies conducted on apple (14, 25, 38, 66), peach (35), raspberry (1) and strawberry (44) indicate that the inheritance of cold hardiness is quantitative. Frequency distributions of the progenies when given are similar to those in Figure 2. The superiority of certain hardy cultivars for transmitting cold hardiness to offspring was evident in all studies. Maternal influence on progenies has been found in certain apple progenies (19, 72) but there were no reciprocal differences detected in progenies presented in Figure 2. In one study the components of genetic variance were determined and a high level additive variance was found to be present (66). The presence of additive variance allows breeders to select and mate parents on the basis of phenotype. Genetic progress can be made by repeated mating of the most hardy phenotypes in recurrent generations.

TABLE I. The Comparison of Cold Hardiness of Commercial Cultivars of Common Fruit Species with that of the Hardiest Clones Within the Species, Species Used in Hybridization with the Commercial Species and Interspecific Hybrids

Species	Common commercial cultivars grown in northern regions[a]	Cold hardiness of commercial cultivars based on hardiness zone[b]	Typical cultivars representing the hardiest clones within the species	Cold hardiness of the hardiest clones within the species based on hardiness zone	Species used in hybridization with commercial species and typical species hybrids	Cold hardiness of hardy species and species hybrids based on hardiness	Authority
Malus pumila (apple)	Cortland McIntosh Red Melba Wealthy	4 (-34.4 to -28.9° C)	Battleford Dr. Bill Heyer 12 Hibernal	2 (-45.6 to -40.0° C)	M. baccata M. baccata x M. pumila Osman Dolgo Rescue	2 (-45.6 to -40.0° C)	Strang and Stushnoff, 1975
Pyrus communis (pear)	Anjou Bartlett Bosc	5 (-28.9 to -23.3° C)	Moe Patten Parker	4 (-34.4 to -28.9° C)	P. ussuriensis P. ussuriensis x P. communis John Philip Olia	3 (-40 to -34.4° C)	Brooks and Olmo, 1972; Quamme, 1976
Prunus armeniaca (apricot)	Alfred Farmingdale Goldcot Veecot	6 (-23.3 to -17.9° C)	Haggith	5 (-28.9 to -23.3° C)	P. besseyi P. mandshurica P. sibirica P. mandshurica x P. armeniaca Morden 604 P. besseyi x P. armeniaca Yuksa	3 (-40 to -34.4° C)	Bradt et al., 1970; Hammel, 1976; Kerr, 1937; R.E.C. Layne, personal communication
P. avium (sweet cherry)	Bing Hedelfingen Windsor	6 (-23.3 to -17.9°)			(P. avium x P. Pennsylvanica x P. besseyi) Zumbra	4 (-34.4 to -28.9° C)	Bradt et al., 1970; Hammel, 1970
P. cerasus (sour cherry)	Montmorency	5 (-28.9 to -23.3° C)	North Star Meteor Coronation	4 (-34.4 to -28.9° C) 3 (-40 to -34.4° C)	P. cerasus x P. Pennsylvanica Dropmore	2 (-45.6 to -40° C)	Bradt et al., 1970; Hammel, 1976

Genus (common name)	Cultivars[a]	Rating[b]	Cultivars[a]	Rating[b]	Related species/hybrids[a]	Rating[b]	Authorities
P. domestica (European plum)	Damson, Italian, Stanley	6 (-23.3 to -17.9° C)	Dietz, Russian Green, Gage	4 (-34.4 to -28.9° C), 3 (-40 to -34.4° C)			*Bradt et al., 1970; Hammel, 1976*
P. Persica (peach)	Elberta, Golden Jubilee, Red haven	6 (-23.3 to -17.9° C)	Siberian C, Harrow Blood, Veteran	6 (-23.3 to -17.9° C), 3 (-40 to -34.4° C)	*P. besseyi*, *P. besseyi* x *P. Persica* Kerr Hybrids	3 (-40 to -34.4° C)	*Bradt et al., 1970; Hammel, 1976*
P. salicina (Japanese plum)	Burbank, Shiro	6 (-23.3 to -17.9° C)	Brooks 41, Ivanovka, Ptitsin 5, 9, 10 and 12	3 (-40 to -34.4° C)	*P. americana*, *P. besseyi*, *P. nigra*, *P. salicina*, x *P. americana* Radisson, *P. nigra*, x *P. salicina* Eclipse Parkside, *P. besseyi* x *P. salicina* Sapa Manor	3 (-40 to -34.4° C)	*Bradt et al., 1970; Hammel, 1976; Morrison et al., 1954*
Vitis vinifera (European grape)	Chenin Blanc, Cabernet Sauvignon, Gamay Beaujolais	7 (-17.9 to -12.2° C)	*V. riparia*, *V. labrusca*, American and French hybrids derived from *V. vinifera*, *V. riparia*, and *V. labrusca* and other species	3 (-40 to -34.4° C), 5 (-28.9 to -23.3° C), 5 to 6 (-17.9 to -28.9° C)			*Bradt et al., 1970; Childers, 1954; Pienquet, et al., 1977*

[a] Some common cultivars are listed as being typical of each category. A more complete list is provided by the given authorities.

[b] The cold hardiness rating of cultivars is based on the average annual minimum temperature zone (Figure 3) in which they have been reported to be grown.

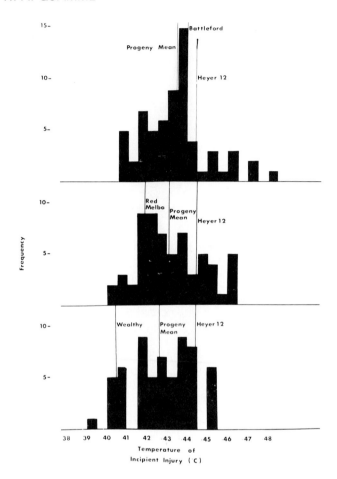

FIGURE 2. Segregation distributions for incipient injury temperatures (as determined in Figure 1) among Red Melba and Wealthy (hardy in Zone 4; see Figure 3) and Battleford (hardy in Zone 3) with Heyer 12 (hardy in Zone 3). The study was carried out cooperatively by Wilbert Ronald, Research Station, Morden and H. A. Quamme, Research Station, Harrow.

IV.   POTENTIAL GERMPLASM FOR IMPROVING COLD HARDINESS OF
       TEMPERATE FRUIT SPECIES BY BREEDING

In Table I the approximate cold hardiness based on the plant hardiness zonation map (Figure 3) of commercial cultivars of different fruit species is compared with that of the most cold hardy cultivars within the species and that of

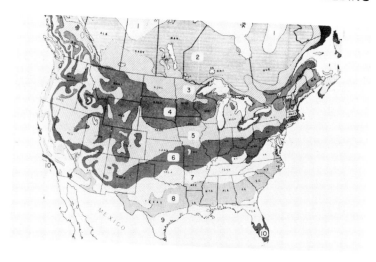

FIGURE 3. Approximate ranges of the average annual minimum temperature zones (^{o}C). Zone 1: below -45.6^{o} C; 2: -45.6 to -40° C; 3: -40 to -34.4^{o} C; 4: -34.4 to -38.9^{o} C; 5: -28.9 to -23.3^{o} C; 6: -23.3 to -17.9^{o} C; 7: -17.9 to -12.2^{o} C; 8: -12.2 to -6.7^{o} C; 9: -6.7 to -1.1^{o} C; 10: -1.1 to 4.4^{o} C. (From Plant Hardiness Zone Map. U.S. Dept. Agr. Misc. Publ. 814, 1960).

species which hybridize with the commercial species and some of the interspecific hybrids. It is evident that there is much potential germplasm for improvement of cold hardiness in many fruit species by both intraspecific and interspecific hybridization. If the cold hardiness present in the hardiest germplasm sources can be combined with other desired traits present in commercial cultivars, it should be possible to extend the range or fruit production and reduce the risk of cold injury in present regions of commercial fruit production.

Considerable progress has already been made in breeding cold hardy apple cultivars. Hardy Russian cultivars and hybrids of *Malus baccata* have been used as a source of cold hardiness to develop cultivars for the home garden and small scale orchards on the Canadian prairies. Size and quality are recovered in the first few backcrosses of *M. baccata* to large apple types (64). Crosses of hardy *M. pumila* types yield progenies with high levels of cold hardiness (Figure 2). Improvement in cold hardiness of apple using cold hardy sources within *M. pumila* has enabled commercial apple production to be extended into the colder regions of Minnesota (61). *Malus baccata* hybrids and hardy clones of *M. pumila*

have also been used as sources of cold hardiness for development of new cold hardy rootstocks (9).

Considerable potential germplasm also exists for the im-
provement in the cold hardiness of pears. Pear production
could be extended into the northern apple growing regions
(Zone 4) if cultivars could be produced equal to the hardiest
Pyrus communis cultivar, e.g. Moe and of better quality (46).
Further increase in cold hardiness seems possible by using
P. ussuriensis Max. in hybridization with *P. communis*. Most
of the size and quality of *P. communis* are recovered in the
first few backcrosses of *P. ussuriensis* with *P. communis*.
Several *P. ussuriensis* hybrids have been developed for home
gardens on the Canadian prairies (34).

In apricot, the lack of tree hardiness does not appear to
limit fruit production as much as the lack of flower bud
hardiness. The flower buds deacclimate earlier than other
fruit species and are susceptible to low temperatures in the
late winter and spring. *Prunus manchurica* L. and *P. sibirica*
L. survive and produce fruit in the northern Great Plains, but
in warmer climates such as the Great Lakes they are more
susceptible to flower bud injury than *P. armeniaca* because
they deacclimate earlier. Selection within *P. armeniaca* has
produced cultivars which are more hardy and more consistant
in cropping (Layne, personal communication). *Prunus besseyi*
Bailey, the Western sandcherry, is a hardy species which can be
hybridized with *Prunus armeniaca*. The hybrids are hardy and
can be backcrossed with apricot and peach (24). No fruit cultivars have been developed from these hybrids, but they have
been used as dwarfing rootstocks for apricot (22).

There appears to be limited potential germplasm for
increasing the hardiness of sweet cherry, *P. avium*, through
breeding (Table I). Few varieties exist which can be grown
beyond Zone 6. The possible exceptions are some cultivars of
P. avium which are reported to survive at Michurinsk in the
U.S.S.R. (41). *P. avium* has been hybridized with *P. besseyi*
and *P. pensylvanica* L. to produce a hardy type of cherry, but
it more resembles *P. cerasus* than *P. avium* (4). Further
interspecific hybridization, apparently, has not been tried.

The breeding and selection of sour cherry within *Prunus
cerasus* using hardy Asiatic sources of cold resistance have
resulted in several commercial cultivars, Meteor and North
Star, which are more hardy than common commercial cultivars,
such as Montmorency. Hardy interspecific hybrids have been
produced between *P. cerasus* and *P. pensylvanica*, but the
fruit is not edible (4).

European plums are not produced commercially much beyond
Zone 6, but some potential appears to exist for selecting
within *P. domestica* to develop hardier cultivars. Some hardy
European plum clones, such as Dietz and Russian Green Gage

survive to Zones 3 and 4.   More potential appears to exist
in *P. salicina* for improvement in cold hardiness.   Clones
exist which are much superior in cold hardiness to common
commercial cultivars.   Furthermore, cold hardy hybrids have
been derived from *P. americana* Marsh, *P. besseyi* and *P.
nigra* Alt. (Table I).   Many of these hybrids have excellent
fruit quality.

Peach appears to be another species in which the possibility of improvement in cold hardiness appears to be limited.
Commercial peach production doesn't extend beyond Zone 6.
Hardy peach germplasm from northern China, e.g. Siberian C,
offer some potential germplasm for cold hardiness improvement
within peach growing areas, but they do not survive much
beyond Zone 6.   The flower buds of peach appear to be less
hardy than other parts of the tree.   Hybrids have been made
between peach and *P. besseyi* which are flower bud and wood
hardy in Zone 3, but they appear to be sterile (Wilbert
Ronald, Research Station, Morden, Man., personal communication).

Grape appears to be a species of considerable potential
germplasm for improvement in cold hardiness through intraspecific hybridization.   Production without protection of
the European grape, *Vitis vinifera*, is risky in Zone 6.   However, the American and French hybrids derived from *V. riparia*
Michx., and *V. labrusca* L. are commercially grown in Zone 6
and parts of Zone 5.   *V. riparia* is one of the hardiest species of grape.   This species survived to -40° C at Morden,
Manitoba, without snow cover.   *V. amurensis* Pupr. from
northern China has also been used as a source of cold hardiness in breeding programs (58).

Small fruit species such as strawberries, raspberries and
blueberries have not been included in Table I because they are
protected from low temperature by snow or mulch, and it is
difficult to compare hardiness of these species on the basis
of the plant hardiness zonation map.   However, there are examples which indicate that progress has been made in improving
cold hardiness of these crops and further progress might be
possible.

Many of the hardy cultivars of red raspberry grown in
North America have been derived from hybridizing the European
red raspberry, *Rubus ideaus* L. var. *vulgatus* with hardier
forms of the native red raspberry, *Rubus ideaus* var. *strigosus*
(Michx.) Maxim. and the native black raspberry *Rubus occiden-
talis* L.   Hybrids of *R. ideaus vulgatus* and *R. ideaus stri-
yosa vulgatus* have been developed which survive without protection in most of the northern agricultural regions (4).

Strawberry (*fragaria* x *ananassa* Duch.) is the result of
hybridizing tender species, *F. chiloensis* (L) Duch. with a
more hardy *F. virginiana* Duch.   *F. ovalis* (Lehn) Rydb. has

been further used to improve the cold hardiness of *F. ananassa* for the central Great Plains region (44).

The common cultivated form of blueberry is the highbush blueberry which is a complex hybrid of *Vaccinium angustifolium* Aiton, *V. australe* Small and *V. corymbosum* L. Highbush blueberry production extends to about Zone 6. Half high hybrids have been developed using hardy selections of the low bush blueberry, *V. angustifolium* which are hardier and take better advantage of snow cover. It seems possible that cultivated production could be extended into colder regions using these hybrids (48).

V. THE RELATIONSHIP OF DEEP SUPERCOOLING TO THE NORTHERN
    LIMITS OF FRUIT PRODUCTION AND BREEDING COLD HARDY
    CULTIVARS.

Many fruit species avoid freezing in some, but not all of their tissues by deep supercooling to as low as -45° C in the mid-winter (42, 46, 47). Deep supercooling appears to involve mainly the xylem ray parenchyma and the flower bud primordium. The water appears to be in a finely divided state and isolated from the ice in adjacent tissues. In the absence of nucleation by external ice the tissue water supercools to extremely low temperatures. Fine droplets of pure water can be supercooled to as low as -38° C at which they spontaneously freeze in the absence of external nucleators. The addition of solutes further depresses the supercooling point of water to a degree dependent on the solute concentration, but freezing occurs below some critical temperature. When freezing occurs in deep supercooled tissue, it is invariably lethal (5).

The deep supercooling point of the xylem ray parenchyma during the mid-winter of apple, pear (46), and grape (42) appears to be related to the average annual minimum temperature isotherms at the northern limit of survival for each of these species. The freezing damage to the xylem closely resembles the black heart condition observed after severe winters in northern regions of production.

The flower bud primordia of apricot, cherry, peach and plum also appear to avoid freezing by supercooling (47). In common commercial *Prunus* cultivars the extent of supercooling is about -25° C and this temperature approximately coincides with the average annual minimum temperature at the northern range of their production (Zone 6). Flower bud injury in which the flower primordium is killed by mid-winter freezes is common in northern regions. Often only the flower buds are killed, but other parts of the tree survive. Deep supercooling in both the xylem and flower buds appears to be related to northern distribution of native grape, *V. riparia* (42).

Deep supercooling appears to be a major limitation to the northern distribution of many fruit species, but it may also represent a limitation to genetic improvement. There appears to be an ultimate level of deep supercooling in the xylem. The lowest temperature of deep supercooling found in a study of 25 forest species was -47° for *Fraxinus nigra* March. Many species which survive to much lower temperatures and range to the Arctic do not deep supercool (15). This ultimate super-cooling level may be important to improving the hardiness of apple and pear. The lowest temperature at which deep super-cooling has been observed in varietal comparisons of apple was -45° C in Dolgo crab. Bark and bud tissues survive to 20° C below the xylem. In three apple progenies involving hardy parents, the wood of all seedlings was injured above -48° C. An ultimate level of -48° C, however, would allow survival in the coldest agricultural regions.

In *Prunus*, there appears to be more potential for improv-ing flower bud hardiness through breeding. In some *Prunus* species, interspecific hybridization with extremely hardy species such as *P. besseyi* or *P. pensylvanica* may provide a greater degree of deep supercooling or may provide a means of entirely avoiding deep supercooling as a mechanism of sur-vival. *P. besseyi* might be the common linking species which could be used to improve cold hardiness of several of the most important *Prunus* fruit species, apricot, cherry and peach.

REFERENCES

1. Aalders, L. E., and Craig, D. L. *Can. J. Plant Sci. 41*, 466 (1966).
2. Bittenbender, H. C., and Howell, Jr., G. S., *Amer. Soc. Hort. Sci. 99*, 187 (1974).
3. Bradt, O. A., Hutchinson, A., Kerr, E. A., Ricketson, C. L., and Tehrani, G., *Ont. Dept. Agri. and Food Publ. 430*, 80 pages (1970).
4. Brooks, R. M., and Olmo, H. P., "Register of New Fruit and Nut Varieties," 2nd ed. Univ. of Calif. Press, Berkeley (1972).
5. Burke, M. J., Gusta, L. V., Quamme, H. A., Weiser, C. J., and Li, P. H., *Ann. Rev. Plant Physiol. 27*, 508 (1976).
6. Chaplin, C. E., *Proc. Amer. Soc. Hort. Sci. 52*, 121 (1948).
7. Childers, N. F., "Modern Fruit Science." Sommerset Press, Inc., New Jersey (1973).

8. Craig, D. L., Gass. D. A., and Fensom, D. S., *Can. J. Plant Sci. 50*, 59 (1970).
9. Cummins, J. N., and Aldwinkle, H. S., *HortScience 9*, 367 (1974).
10. Darrow, G. M., "The Strawberry. History, Breeding and Physiology." Holt, Rinehart and Winston, New York (1966).
11. Darrow, G. M., *J. of Hered. 9*, 179 (1920).
12. Dexter, S. T., Tottengham, W. E., and Graber, L. F., *Plant Physiol. 7*, 63 (1932).
13. Evert, D. R., and Weiser, C. J., *Plant Physiol. 47*, 204 (1971).
14. Fejer, S. O., *Can. J. Plant Sci. 56*, 303 (1976).
15. George, M. F., Burke, M. J., Pellett, H. M., and Johnson, A. G., *HortScience 9*, 519 (1974).
16. Gill, P. A., and Waister, P. D., *J. Hort. Sci. 51*, 509 (1976).
17. Greenham, G., *Can. J. Bot. 44*, 1471 (1966).
18. Harris, R. E., *Can. J. Plant Sci. 50*, 249 (1970).
19. Harris, R. E., *Can. J. Plant Sci. 45*, 159 (1965).
20. Hildreth, A. C., "Determination of Hardiness in Apple Varieties and Relation of Some Factors to Cold Resistance." Minn. Agr. Exp. Sta. Tech. Bull., 42 pages (1926).
21. Hummel, R., *Fruit Var. J. 31*, 61 (1977).
22. Hutchinson, A., *Ont. Dept. Agri. Food. Publ. 334*, 21 pages (1969).
23. Janick, J., and Moore, J. N., "Advances in Fruit Breeding." Purdue Univ. Press, West Lafayette (1975).
24. Kerr, W. L., *Proc. Amer. Soc. Hort. Sci. 37*, 73 (1937).
25. Lantz, H. L., and Pickett, B. S., *Proc. Amer. Soc. Hort. Sci. 40*, 237 (1940).
26. Lapins, K., *Can. J. Plant Sci. 42*, 521 (1962).
27. Lapins, K., *Proc. Amer. Soc. Hort. Sci. 81*, 26 (1962).
28. Larcher, W., and Eggarter, H., *Protoplasma 41*, 595 (1960).
29. Layne, R. E. C., *Proc. XIX Int. Hort. Cong. 1 A*, 332 (1974).
30. Levitt, J. "The Hardiness of Plants," 278 pages. Academic Press, New York (1956).
31. Levitt, J., "Responses of Plants to Environmental Stresses," 697 pages. Academic Press, New York (1972).
32. Maney, T. J., *Proc. Amer. Soc. Hort. Sci. 40*, 215 (1940).
33. McKenzie, J. S., and Weiser, C. J., *Can. J. Plant Sci. 55*, 651 (1975).
34. Morrison, J. W., Cumming, W. A., and Temmerman, H. J., *Can. Dept. Agri. Publ. 1222*, 31 pages (1965).

35. Mowry, J. B., *Proc. Amer. Soc. Hort. Sci. 85*, 128 (1964).
36. Murawski, J., *Zuchter 31*, 52 (1961).
37. Nelson, S.H., and Blair, D.S., *Progress Report, 1954-58*. Hort. Div. Central Expt. Farm. Can. Dept. Agr. (1959).
38. Nybom, N., Bergendal, P. O., Olden, E. L., Tamas, P., *Eucarpia 182*, 66 (1962).
39. Olien, C. R., *Ann. Rev. Plant Physiol. 18*, 387 (1967).
40. Olien, C. R., *Barley Genetics 2*, 356 (1969).
41. Oratovsky, M. T., *Rep. Soviet Sci. 16th Int. Hort. Congr., Moscow*, p. 67 (1962).
42. Pierquet, P., Stushnoff, C., and Burke, M. J., *J. Amer. Soc. Hort. Sci. 102*, 54 (1977).
43. Pomeroy, M. K., and Fowler, D. B., *Can. J. Plant Sci. 53*, 489 (1973).
44. Powers, L., *J. Agr. Res. 70*, 95 (1945).
45. Proebsting, E. L., and Mills, H. H., *Proc. Amer. Soc. Hort. Sci. 89*, 85 (1961).
46. Quamme, H. A., *Can. J. Plant Sci. 56*, 493 (1976).
47. Quamme, H. A., *J. Amer. Soc. Hort. Sci. 99*, 315 (1974).
48. Quamme, H. A., Stushnoff, C., and Weiser, C. J., *HortScience 7*, 500 (1972).
49. Quamme, H. A., Evert, P. R., Stushnoff, C., and Weiser, C. J., *HortScience 7*, 24 (1972).
50. Quamme, H. A., Layne, R. E. C., Jackson, H. O., and Spearman, G. A., *HortScience 10*, 521 (1975).
51. Reid, W. S., Harris, R. E., and McKenzie, J. S., *Can. J. Plant Sci. 56*, 623 (1976).
52. Sakai, A., and Weiser, C. J., *Ecology 54*, 118 (1973).
53. Scott, D. H., and Cullinan, F. P., *J. Agr. Res. 73*, 207 (1946).
54. Scott, K. R., and Spangelo, L. P. S., *Proc. Amer. Soc. Hort. Sci. 84*, 131 (1964).
55. Simons, R., *HortScience 7*, 401 (1972).
56. Siminovitch, D., and Briggs, D. R., *Plant Physiol. 28*, 15 (1953).
57. Siminovitch, D., Therrien, H., Gfeller, F., and Rheaume, B., *Can. J. Bot. 42*, 637 (1964).
58. Snyder, E., *Proc. Amer. Soc. Hort. Sci. 34*, 426 (1937).
59. Steponkus, P. L., and Lanphear, F. O., *Plant Physiol. 42*, 1423 (1967).
60. Strang, H. G., and Stushnoff, C., *Fruit Var. J. 29*, 78 (1975).
61. Stushnoff, C., *HortScience 7*, 10 (1972).
62. Svejda, F., *Can. J. Plant Sci. 46*, 441 (1966).
63. United States Department of Agriculture, *Misc. Publ. 814*, (1960).
64. Ure, C. R., *Canada Agr. Instit. Rev., Jan-Feb.*, 20 (1960).

65. van Adrichem, M. C. J., *Can. J. Plant Sci.* *50*, 181 (1970).
66. Watkins, R., and Spangelo, L. P. S., *Theor. Appl. Gen.* *40*, 195 (1970).
67. Weaver, G. M., and Jackson, H. O., *Can. J. Plant Sci.* *49*, 459 (1969).
68. Weaver, G. M., Jackson, H. O., and Stroud, F. D., *Can. J. Plant Sci.* *48*, 37 (1968).
69. Weiser, C. J., *Science 169*, 1269 (1970).
70. Wilner, J., *Can. J. Plant Sci.* *40*, 630 (1960).
71. Wilner, J., *Can. J. Plant Sci.* *41*, 309 (1961).
72. Wilner, J., *Can. J. Plant Sci.* *45*, 67 (1964).

BREEDING FROST-RESISTANT POTATOES
FOR THE TROPICAL HIGHLANDS

R. N. Estrada

Breeding & Genetics Department
International Potato Center
Lima, Peru

I. INTRODUCTION

The tropical Andean highlands, where potatoes are cultivated, are characterized by low average temperatures ($12°$ C or
less), short day length (11 to 13 hours), high intensity of
light (10,000 to 12,000 foot candles) and 300 to 500 mm (12
to 20 inches) of rainfall during the growing season. Frequent
frosts which may occur during the early growth of potatoes,
in mid-season, or toward the end of the growing period cause
severe reduction in tuber yields. Temperatures between $-2°$ C
and $-6°$ C most frequently occur one to three hours in the
early morning, depending on environmental conditions (5).
Wild and primitive potato species frequently survive such
environmental stress and farmers are encouraged to cultivate
clones of primitive species despite relatively low yields and
the bitter flavor produced by alkaloids. These bitter potatoes have been used since Inca times for producing "chuño", a
dehydrated freeze dried potato suitable for prolonged storage.
For new potato hybrids to succeed in such environmental
stress areas, the following four characteristics are
necessary:

(1) Fast and vigorous growth under low average temperatures and short day length.
(2) Relatively early maturity, four months.
(3) Fast and abundant tuberization (bulking).
(4) High degree of frost resistance in the foliage.

This research was initiated in 1968 in Colombia (ICA) by a grant from the Rockefeller Foundation which provided the scientific support of the Plant Hardiness Laboratory, University of Minnesota. Additional scientific input and equipment by CIP after 1972, in addition to a research contract between CIP and the University of Minnesota, have permitted a wide and successful development of this study during the last four years.

II.   METHODOLOGY

Four steps were undertaken to assure success in breeding for resistance to environmental stress:

(A) Genetic resources with frost resistance were obtained for use as parents.

(B) Successful cross-breeding with improved cultivated clones was accomplished.

(C) Artificial methods of testing were developed to screen mass populations in early stages.

(D) Adequate field testing sites were selected to reconfirm results of artificial tests of seedlings and to evaluate the yield of selected clones under environmental stress.

An elaboration of these four steps follows.

In reference to step A, the following genetic resources were used to produce improved genetic stocks:

(1) Frost-resistant, wild species collected in Peru, Bolivia, and Mexico were crossed with diploid and tetraploid cultivated species. Resistant species and diverse interspecific crosses were obtained from the Potato Introduction Station, Sturgeon Bay, Wisconsin, USA.

(2) Primitive cultivated species with high resistance to frost such as S. ajanhuiri (2n = 24), S. juzepczukii (2n = 36) and S. curtilobum (2n = 60) were used for crosses.

(3) Highly selected clones with commercial characters and other desirable resistances from the species S. andigena, S. stenotomum, S. phureja, were used for quality, yield, and some degree of frost tolerance. These species, as well as improved tuberosum x andigena hybrids developed in Peru and Columbia and some tuberosum varieties obtained primarily from Mexico, were necessary for the gradual nobilization of the wild and primitive cultivated species. Tables I and II illustrate the species used for cross-breeding.

TABLE I.   *Wild Tuber Bearing Solanum Species Used in Breeding for Frost Resistance*

Species name	Abbreviation	Chromosomes, 2n
S. bukasovii	buk	24
S. boliviense	blv	24
S. brevicaule	bvc	24
S. chacoense	chc	24
S. chomatophilum	chm	24
S. commersonii	cmm	24
S. megistacrolobum	mga	24
S. multidisectum	mlt	24
S. raphanifolium	rap	24
S. sanctae rosae	sct	24
S. venturi	vnt	24
S. vernei	vrn	24
S. acaule	acl	48, 72
S. demissum	dms	72

TABLE II.   *Cultivated Potato Species Used in Breeding for Frost Resistance*

Species name	Abbreviation	Chromosomes, 2n
S. phureja	phu	24
S. stenotomum	stn	24
S. ajanhuiri	ajh	24
S. juzepczukii	juz	36
S. tuberosum	tbr	48
S. andigena	adg	48
S. curtilobum	cur	60

TABLE III.   Crosses and Selections Made During 1974,
              1975, 1976, and 1977 for Frost Resistance

Species used	11	16	9	12	21
Families obtained	31	117	232	220	600
Seeds obtained	5050	10370	31500	27300	74220
Seedlings obtained	1513	4300	17860	15000	38670
Seedlings resistant (artificial tests)	60	100	410	320	890
Clones selected	25	40	110	--	175

In step B, numerous seeds, seedlings and diverse families
were obtained through cross-breeding during 1974, 75, 76 and
77. Details can be seen in Table III.

With step C, the following methods were applied when
screening mass populations of seedlings:

(1) Growth chamber testing which was conducted when seedlings, grown in the greenhouse, were five to eight cms high.
The seedlings had a two to three day pre-hardening period and
received the last day, temperature shocks of -4° C to -5° C
for about two hours. From 80 to 90 percent were killed by
this test.

(2) The low temperature bath test. Here excised leaflets
were tested under controlled, yet similar temperatures, but
using different instruments. This method, first described by
Sukumaran and Weiser (7) was applied and later the modifications suggested by Chen et al. (2) were added. This method
complimented the growth chamber system as some seedlings which
there escaped damage, showed considerable injury in the leaflet test (Figure 1). About 95 percent of the population was
eliminated after both tests were performed.

Field tests have indicated that about 50 percent of the
potato material showing resistance to artificial tests will
also display resistance in the field. This finding is supported by the studies of Alvarado et al. (1) and by
Murillo (6). Figure 2 illustrates this point.

Such artificial tests have displayed a potential for
screening field resistant material and thus form a tool for
early elimination of a great majority of frost susceptible
populations.

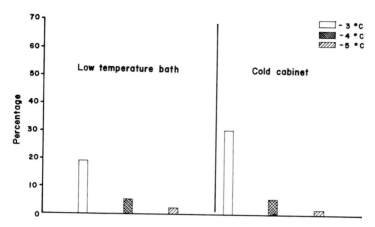

FIGURE 1. Percentage of clones with resistance to -3° C,
-4° C and -5° C tested by artificial methods. (After
Murillo, 1977).

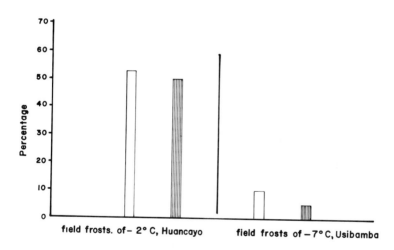

FIGURE 2. Percentage of clones with resistance to field
frosts and which were selected as resistant to -3° C in
cold cabinet ☐ and in low temperature bath ▥. (After
Murillo, 1977).

Concerning step D, three main locations are now being used to test selected materials in the field:

(1) Usibamba, a Peruvian highland area, south-west of Huancayo, 3,700 mts.
(2) Belen Experimental Station, a Bolivian high plateau area, near Lake Titicaca, 3,900 mts.
(3) Atizapan experimental field in Toluca Valley, Mexico, 2,650 mts. during winter season.

Nearly 38,000 seedlings have been tested. From this group about 200 frost-tolerant selected clones are being checked for such characters as yield, tuber quality and disease resistance.

The following scale (4) was used to measure the degree of frost damage in the field:

(1)   0 - 10% foliage damaged
(2)   11 - 20% foliage damaged
(3)   21 - 30% foliage damaged
(4)   31 - 40% foliage damaged
(5)   41 - 50% foliage damaged
(6)   51 - 60% foliage damaged
(7)   61 - 70% foliage damaged
(8)   71 - 80% foliage damaged
(9)   over 80% foliage damaged

III. RESULTS

Tables 4, 5, and 6 summarize field frost resistance observations and the yield of 14 hybrid types at various ploidy levels and from several cross-breeding methods. These hybrid types are represented by 574 clones which were obtained from approximately 20,000 seedlings (1974, 75, 76). This field evaluation was made in Usibamba, Peru, during the 1976-1977 planting season. Two controls were used as a reference: S. *curtilobum* clones of a potato species commonly cultivated in the highlands of Peru and Bolivia because of frost-resistance and the relative high yield; and S. *tuberosum* clones which represent frost susceptible clones.

TABLE IV.   *Yield and Foliage Reaction Obtained in Hybrids Originated from Tetraploid and Pentaploid Parents After -4° C Field Frosts, Usibamba, Peru*

Type of cross	Clones tested	Clones selected	Kgs/ pl.	Average fr. damage 1-9 scale
(tbr x acl) x tbr x adg x adg	56	22	0.90	2.8
(tbr x adg) x adg	94	25	0.79	2.1
cur x tbr	26	7	0.74	2.3
cur x adg	50	16	0.70	2.1
adg x adg	34	15	0.63	3.1
tbr x adg	79	28	0.51	2.8
cur (CONTROL)	23	7	0.58	1.0
tbr (CONTROL)	9	7	0.34	5.1

TABLE V.   *Yield and Foliage Reaction Obtained in Hybrids Originated from Tetraploid x Diploid Parents After -4° C Field Frosts, Usibamba, Peru*

Type of cross	Clones tested	Clones selected	Kgs/ pl.	Average frost damage
adb x (blv x phu)	18	15	0.85	2.6
(adb x adb) x phu	73	19	0.75	2.3
adg x phu	5	2	0.75	2.5
adg x stn	15	3	0.74	1.3
(adg x phu) x stn	5	2	0.73	2.5
cur (CONTROL)	23	7	0.58	1.0
tbr (CONTROL)	9	7	0.34	5.1

TABLE VI.   *Yield and Foliage Reaction Obtained in Hybrid*
Originated from Diploid Parents After -4° C
Field Frosts, Usibamba, Peru

Type of cross	Clones tested	Clones selected	Kgs/ pl.	Average frost damage
ajh x stn	16	8	0.71	1.6
phu x stn	25	8	0.63	2.5
(ajn x stn) x phu	13	10	0.58	2.5
wild sp. x stn (five different)	65	5	0.49	2.2
cur (CONTROL)	23	7	0.58	1.0
tbr (CONTROL)	9	7	0.34	5.1

IV.   CONCLUSIONS

From the results observed in Tables 4, 5, and 6, several conclusions were drawn:

(1) There is low correlation between yield and frost resistance in the foliage when frosts are moderate (-4° C). However, a minimum degree of resistance is required, with no more than 30 percent damage, to obtain acceptable yields.

(2) Early tuberizing materials which thrive under cold temperatures, but are not highly resistant to freezing temperatures, may be adaptable.

(3) A representative degree of frost resistance may be found in hybrids obtained between some cultivated selected clones, both tetraploids and diploids.

(4) Tetraploid-diploid crosses could be a useful approach to breeding frost-resistant potato materials.

(5) A build up of frost resistance has been observed after intercrosses with materials of moderate resistance.

The yields expressed in tons/ha varied among groups from 12.5 ton/ha to 22.5 ton/ha. However, there were outstanding clones in several groups which produced the equivalent of 31 ton/ha. Some of the clones selected for frost resistance have been analyzed and reveal high content of total solids (30

percent) and protein (over ten percent on dry basis), CIP
Annual Report (3). The flavor of many clones has also been
found to be fair to good.

REFERENCES

1. Alvarado, L. F., Estrada, R. N., and Riveros, G., *in*
    "Evaluación de papas resistentes a heladas en estado de
    plántulas" 7a, p. 123. Reunión Sociedad Latinoamericana
    de Papa, Bogotá, Colombia (1972).
2. Chen, P. M., Burke, M. J., and Li, P. H., *Bot. Gaz. 137*,
    313 (1976).
3. International Potato Center, *Annual Report 1976*, 129 pp.
    (1977).
4. Mendoza, H. A., and Estrada, R. N., *in* "Proceedings of
    the International Conference on Stress Physiology in
    Important Crop Plants." Yonkers, New York (in press).
5. Muñoz, C., and Sanchez, W. *in* "Estudio de la distribución
    espacial y temporal de las heladas en la Hoya del Lago
    Titicaca," 63 pp. I - Seminar of Natural Resources,
    Environment and Ecologic Systems, Lima, Peru (1974).
6. Murillo, V., *M. S. Thesis*, p. 65. Universidad Nacional
    Agraria, Lima, Peru (1977).
7. Sukumaran, N. P., and Weiser, C. J., *HortScience 7*, 467
    (1972).

Part VI
Cryopreservation and Cryoprotection

SURVIVAL OF PLANT GERMPLASM
IN LIQUID NITROGEN[1]

A. Sakai
Y. Sugawara

The Institute of Low Temperature Science
Hokkaido University
Sapporo

ABSTRACT

Suspension-cultured cells of sycamore, which were immersed in liquid nitrogen after prefreezing to the temperatures from -30 to -50° C in the presence of cryoprotectants,
could proliferate vigorously when rewarmed rapidly. It was
also demonstrated that less hardy cells could survive immersion in liquid nitrogen by rapid passage through the growth
zone of intracellular ice crystals when cells were partially
dehydrated by extracellular freezing in the presence of
cryoprotectants. It appears that prefreezing and cryoprotectants enhance innocuous intracellular freezing by slowing
down the growth rate of intracellular ice crystals. This
study provides a possibility for long-term preservation of
germplasm in liquid nitrogen.

[1]*Contribution No. 1871 from the Institute of Low Temperature Science.*

345

I. INTRODUCTION

Genetic gene preservation is becoming increasingly urgent
(2). Freezing storage in liquid nitrogen was found to be a
suitable method for animal cell lines. Liquid nitrogen is
readily available commercially and relatively cheap to buy
in countries where supplies of liquid nitrogen are obtainable.
It has been well known that plant materials such as seeds,
pollen, tubers, etc., which resist an intensive dehydration
retain their viability after direct immersion in liquid nitrogen from room temperature when rewarmed slowly or rapidly
(23). Fully-hydrated seeds, however, except for those that
over-winter in cold climates, are generally killed by slight
freezing (23).
Sakai (14, 15) reported that very hardy twigs and fruit buds
(24) could survive immersion in liquid nitrogen after subsequent slow rewarming in air at 0° C, if they were sufficiently
dehydrated by extracellular freezing. In partially freeze-
dehydrated cells, however, a rapid rewarming following immersion in liquid nitrogen is required for maintaining their
viability (16). This prefreezing method is applicable to less
hardy materials, if any cryoprotectant is previously used on
them, so that freezing to about -70° C can be resisted (14).
Another method for maintaining viability at super-low
temperatures with ultrarapid cooling and rewarming was presented by Luyet (7). This method consists of preventing the
growth of the intracellular ice crystals formed during rapid
cooling to super-low temperatures by rapid passage through
the zone of growth. It is, however, well known that less
hardy cells which cannot survive freezing -10° C or above are
quite apt to cause lethal intracellular freezing during rapid
cooling (16, 18, 21). Thus, this method might not be applicable to less or non-hardy plant materials. Sakai *et al.*,
however, demonstrated that even less hardy cells could survive
immersion in liquid nitrogen, provided that they were partially dehydrated by slow freezing in the presence of cryoprotectants (16, 19), and were then rapidly cooled to and rewarmed from liquid nitrogen (20).
This study was designed to examine some factors contributing to the survival of less or non-hardy cells cooled to the
temperature of liquid nitrogen, and also to list up the results on survival of tissue cultures cooled to the temperature
of liquid nitrogen, to find methods applicable to long-term
genetic preservation.

II. MATERIALS AND METHODS

Materials: an unicellular marine red alga (*Porphyra yezoensis*) in the leafy stage, cortical parenchymal cells from mulberry twigs (*Morus bombycis*), suspension cultures of sycamore (*Acer pseudoplatanus*) were mainly used as the experimental material. Thin tangential tissue sections (20 to 30 μ thick, unicellular, 1 to 2 mm wide and 2 to 3 mm long) were cut with a razor from the cortical regions of a 1-year-old twig. Ten tissue sections were used in each experiment. Cortical tissue sections from mulberry twigs and pieces of red alga were mounted between cover glasses (18 x 18 mm) with about 0.03 ml of water or other solutions. The rapidly cooled cells were rewarmed, either rapidly in a water bath at 35° C (about 500° C/sec) or slowly in air at 0° C (2° C/sec). The rewarming rate was represented as the time required for the temperature to rise from the final temperature to −2° C. Tissue sections or pieces of alga mounted between cover glasses were cooled rapidly by immersion in liquid nitrogen following prefreezing at different temperatures (cooling rate: about 170° C/sec). The temperature was determined with a 0.1 mm copper-constantan thermocouple and was recorded by an oscilloscope. The cooling rate was determined by placing a thermocouple between cover glasses.

The viability of cortical cells was determined by vital staining with neutral red and the plasmolysis test. In this test plasmolysis and deplasmolysis were repeated twice with a twofold isotonic balanced salt solution and water. Normally stained and plasmolyzed cells were regarded as viable ones. The viability of alga cells was determined by a vital staining test.

Tissue-culture cells used in this experiment were originally derived from the tissue slice isolated from the cambial area of twigs of sycamore. Callus was cultured on a modified White's agar medium containing 3 mg/l of 2.4-D and 10% (v/v) coconut milk essentially according to the method described by Lamport (5). Agar-grown callus was transferred into 0.5 or 1-liter of Sakaguchi's flask containing 100 or 300 ml of modified White's medium, respectively, which were supplemented with 6 mg/l of 2.4-D and 10% coconut milk.

Cultures were agitated at 120 cycles/min on a reciprocal shaker at 26° C under continuous illumination of about 280 ft-c at culture level. Continuous subcultures were performed by pipetting aliquots into fresh liquid medium at 3- to 4-week intervals for 1 year or more. It was confirmed in the preliminary experiment on cryoprotectants that the highest survival was obtained in the frozen-thawed cells in the presence of 10% DMSO and 5% glucose. Cell suspensions to be frozen were collected in a graduated spitz tube (15 mm diameter, 110 mm long). To the cell suspensions from which excess medium was removed, cryoprotectants were gradually added and sufficiently mixed. The opening of the tube was covered with two layers of aluminum foil. After standing at about 22° C for 15 min, the cell suspensions were immersed in an ethanol bath at -10° C. Then, frozen cell suspensions were successively cooled down at 5 min intervals to the test temperatures of -15, -23, -30, -40, -50, and -70° C, respectively. Some cell suspensions frozen at these temperatures for 5 min were then immersed in liquid nitrogen (cooling rate: 420° C/min). Cell suspensions held at each test temperature or immersed in liquid nitrogen for 15 or 20 min after prefreezing to these temperatures, were rewarmed rapidly in water at 40° C (540° C/min) or slowly in air at 0° C (10° C/min), respectively. Immediately after thawing, the frozen-thawed and unfrozen control cell suspensions were diluted to 10 ml with fresh cultural medium. This was repeated four to five times.

The cell viability was determined with TTC reduction rate or regrowth rate. The TTC test was performed according to the method of Steponkus and Lanphear (26). To determine the growth rate, frozen-thawed cells suspended in 10 ml of fresh medium were cultured in 100 ml Erlenmeyer flasks, under previously described conditions. The growth rate was estimated photometrically at 610 nm using a Bausch-Lomb colorimeter according to the method described by Eriksson (3).

III. RESULTS AND DISCUSSION

At first, an unicellular marine red alga (*Posphyra yezoensis*) in the leafy stage was used as the experimental material. Pieces of the alga mounted in sea water between cover glasses were observed to resist slow freezing to -20° C. To investigate the effect of the degree of prefreezing on the survival of ultra-rapidly cooled alga cells, pieces of red alga (10 mm^2) prefrozen slowly to various temperatures between -5 and -30° C were immersed in liquid nitrogen (cooling rate: 170° C/sec), then they were rewarmed rapidly in sea water at 35° C (250° C/sec) or slowly in air at 0° C (1° C/sec). Results are summarized in Figure 1. Algal cells prefrozen between -10 and -20° C, all survived immersion in liquid nitrogen when rewarmed rapidly, while all were killed when rewarmed slowly. Cells prefrozen at -5° C might be killed as a result of prompt formation of large ice grains within the cells during cooling to the temperature of liquid nitrogen. These results suggest that some freezable water remains in the cells prefrozen at the temperatures between

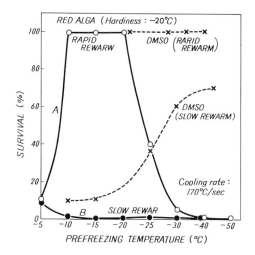

FIGURE 1. *Survival of red algal cells immersed in liquid nitrogen after prefreezing to different temperatures and rewarmed slowly in air at 0° C or rapidly in sea water at 30° C. Material: Red alga cells (Porphyra yezoensis) in the leafy stage. Pieces of alga (10 mm²) were mounted between cover glasses with sea water, frozen slowly to different temperatures, then immersed in liquid nitrogen. A. B.: without DMSO. Cooling rate: 170° C/sec; rewarming rate: 250° C/sec in water at 40° C, 1° C/sec in air at 0° C.*

-10 and -20° C, but the ice crystals formed in the cells during ultra-rapid cooling may be small enough to be innocuous. Thus, they may melt before they have time to grow to a damaging size when rewarmed rapidly, and cells remain viable (16-18, 20). Actually, we observed that the survival of algal cells prefrozen to -10° C abruptly decreased at temperatures between -50 and -30° C during rewarming in air at 0° C following removal from liquid nitrogen. This indicates that small innocuous intracellular ice crystals have grown to a damaging size in this temperature range during rewarming. The survival percentage in the cells immersed into liquid nitrogen following prefreezing to -25 or -30° C, abruptly decreased as a result of frost injury during slow freezing to -25 or -30° C. Red algal cells treated with 1.5 M DMSO solution resisted slow freezing to -40° C at least. Most of the cells immersed in lqiuid nitrogen after prefreezing to -40° C survived even when rewarmed slowly in air (Figure 1).

Cortical cells taken from mulberry twigs in spring immediately before bud opening survived slow freezing to only -10° C. However, these cells became resistant to slow

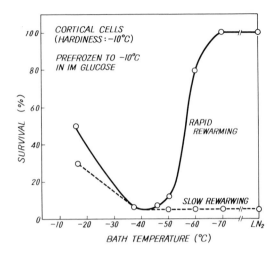

FIGURE 2. Survival of cortical cells immersed in baths at different temperatures and rewarmed rapidly. Material: cortical cells of mulberry twigs immediately before bud opening, which resisted slow freezing to -10° C. Cortical tissue sections mounted in 1 M glucose solution between cover glasses were prefrozen to -10° C, then cooled in baths at different temperatures for 5 min before being rewarmed rapidly.

freezing to -20° C in the presence of slightly hypertonic glucose of DMSO solution. Cortical tissue sections mounted in 1 M glucose or DMSO solution between cover glasses were prefrozen to -10° C, then immersed in baths held at various temperatures from -15 to -70° C, or liquid nitrogen. These ultra-rapidly cooled cells (150° C/sec) were rewarmed slowly or rapidly. As shown in Figure 2, in the cells cooled rapidly by immersion in baths at -70° C or liquid nitrogen, viability was retained when rewarmed rapidly. When the cells immersed into liquid nitrogen after prefreezing at -10° C were held at -35 or -70° C for 5 min following removal from liquid nitrogen before being rewarmed rapidly in water at 35° C, all cells held at -70° remained alive, but were all killed at -35° C. These facts suggest that in the cells prefrozen at -10° C, some freezable water remained in the cells, but the ice crystals formed in the cells during a rapid cooling by immersion into baths below -70° C may be small enough to be innocuous. As shown in Figure 2, the survival curve abruptly rose with the decrease in temperatures from -50 to -70° C, which is characteristic of rapid rewarming. However, in the cells cooled in baths between -30 and -50° C, survival was very low, even when rewarmed rapidly. This suggests that intracellular ice crystals formed during rapid cooling are large enough to be immediately damaging in this temperature range (16, 20). To check this possibility, effect of pre-freezing on the growth rate of intracellular ice crystals formed during rapid cooling was determined with very hardy cortical cells which survived slow freezing below -70° C. These hardy cells prefrozen at -5, -8, and -10° C respectively were immersed in baths held at different temperatures for different lengths of time, before being rewarmed rapidly in water. As shown in Table I, length of time held without causing injury in the cells immersed in baths at different lengths of time was observed to increase with more prefreezing. Sakai (16, 18, 20) also previously demonstrated that prefreezing and the use of cryoprotectants contributed to inducing innocuous intracellular freezing by slowing down the growth rate of intracellular ice crystals and by preventing lethal intracellular freezing from occurring following freezing of the surrounding medium. Thus, less hardy cells could survive immersion in liquid nitrogen by rapid passage through the growth zone of intracellular ice crystals, when cells were partially dehydrated by slow freezing in the presence of cryoprotectants, especially DMSO. The cooling and rewarming rates necessary to survive with rapid cooling to and rewarming from liquid nitrogen vary considerably with or without

TABLE I. *Length of Time Held without Causing Injury in the Cells Immersed in Baths at Different Temperatures*

Prefreezing temperature (°C)	Bath temperature (°C)		
	-50	-60	-70
- 5	Below 0.5[a]	2	—
- 8	3.5	10	2 min
-10	10	30	Above 10 min

Material: Very hardy mulberry cortical cells which survived slow freezing below -70° C. Tissue sections prefrozen at -5, -8 and -10° C were immersed in baths at -50, -60 and -70° C respectively for different times before being rewarmed rapidly in water at 35° C.

[a]*Time (sec) held in baths without causing injury before being rewarmed rapidly.*

suspending mediums, the amount of the medium, the degree of prefreezing and the freezing resistance in the presence of cryoprotectants of the cells used. In general, lower cooling and rewarming rates, require much prefreezing.

Freezing storage of tissue cultures in liquid nitrogen might be valuable for the preservation of plant genetic resources. Usually 0.5 to 1.0 ml suspension cultures in a spitz tube or ampoules are used for freezing storage. However, it seems difficult for cultures to cool ultra-rapidly to and rewarm from liquid nitrogen, unlike red alga cells and twig cortical cells. Suspension cultures of 0.5 ml in a spitz tube were slowly frozen to each test temperature down to -76° C in the presence or absence of cryoprotectants (12% DMSO and 5% glucose). These frozen cells were rapidly rewarmed in water at 40° C. Figure 3 shows TTC reduction rate of frozen cells decreased with decreasing temperatures down to -76° C. In the cells frozen in distilled water, only a negligible reduction was observed even after freezing to -10° C. Also, in the cells frozen to -30° C in the presence of cryoprotectants, little or no difference was observed in TTC reduction between the cells rewarmed in water at 40° C or in air at 0° C. Cultures of 0.5 ml in the presence of cryoprotectants were prefrozen to different temperatures ranging from -23 to -76° C. These prefrozen cells were immersed in liquid nitrogen (420° C/min) and were subsequently rewarmed rapidly

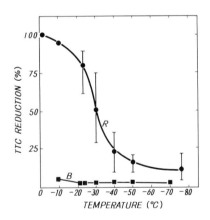

FIGURE 3. *TTC reduction rate of cultured cells frozen slowly to each test temperature down to -76° C and subsequently rewarmed rapidly in the presence or absence of cryoprotectants. R. B.: In the presence (R) or absence (B) of cryoprotectants (12% DMSO plus 5% glucose). Material: Suspension cultures of sycamore. These frozen cells were rewarmed rapidly in water at 35° C.*

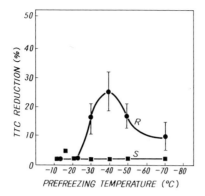

FIGURE 4. TTC reduction rate of the cells immersed in liquid nitrogen after prefreezing to various temperatures in the presence of cryoprotectants. These cells immersed in liquid nitrogen for 10 min were rewarmed rapidly in water at 35° C (R) or slowly in air at 0° C.

in water at 40° C (540° C/min) or slowly in air at 0° C (10° C/min). As shown in Figure 4, the TTC reduction rate of the cells immersed in liquid nitrogen after prefreezing to -23° C was less than 5% regardless of rewarming methods. This indicates that cell injury occurs during rapid cooling (18, 20). TTC reduction rates of the cells prefrozen at the temperatures from -30 to -50° C were much higher than those above -23° C or at -70° C (Figure 4). The cells immersed in liquid nitrogen after prefreezing at -30, -40 and -50° C, respectively, grew vigorously while this was not observed in the cells prefrozen above -23° C or at -70° C.

Recently we observed that cultures of *Ruta graveolens* prefrozen to -45 or -50° C, *Daucus carrota*, prefrozen below -30° C in the presence of cryoprotectants proliferated vigorously after immersion in liquid nitrogen when rewarmed rapidly. The cultures of *Nicotiana tabacum* and *Broussonetia kazinoki* were also observed to grow vigorously after slow freezing to -30° C, but not to -40° C or to the temperature of liquid nitrogen after prefreezing to -30° C. Further investigation is required to enable other species to survive prefreezing to -40 or -50° C prior to immersion in liquid nitrogen.

The survival of frozen cultured cells depends on a number of factors—the most critical being the rate of cooling and rewarming, cryoprotectants, the age and nature of suspensions and cell density from which growth occurs.

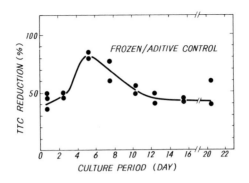

FIGURE 5. *Changes in TTC reduction rate of the frozen-thawed and unfrozen additive control cells during a passage of suspension cell culture of sycamore. Frozen: Cells frozen slowly to -30° C in the presence of cryoprotectants (12% DMSO plus 5% glucose). Additive control: Unfrozen cells treated with the same cryoprotectants as frozen cells.*

In the present study, we observed that highest survival is recorded for cells at lag phase or early exponential phase (Figure 5). Correlation between stage of culture growth and survival has also been reported (6, 13, 29). The suspension cultured sycamore cells corresponding to these phases were observed electronmicroscopically to be filled with a highly dense cytoplasm resulting from new protein synthesis and to have a small cell size as compared with the cells in the advanced phases by Shutton-Jones and Street (27). Nag and Street (8) also observed that reduction in survival percentage during freezing and thawing is primarily due to death of the larger more highly vacuolated free cells of the suspension which do not give rise to colonies even with unfrozen suspensions, and that in large nodules of *Linum* grown in liquid cultures, localized groups of meristematic cells at intervals along the periphery, which are non-vacuolated, green and contain organelle-filled cytoplasm have the greatest resistance to freezing in the presence of DMSO (13). He also observed that if large nodules are frozen, only certain areas, presumably meristematic, are green and begin growth rather than the whole tissue mass.

These results indicate the importance for successful freeze-preservation of the stage in the growth cycle of suspensions and of the effective cell density from which growth occurs. Withers and Street (29) clearly indicated that the capacity of the surviving cells to embark upon cell division is significantly affected by the stage in the batch culture growth cycle at which they are frozen. Latta (6) reported that samples for cultures of *lpomea* containing 44 g/1 sucrose did not survive freezing to -40° C in the presence of cryoprotectants, but those from cultures containing 65 g/l did survive. Withers and Street (29) also observed that the same pregrowth conditions enhances resistance to injury by cryoprotectants. It appears that pregrowth conditions in the high level of sugars or mannitol seem to be a prominent method of improving sensitivity to cryoprotectants and freezign of cultures.

Cytological transition from a thin cytoplasm surrounding a large central vacuole in summer to more dense and extensive cytoplasm in winter has been observed (11, 12), although the role in increasing hardiness is obscure. Thus, attention has been directed to the importance of cell size, and degree of vacuolation and the stage in the growth cycle of the cell suspension cultures subjected to freezing and thawing. However, further investigation is required to more fully explain differences between the cells of different species in their susceptibility to injury during the freeze-thawing cycle and to optimize conditions within each species.

Experience with suspension cultured cells and callus is sufficiently promising to suggest that it should be possible to preserve at least some plant species by storing them as tissue cultures in liquid nitrogen, if these cells are inherently stable (28). Nag and Street (8) first demonstrated that cultured cells from high embryogenic cell lines of carrot stored in liquid nitrogen for 45 days began to initiate embryo and then developed into healthy plants of normal morphology; no decline in cell survival was observed during storage in liquid nitrogen over 100 days (8). They also reported that no morphogenic potential was impaired and that examination of large numbers of metaphase plants in squash preparations has yielded no evidence of any effect of freezing preservation on the karyotype (8).

The value of the technique for the conservation of genetic resources would depend upon the ease with which tissue cultures possessing high propagation potential could be isolated from those species which are not readily preserved by orthodox methods. Since regeneration and morphogensis of plants derived from culture cells seem to be difficult except to some species, meristematic tissues should be used for a long-term preservation (24, 25).

Cryoprotectants, culture stage and cooling and rewarming conditions which enabled cultures and meristematic tissues to survive freezing to the temperature of liquid nitrogen are listed in Table II.

TABLE II. *Plant Cultures and Meristematic Tissues which Survived Freezing to the Temperature of Liquid Nitrogen*

Species	Pre-treatment	Cryopro-tectants	Culture stage or condition	Cooling rate, prefreezing temperature	Re-warming rate	Reference
SUSPENSION CULTURES						
Daucus catota	--	DMSO (10%) or Glycerol (5%)	7-day-old	2 to 4° C/min -40 LN	Rapid	Latta (6)
Daucus catota	--	DMSO (5%)	5~15-day-old	20 C/min -40 or -100 LN	Slow or Rapid	Nag and Street (8)
Daucus catota	--	DMSO (5~7%)	5~8-day-old	20 C/min -100 LN	Rapid	Bajaj (1)
Atropa belladonna	--	DMSO (5%)	8-day-old	20 C/min -40 or -100 LN	Rapid	Nag and Street (9)
Acer pseudo-	--	DMSO (10%) plus Glucose (5%)	5-day-old	Slow freezing -30 to -50 LN	Rapid	Sugawara and Sakai (1974)
Ruta graveolens	--	DMSO (7%) plus Glucose (5%)	7-day-old	Slow freezing -40 to -50 LN	Rapid	Sakai (1974)
Nicotiana tabacum (Haploid)		DMSO (5 7%)	5~8-day-old	20 C/min -100 LN	Rapid	Bajaj (1)

356

	4° C, 3 days hardening	DMSO (5%)	Shoot apices slices	50° C/min	Rapid	Seibert and Wetherbee (25)
MERISTEMATIC TISSUES						
Dianthus caryophyllus	4° C, 3 days hardening		Shoot apices slices		Rapid	Seibert and Wetherbee (25)
Malus pumila (McIntosh red)	-3° C, 20 days hardening	--	Winter bud	Slow freezing -30 to -40 LN	Slow	Sakai and Nishiyama (24)
Very hardy twigs	-3° C, 20 days hardening	--	--	Slow freezing -30 LN	Slow	Sakai (15)
TISSUE CALLUS						
Machantia Polymorphoa (moss)	0° C, 33 days (hardening)	DMSO (5%) plus Glucose (10%)	17-day-old	Slow freezing -20 to -40 LN	Rapid	Noomi and Takeuchi (1976)
Populus gelrica	10° C, 5 days; 0° C, 20 days (hardening)	--	10-day-old	Slow freezing -20 to -40 LN	Slow	Sakai and Sugawara (22)
Sambucus racemosa	10° C, 5 days; 0° C, 33 days; 10% sugar medium (hardening)	--	10-day-old	Slow freezing -30 LN	Slow	Yoshida (1977)

a -40 LN: Indicated cells prefrozen at -40 C prior to immersion in liquid nitrogen.

REFERENCES

1. Bajaj, Y. P. S., *Physiol. Plant. 37*, 263 (1976).
2. Bennett, E., *in* "Plant Induction and Genetic Conservation: Genecological Urgent Problems," p. 113. Scottisch Plant Breeding Station, William Blackwood and Sons Ltd., Edinburgh (1965).
3. Ericksson, T., *Physiol. Plant. 18*, 976 (1965).
4. Finkle, B. J., Sugawara, Y., and Sakai, A., *Plant Physiol. (Suppl.) 56*, 80 (1975).
5. Lamport, D. T. T., *Exp. Cell. Res. 33*, 195 (1964).
6. Latta, R., *Can. J. Bot. 49*, 1253 (1971).
7. Luyet, B. J., *Biodynamica 1*, 1 (1937).
8. Nag, K. K., and Street, H. E., *Nature 245*, 270 (1973).
9. Nag, K. K., and Street, H. E., *Physiol. Plant. 34*, 254 (1975).
10. Nag, K. K., and Street, H. E., *Physiol. Plant. 34*, 261 (1975).
11. Ōtsuka, K., *Low Temp. Sci., B. 30*, 33 (1972).
12. Pomeroy, M. K., and Siminovitch, D., *Can. J. Bot. 49*, 787 (1971).
13. Quatrano, R. S., *Plant Physiol. 43*, 2057 (1968).
14. Sakai, A., *Nature 185*, 393 (1960).
15. Sakai, A., *Plant Physiol. 40*, 882 (1965).
16. Sakai, A., *Plant Physiol. 41*, 1050 (1966).
17. Sakai, A., and Ōtsuka, K., *Plant Physiol. 42*, 1686 (1967).
18. Sakai, A., and Yoshida, S., *Plant Physiol. 42*, 1695 (1967).
19. Sakai, A., and Yoshida, S., *Cryobiology 5*, 160 (1968).
20. Sakai, A., *Cryobiology 8*, 225 (1971).
21. Sakai, A., and Ōtsuka, K., *Plant and Cell Physiol. 13*, 1129 (1972).
22. Sakai, A., and Sugawara, Y., *Plant and Cell Physiol. 14*, 1201 (1973).
23. Sakai, A., and Noshiro, ., *in* "Crop Genetic Resources for Today and Tomorrow" (O. H. Frankel and J. G. Hawkes, eds.), p. 317. Cambridge Univ. Press, London (1975).
24. Sakai, A., and Nishiyama, Y., *HortScience*, in press (1978).
25. Seibert, M., and Wetherbee, P. J., *Plant Physiol. 59*, 1043 (1977).
26. Steponkus, P. L., and Lanphear, F. O., *Plant Physiol. 42*, 1423 (1967).
27. Sutton-Jones, B., and Street, H. E., *J. Exp. Bot. 19*, 114 (1968).

28. Torrey, J. G., *Physiol. Plant.* *20*, 265 (1967).
29. Withers, L. A., and Street, H. E., *Physiol. Plant.* *39*, 171 (1977).

Plant Cold Hardiness and Freezing Stress

ULTRACOLD PRESERVATION OF SEED GERMPLASM

P. C. Stanwood
L. N. Bass

National Seed Storage Laboratory
USDA-ARS
Fort Collins, Colorado

I. INTRODUCTION

Preservation of plant germplasm is not new. From the beginning of agriculture, plant materials have been saved from
year to year for planting the next year's crop. As cropping
systems advanced and man became more mobile, it became necessary to store plant germplasm longer than from one growing
season to the next. Within the past century, plant scientists have recognized the necessity of preserving plant germplasm not only for planting the next crop but also to retain
the broad genetic base needed by future generations of plant
breeders. Germplasm preservation makes gene complexes
available that would otherwise be lost. To preserve the
broad gene base needed to meet all future demands, a wide
range of plant materials must be preserved indefinitely.
Techniques are available for extending the storage life of
seeds, and a program of preserving plant germplasm as seeds
is well established. However, a system is just being initiated for preservation of materials that must be propagated
vegetatively.
World-wide introduction and utilization of improved
cultivars and new cultural practices (1, 4, 14, 22) has developed a universal need to collect and preserve indigenous
germplasm from the centers of origin of crop plants. Aware-
ness of this need has prompted many countries to develop systems for the exploration, importation, multiplication,

evaluation, distribution, and preservation of these vanishing plant resources. Skrdla (19) has discussed the U.S. Plant Introduction System in detail.

A majority of the plant materials in these systems are preserved as seeds. Consequently, this paper is concerned with the long-term preservation of seeds, and it concentrates on the use of ultracold (-100° C and below) storage in general and specifically on the use of liquid nitrogen (LN_2).

II.   SOURCES OF SEED GERMPLASM LOSS

The depletion of the seed supply of an accession held in long-term storage repository can be attributed to three factors: 1) germination testing; 2) loss of viability; and 3) distribution.

Testing for viability usually is the primary source of accession depletion. For long-term preservation, viability testing is necessary to identify accessions that are in danger of being lost because of low germination. Over the life of a given stored seed sample, viability testing can amount to 20 to 40% of the sample depletion, depending upon the frequency of testing and the stored sample size.

Seed distribution usually is not as significant as viability testing in reducing the quantity of stored seeds. At the National Seed Storage Laboratory, Fort Collins, Colorado, it is not uncommon to distribute only 20 thousand seeds annually, compared to the nearly three million seeds used in viability testing.

The most variable factor affecting seed preservation is the rate of viability decline. Viability decline can be measured in hours for orchids or decades for the hard-seeded legumes (5). Obviously, the slower the rate of decline in viability, the longer an accession can be held without the need for a seed increase.

Long-term seed preservation must emphasize a reduction in the rate of viability loss during storage. If viability drops rapidly, seeds must be tested frequently, which rapidly depletes the existing seed stocks. With rapid depletion, an accession must be replenished more often, which introduces additional problems associated with field growing and handling. The risk of genetic changes is high during the regrowing process; therefore, seed increase should be reduced to a minimum (17).

III. SEED VIABILITY DECLINATION

Seed moisture content and storage temperature have been identified as the two most important controllable factors affecting the longevity of seeds (5, 6, 15, 20). Numerous experiments using many kinds of agricultural seeds stored at temperatures from -18 to 40° C have suggested that optimum seed moisture content ranges from 4 to 7% for orthodox (easily preserved) species. These seed moistures are easily obtained without damage to the seeds.

A class of seeds called "recalcitrant types" cannot be easily dried to a low moisture content, and they are thus difficult to preserve. Presumably, their high seed moisture content accelerates viability loss during storage.

Assuming that the optimum seed moisture content for orthodox seeds has been determined and that recalcitrant seeds cannot be dehydrated without damage, modifications in storage temperature appear to offer the greatest potential for extending seed viability. According to Roberts (16), the colder the storage temperature, the longer the seeds remain viable. Therefore, ultracold storage should considerably lengthen seed storage life.

IV. ULTRACOLD PRESERVATION

There are many examples of long-term preservation of biological materials at subfreezing temperatures, particularly at LN_2 temperature (-196° C). LaSalle (7) maintained cells of micro-organisms cryogenically for 10 years with 96% viability. Dietz (3) maintained viability of unicellular plant material for up to 11 years in LN_2 without problems. Cryogenic preservation of bull semen with LN_2 is a common practice throughout the world, and Picket and Berndtson (12) indicated that loss of viability was nonsignificant for several years. Other examples of preservation of micro-organisms in LN_2 are abundant in the literature.

There is little information on the storage of seeds of higher plants for long periods at ultracold temperatures. Therefore, it is not known whether seeds would be damaged by long exposure to LN_2 temperature. However, the examples with other biological materials cited above suggest that loss of seed viability would be minimal at LN_2 temperature.

Depletion of food reserves (starvation of meristematic cells), accumulation of toxic compounds, breakdown in mechanisms for triggering germination, inability of ribosomes to disassociate, enzyme degradation and inactivation, lipid autooxidation, formation and activation of hydrolytic enzymes, degradation of functional structures, and genetic degradation have all been suggested as possible causes of seed deterioration (2, 6, 16). Most of these are biochemical in nature, and their activity should decrease with reduced temperature (8). Presumably, most of these reactions would be stopped or greatly reduced at or below LN_2 temperature.

One of the greatest problems of storing biological material in LN_2 is to cool the material to the ultracold temperature and return it to ambient temperature without damage (11). The length of time the material is held in LN_2 does not appear to be a problem biologically. Cryobiological literature suggests that during the cooling and warming process, the more hydrated a biological material is, the more susceptible it is to injury. Orthodox seeds would, in general, have an advantage over other biological materials in that they would easily be dried to a relatively low moisture content (4 to 7%), thus effectively eliminating many moisture-associated problems during the freezing and thawing processes. Conversely, recalcitrant seeds probably would have the same freezing and thawing problems as other cryogenically stored biological materials and would thus be more difficult to handle than orthodox seeds.

Thus, the biological feasibility of storing orthodox and recalcitrant seeds at LN_2 temperature, except for unknown long-term implications, is reduced to determining which species can survive LN_2 freezing temperature and return to ambient temperature without damage. Lipman and Lewis (9) stored seeds of sugarcane, spinach, cucumber, sugar beet, buckwheat, barley, purple vetch, oat, onion, mustard, and *Melilotus* ssp. for 30 days and seeds of pea, corn, squash, alfalfa, and sunflower for 60 days at -196° C. After rewarming these seeds, laboratory and greenhouse evaluation indicated no decline in seed germination or vigor as compared to germination and vigor of the controls held at room temperature. Lipman (10) subjected seeds of vetch, wheat, barley, tobacco, flax, buckwheat, spinach, milo maize, and *Melilotus* ssp. to 1.35 K (-272° C) for several hours. He again found no differences between ultracooled seeds and controls when the seeds were germinated and grown in a greenhouse. No information concerning recalcitrant seeds being exposed to LN_2 was found.

Seeds of rice (*Oryza sativa* L. cv. Horya), winter wheat (*Triticum aestivum* L. cv. Mukakomugi), soybean (*Glycine max* cv. Horei), alfalfa (*Medicago sativa* L. cv. DuPuits), and Italian ryegrass (*Lolium multiflorum* Lam. cv. Billion) at moistures from 5 to 26% (dry-weight basis) were exposed to LN_2 and then rewarmed at different rates (18). Generally, seeds below 8% in moisture were not damaged by the LN_2

TABLE I. *Kinds of Seeds for which Germination was not Adversely Affected by Storage in Liquid Nitrogen for Periods up to 6 Months (21)*

Alfalfa (Medicago sativa L.)
Barley (Hordeum vulgare L.)
Bean (Phaseolus vulgaris L.)
Bluegrass (Poa pratensis L.)
Cabbage (Brassica oleracea
  var. capitata L.)
Cantaloupe (Cucumis melo L.)
Carrot (Daucus carota L.)
Castorbean (Ricinis communis
  L.)
Clover (red) (Trifolium
  pratense L.)
Corn (Zea mays L.)
Cotton (Gossypium hirsutum
  L.)
Cowpea (Vigna unguiculata
  {L.} Walp. Subsp.
  unguiculata)
Cucumber (Cucumis sativus L.)
Datura (Datura metel L.)
Eggplant (Solanum melongena
  L.)
Fescue (red) (Festuca rubra
  L.)
Fir (White) (Abies concolor
  {Gord. & Glend.} Lindl. ex
  Hildebr.)
Lettuce (Lactuca sativa L.)
Millet (Pearl) (Pennisetum
  americanum {L.} Leeke)
Nasturtium (Tropaeolum majus
  L.)

Oat (Avena sativa L.)
Onion (Allium cepa L.)
Orchardgrass (Dactylis
  glomerata L.)
Pea (Pisum sativum L.)
Peanut (Arachis hypogaea L.)
Pepper (Capsicum annuum L.)
Pine (Ponderosa) (Pinus
  ponderosa Dougl. ex P. & C
  Lawson)
Poppy (Papaver somniferum L.)
Rice (Oryza sativa L.)
Rye (Secale cereale L.)
Safflower (Carthamus
  tinctorius L.)
Sorghum (Sorghum bicolor {L.}
  Moench)
Soybean (Glycine max {L.} Merr.)
Squash (Cucurbita maxima Duch.)
Sudangrass (Sorghum sudanense
  {Piper} Stapf.)
Sugar Beet (Beta vulgaris L.)
Timothy (Phleum pratense L.)
Tobacco (Nicotiana tabacum L.)
Tomato (Lycopersicon esculentum
  Mill.)
Trefoil (Lotus corniculatus L.)
Vetch (Vicia ssp.)
Watermelon (Citrullus lanatus
  {Thunb.} Matsum. and Nakai)

temperature and the rate of rewarming was not important in their survival. A notable exception was soybean, for which the optimum seed moisture for slow (0.5° C/sec) rewarming was 14.6% and for fast (9.0° C/sec) rewarming was 15.8%.

Rajewsky (13), working with *Arabidopsis* ssp., noticed no detrimental effect on seed germination from exposure to LN_2 temperature. He also noted that seeds exposed to 400,000 of X-radiation at LN_2 temperature germinated at 58%, whereas those irradiated at room temperature had only 39% germination. The implication was that lethal radiation damage was substantially reduced at the LN_2 temperature.

Stanwood (21) screened commonly cultivated plants for susceptibility of seeds to damage from exposure to LN_2 temperature. At moistures below 13% (wet-weight basis) the 42 species tested showed no damage to germination (Table I).

V.   PHYSICAL CONSIDERATIONS--LIQUID NITROGEN VS. MECHANICAL REFRIGERATION

Conventional refrigeration of seed storage facilities has serious drawbacks in terms of the reliability of continuous operation. Compressors, refrigerant lines, circulating fans, and cooling water systems all require continuous maintenance and can have high failure rates, resulting in discontinuous operation. Many of the operational aspects of mechanical refrigeration are eliminated with LN_2 storage systems.

For example, some biological LN_2 storage containers can hold up to 50,000 1.2-ml vials at $-196°$ C for up to 130 days without recharging. These devices are essentially passive and require no maintenance except periodic refilling with LN_2. Furthermore, the failure rate of LN_2 storage containers is very low. When it does occur, usually through loss of vacuum, it does so gradually and is easily detected and corrected. Therefore, LN_2 storage can show substantial advantage over mechanical refrigeration because of this high reliability factor.

Estimates of relative costs of conventional storage at $-20°$ C and LN_2 storage are difficult to compare. A reasonable cost estimate for conventional storage (including maintenance, electricity, and equipment deterioration) of sealed cans (75 cm^3) at $-20°$ C may range from $0.16 to $0.50 per sample-year. For seeds the size of sorghum (30 seeds/gram) and smaller, the comparable LN_2 storage cost for an equal number of seeds may range from $0.03 to $0.30 per sample-year. It should be noted that any storage cost figure must be used with caution, because it depends upon seed size and the number of seeds being stored. The effect of seed size on maintenance cost is illustrated by the cost estimates for various agricultural crops given in Table 2.

TABLE II.   *Estimated cost of liquid nitrogen storage for at least 1,000 seeds of various agricultural crops[a]*

Species	Minimum number of seed	Estimated LN_2 storage cost per year[b]
Alfalfa	2400	$ 0.03
Barley	1000	0.48
Bean	1000	1.50
Bluegrass	4500	0.03
Carrot	2500	0.03
Clover	2600	0.03
Corn	1000	1.41
Cucumber	1000	0.36
Eggplant	1000	0.03
Fescue	1500	0.03
Flax	1000	0.06
Impatiens	4200	0.03
Oat	1000	0.45
Onion	1600	0.06
Orchardgrass	1500	0.03
Rice	1000	0.30
Rye	1000	0.15
Ryegrass	1000	0.18
Safflower	1000	0.33
Sorghum	1000	0.30
Soybean	1000	1.26
Sugar Beet	1000	0.60
Tobacco	4500	0.03
Tomato	1000	0.09
Wheat	1000	0.21

[a]*Based on LN_2 at $0.40/1.*
[b]*Cost includes LN_2, equipment depreciation, maintenance, and inventory system.*

VI.   CRYOGENIC SEED PRESERVATION INVESTIGATIONS

Several questions must be answered before LN$_2$ storage can be used for routine long-term preservation of seeds. The most important questions are: 1) Can seeds of a given plant species withstand being subjected to LN$_2$ temperature without loss of viability? 2) Does seed moisture content at the time of subjection to LN$_2$ temperature affect seed response? 3) How long will seeds remain viable in LN$_2$? 4) How do costs of LN$_2$ storage compare with costs of conventional refrigerated storage?

VII.   RESEARCH OBJECTIVES

Research projects are being developed:

A.   *To Determine the Effect of Various Periods of Exposure to LN$_2$ Temperature on Seed Viability, Seedling Vigor, and Genetic Stability*

Seeds of various moisture levels stored at various subfreezing temperatures, including LN$_2$ temperature, will be tested periodically for germination and vigor; if possible, seedlings will be grown to maturity to determine any genetic changes that may have occurred during storage.

B.   *To Determine Appropriate Methods for Freezing and Thawing Recalcitrant Type Seeds*

Various procedures and cryoprotectants will be investigated to determine the maximum allowable dessication for such seeds before they are subjected to LN$_2$ and other low-temperature storage conditions.

C.   *To Determine Variations in the Rate of Viability Decline between Cultivars of a Given Crop Species and between Samples of the Same Cultivar when Stored under the Same Condition*

Seeds from two or more sources of the same cultivar will be stored under the same conditions to determine whether real differences in viability loss exist between sources of the same cultivar and between lots of the same source. Seeds having a range of moisture contents will be used.

D. *To Determine, If Possible, Any Source of Damage to Seeds Stored at LN$_2$ Temperature for Both Short- and Long-Term Periods*

Evaluations will be made of possible membrane damage, biochemical changes, and chromosomal mutations.

E. *To Evaluate the Efficacy of Various LN$_2$ Containers, Associated Equipment, and Inventory Systems as They Relate to Operational Costs for Use in Routine Germplasm Preservation*

Undoubtedly some LN$_2$ storage containers operate more efficiently than others. Some associated equipment and inventory control systems may be better suited to germplasm preservation than others. Studies will be conducted to identify the most efficient and least costly LN$_2$ storage containers, the most efficient inventory control systems, and the most suitable associated equipment.

VIII. SUMMARY

Preservation of valuable plant germplasm, as seeds, has received considerable emphasis in recent years. Long-term storage techniques for many seeds are available; however, viability loss still occurs, necessitating the allocation of scarce resources for monitoring viability and for periodic growing of new seed crops. Additionally, the seed increase process includes a high risk of contamination and of changes in the genetic composition of the gene pool contained in the original stock. Reducing storage temperature into the ultracold range apparently can extend seed life considerably, thus reducing maintenance requirements of individual accessions and assuring a better quality of seeds, both genetically and physiologically.

REFERENCES

1. Anonymous, "Genetic Vulnerability of Major Crops." Nat. Acad. Sci., Washington, DC (1972).
2. Copeland, L. O., *in* "Principles of Seed Science and Technology," p. 201. Burges Press (1976).
3. Dietz, A., *in* "Nitrogen Preservation of Stock Cultures of Unicellular and Filamentous Microorganisms" (A. P. Rinfret, and B. LaSalle, eds.), p. 22. Round Table Conference on the Cryogenic Preservation of Cell Cultures, Nat. Acad. Sci. (1975).
4. Harlan, J. R., *Science 188*, 618 (1975).
5. Harrington, J. F., *in* "Seed Biology," Vol. III (T. T. Kozlowski, ed.), p. 145 (1972).
6. James, E., *Adv. Agron. 19*, 87 (1967).
7. LaSalle, B., Introduction *in* "Cryogenic Preservation of Cell Cultures," Round Table Conference on the Cryogenic Preservation of Cell Cultures, Nat. Acad. Sci. (1975).
8. Lehninger, A. L., *in* "Biochemistry," p. 147. Worth Publishing, Inc. (1970).
9. Lipman, C. B., and Lewis, G. N., *Plant Physiol. 9*, 392 (1934).
10. Lipman, C. B., *Plant Physiol. 11*, 201 (1936).
11. Merryman, H. T., *in* "Cryobiology" (H. T. Merryman, ed.), p. 40. Academic Press (1966).
12. Pickett, B. W., and Berndtson, W. E., "Principles and Techniques of Spermatozoan Preservation by Freezing," unpublished ch. 16. Animal Reproduction Laboratory, Colorado State University, Fort Collins, Colorado (1975).
13. Rajewsky, B., *British J. Radiology 25*, 550 (1952).
14. Reitz, L. P., *in* "Agronomic Research for Food" (F. L. Patterson, ed.), p. 85. Am. Soc. of Agron. Spec. Publ. 26 (1976).
15. Roberts, E. H., *Annals of Botany 25*, 381 (1961).
16. Roberts, E. H., *in* "Viability of Seeds" (E. H. Roberts, ed.), pp. 14, 253. Syracuse University Press (1972).
17. Roos, E. E., *Personal Communication*. National Seed Storage Laboratory, Colorado State University, Fort Collins, Colorado (1977).
18. Sakai, A., and Noshiro, M., *in* "Crop Genetic Resources for Today and Tomorrow" (O. H. Frankel, and J. G. Hawkes, eds.), p. 217. International Biological Program Publication 2 (1975).
19. Skrdla, W. H., *HortSci. 10*, 570 (1975).
20. Stanwood, P. C., and Bass. L. N., *Agron. Abstracts*, 96 (1976).

21. Stanwood, P. C., *Unpublished Data*. National Seed
     Storage Laboratory, Colorado State University, Fort
     Collins, Colorado (1977).
22. Wilkes, G. H., *Bull. Atomic Scientists*, Feb., 8 (1977).

EFFECTS OF COMBINATIONS OF CRYOTPROTECTANTS
ON THE FREEZING SURVIVAL
OF SUGARCANE CULTURED CELLS

B. J. Finkle
J. M. Ulrich

Western Regional Research Center
U. S. Department of Agriculture
Berkeley, California

ABSTRACT

Suspension cultures of sugarcane cells (*Saccharum* sp.)
were treated with glucose, dimethylsulfoxide, ethylene glycol,
and polyethylene glycol, at various concentrations, alone and
in combinations, before freezing to various temperatures.
Viability of the thawed cells was assessed both by triphenyltetrazolium chloride reduction as determined colorimetrically,
and by cell growth. Mixtures of glucose and dimethylsulfoxide (optimum at about 1.9M total concentration. 1:3 molar
ratio, respectively) were found to give up to a 100% increase
in cryoprotection, compared with the effect of either compound alone at 1.9M concentration. Much of this effect was
from a decrease in toxicity from the mixing of cryoprotectants. Polyethylene glycol alone had no cryoprotective effect
but when added as a 10% solution in addition to glucose and
dimethylsulfoxide, enhanced the cryoprotection. Combinations
with ethylene glycol were less effective and more toxic. Use
of the glucose-dimethylsulfoxide combination allowed freezing
of these cells to -23° C with little decrease in viability,
or to -40° C still with the capability for delayed growth.

I.   INTRODUCTION: SIGNIFICANCE AND BENEFIT OF STABLE, LONG-
     TERM PRESERVATION OF PLANT TISSUES

     One hardly needs to convince this audience about the
benefits of being able to preserve plant tissues for future
plant breeding purposes. The requirements are that the tissues be completely stable for long periods of time and maintain a capability for being regenerated into whole plants. A
promising way to do this would be by deep freezing.
     A preponderance of the cultivars in world agriculture
may soon disappear from the earth if reliable, inexpensive,
long range storage methods are not put into practice very
soon. For instance, low-yield cultivars (but which to the
plant breeder may have valuable character traits) are under-
standably losing out to high-yield varieties throughout the
world (4, 27).
     Dr. Stanwood will be describing how, with many crops,
seeds can be relied upon for preservation. But preserving
those many species where vegetative propagation is the method
of choice, presents perhaps an even greater challenge.
     Successful methods for viable freezing could be an answer
to this problem of vegetative tissue storage.

II. ARTIFICIAL IMPROVEMENTS IN FREEZE-TOLERANCE OF PLANT
     TISSUES

     In attempting to freeze-preserve plant tissue, we wish we
knew better (and could imitate) how plants protect themselves
in surviving cold. (An extreme example is the -253° C temperature of liquid hydrogen to which Tumanov et $al.$, subjected
their twigs of birch and black currant {26}). Lacking this
comprehension, plant cryobiologists have tried various methods
and rates of freezing and thawing, and the addition of chemical substances to preserve plants in the living state through
the freezing and thawing process.
     In 1912, the plant physiologist N. A. Maximov took the
first perceptive steps (8). He added an extract from freeze-
hardy cabbage (and, later, synthetic mixtures) to freeze-
tender, in fact tropical, $Tradescantia$ $discolor$ leaves; they
survived freezing. Much more recently, Olien, with polysaccharides from cereal cell walls (16), and Heber, with protein from spinach leaves (5), have used purified plant fractions in the same way. Many useful leads towards freezing
plant tissues without excessive damage have come from still
other subjects of study: animal tissues such as blood cells

and sperm, insects, and microorganisms, including yeast cells (9). The presence of cryoprotective compounds has been found essential or helpful when freezing to low temperatures.

A. *Effects of Cryoprotectants*

You will find many interpretations of how cryoprotective substances protect against freezing damage. The very large *variety* of interpretations, itself, indicates that we really don't understand the protective effect, partly because we don't know the primary site of freezing damage.

The many types of cryoprotective compounds can be very different in structure. Examples are: polyhydroxy compounds (as monomers, and in varying degrees of polymerization), dimethylsulfoxide, polyvinylpyrrolidone, and amines, to mention but a few. The compounds have *in common* a high degree of chemical polarity and great attraction for water molecules.

It is believed that the primary site of freezing injury may be at some part of the very complex membranes of cells, as discussed in previous papers. The membrane model proposed by Singer and Nicolson (see Figure 4 in the article by Palta and Li) exhibits a fluid mosaic of constituent structural compounds that may be individually accessible to reactants, or to injury. Whether the primary freezing injury is biochemical (from modification of membrane constituents) or physical (from osmotic water flow) is, unfortunately, still not understood.

There are at least three locations where a cryoprotectant may act to avoid freezing damage: on the exterior solution, on the membrane itself and on the interior solution. In the exterior solution there may be lowering of the freezing point, increase in viscosity, etc., depending on the cryoprotectant used. With respect to the membrane, there can occur modification of its molecular configuration and/or its water structure, with consequent effects on its physical strength, solute selectivity or leakiness, etc. A cross-sectional drawing of the fluid mosaic membrane model (Figure 1) emphasizes the many sites with which a cryoprotectant could react to stabilize the intra- and inter-molecular arrangements of structural compounds and of water association in the membrane. In the interior solution (really, in turn, a very complex suspension of membrane-bound suspensions, gels and solutions) there may, again, be marked effects on the freezing point and viscosity—especially if the cryoprotectant enters the cell—and also osmotic changes in cell size and cell sol concentration, as well as effects on enzyme systems.

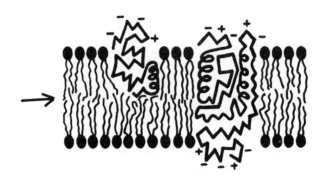

FIGURE 1. *The lipid-globular protein mosaic model of membrane structure: schematic cross-sectional view. The phospholipids are arranged as a discontinuous bilayer with their ionic and polar heads in contact with water. The integral proteins, with the heavy lines representing the folded polypeptide chains, are shown as globular molecules partially embedded in, and partially protruding from, the membrane. From Singer and Nicolson (22).*

B. *Effective Compounds and Their Interactions*

Of the many effective compounds, of which a selection of low molecular weight compounds is shown in Table I, some penetrate quickly, others more slowly or not at all. Large polymers, not shown, such as polyvinylpyrrolidone, dextran, and hydroxyethylstarch--compounds that are too large to enter the cell--are also effective cryoprotectants (11, 18). How do these many compounds act? Could *all* of these various compounds act similarly? Could they affect only the exterior solution or the membrane, and not the interior? This is rather doubtful, judging from a great deal of work that has been done (12, 25). What is more, some compounds are effective even after freezing (13, 19). Enhancement of repair processes may be involved here, in addition to an avoidance of delayed freezing damage (14, 19). Pribor, while studying the effects of polyvinylpyrrolidone and a dextran, of similar

TABLE I. *Comparative Protective Action of Various Solutes on Freezing Injury*[a]

-7°C: Slight Protection	-10°C: Medial Protection	-15°C: Medial Protection	-20°C: High Protection	-25°C: Highest Protection
Raffinose	D-Glucuronic acid	Ascorbic acid	Sucrose	Glucose
Mannitol	Hydroxyproline	Acetoamide	Sorbitol	β-Methyl-D-glycoside
Glucosamine	Glutamic acid	Methyl aceto-amide	Lactate	Xylose
Galacturonic acid	Methyl formamide	Dimethyl forma-mide	Pantothenic acid	Rhamnose
Dimethyl aceto-amide	Formamide	Betaine aldehyde	Methanol	Erythritol
Acetyl glycine	Dimethyl urea	(Cabbage sap)	Ethylene glycol	Xylitol
α-Aminobutyrate	N-Methyl urea	(Pseudoacacia: bark sap)	Glyceraldehyde	Glycerol
α-Glycerophos-phate	Dimethyl thetine		Dimethyl sulfox-ide	Betaine
β-Glycerophos-phate	Acetyl choline		1,3-Dimethyl urea	Trimethylamine N-oxide
Water (N.C.): -5° C	N-Methyl-pyrrolidone		Methyl glycine	
	Triethylene glycol		Dimethyl glycine	
			Methionine sulf-oxide	
			Lactoamide	

[a]Material: cabbage cells, 0.5 M aqueous solutions. (From Sakai and Yoshida {9}).

377

molecular weights (40,000 daltons), on the freezing of red
blood cells, concluded that the compounds behave similarly in
several respects, but they show, also, nine important differences in cryoprotective properties (18). In that paper, and
in another with the compound-pair, glycerol and dimethylsulfoxide (17), he stressed the probability of a multifactor biological interaction of reactants.

It would seem to follow then, that since all the effective
compounds do not act similarly, certain combinations may be
more effective than single agents, used alone. This would
seem particularly true because of still another effect of such
compounds, easily dismissed: in addition to their cryoprotective effect, many are toxic (11). This is, possibly, an
effect that is independent of their cryoprotective action.
Quite possibly, then, the toxic effect from a multiple-compo-
nent cryoprotectant could be *less than* the arithmetic sum of
the individual compound toxicities, particularly because so
many complex interactions are taking place; and possibly, too,
cryoprotection itself could be *more than* additive (synergistic) because so many complex interactions are taking place.
To look at this in another way, increasing the concentration
of a given cryoprotectant by a certain amount could bring it
into a highly toxic range, whereas superimposing a similar
amount of a cryoprotective substance of different type might
not produce a similarly large toxic effect.

Using model systems, Ruwart *et al.* were able to test the
interactive effects of cryoprotectants, or their combinations,
by the manner in which they affected the fluorescence of serum
albumin or tryptophane (20). Some combinations were strictly
additive, others quite different from this, indicating molecular interactions.

In the few instances where combinations of cryoprotectants
have been tested with living cells, there have been some positive and some negative results. McGann and Kruuv, adding to
Chinese hamster lung fibroblasts solutions made 0.5 M in glycerol and DMSO, came to the bold negative conclusion that the
search for "complementary" cryoprotective action "now appears
to be fruitless since two of the most effective agents show
relatively flat age-survival responses to the freeze-thaw
stress" and show no significant additional protection "beyond
that conferred by....one of these agents acting alone" (10).
There have been but few attempts to examine combinations of
cryoprotective compounds in a systematic and quantitative manner. Djerassi and Roy did so, studying the effects of dimethylsulfoxide and several sugars, on the freezing of platelets
from rat blood (2). Figure 2 shows that addition of one compound, alone, gave little or no cryoprotective effect. But
combining dimethylsulfoxide with a sugar, like sucrose, was
substantially protective, and even more so (protection to

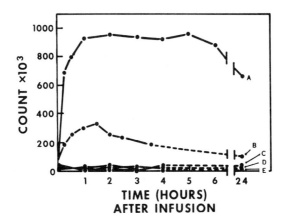

*FIGURE 2. In vitro survival of platelets (average count)
stored 24 hours at -79° C. A: Control, non-frozen platelets.
B: Platelets frozen in 10% sucrose and 10% dimethylsulfoxide.
C, D, E: Platelets frozen in 10% sucrose, in plasma only, or
in 10% dimethylsulfoxide. From Djerassi and Roy (2).*

-196° C) with the sugars xylose or glucose than with the sucrose combination illustrated. Diamond reported a similar ef-
fect--that a combination of DMSO and glucose is required to
preserve the protozoan, *Entamoeba invadens*, to -170° C (1).

Closer to our own work with plant tissue cultures, Latta
(7) found improved cryoprotection of *Ipomoea* cells from a
ternary mixture of DMSO-glycerol-sucrose (2.5%-2.5%-3% w/v).
Sugawara and Sakai selected DMSO-glucose (12%-5%) to freeze
sycamore cell suspensions to -196° C (24) while, with the same
coworkers, Finkle used a combination of glucose-DMSO-ethylene
glycol (4%-3%-2.5%) to increase the survival of tobacco and
carrot cells (3). On the other hand, when Nag and Street used
a DMSO-glycerol combination (5%-10%) for sycamore, belladonna,
and carrot cells, the cryoprotective effect was no greater
than from adding one of the compounds alone (15). More systematic work needs to be done to establish a more general understanding about the action of cryoprotective combinations.

III. USE OF COMBINATIONS OF CRYOPROTECTANTS ON SUGARCANE
      CELLS

In attempting a more systematic approach we used glucose
(G), dimethylsulfoxide (D), ethylene glycol (E), and polyethylene glycol (P), at varying concentrations, to establish a

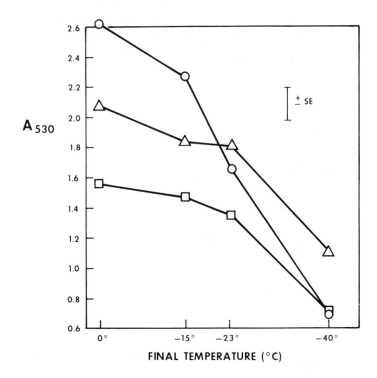

FIGURE 3. Cryoprotectant concentration effects on cane cell temperature-viability curves. (o) 4%-6% GD; (△) 8%-12% GD; (□) 12%-18% GD. Error bar indicates ± SE of the mean (from five sets of duplicate samples for each point), except for the -15° C 4%-6% GD (o) point (from two to six sets of duplicate samples, having a 63% greater SE).

range of freezing protection by each of them while freezing sugarcane cells; we then used *combinations* of the compounds. The experiments also included cells that were treated with protective compounds, but kept at 0° C (unfrozen) as a measure of reagent toxicity.

A. Methods

The sugarcane cells (*Saccharum* sp. hybrid cv. H50-7209), were obtained from P. Moore of the USDA Experiment Station in Hawaii. We grew the cells in medium containing coconut water (6), for 6 to 9 days. Before freezing we added the cryoprotectant, gradually, at 0° and then froze the cells in graduated centrifuge tubes through a stepwise cooling regime (or kept them at 0° as a test of toxicity). Thawing was carried out by swirling in a 40° bath, just to the point of melting, followed by two washings. Then cell viability was tested by the Steponkus and Lanphear triphenyltetrazolium chloride (TTC) method (23), or else the cells were transferred aseptically into tubes containing culture medium and placed on shakers to determine growth, measured turbidimetrically. The TTC viability method measures the activity of a dehydrogenase enzyme system and is quantitated through measuring the amount of a red TTC-reduction product (formazan) extracted into hot alcohol. The absorption of the extract is read spectrophotometrically at 530 μm.

B. *Results*

At first we attempted to optimize the cryoprotective action of G and D, alone, and then in combination. The best ratio of G and D was found to be about 8%:12% G:D. The effects of adding multiples of 4%-6% GD to the cane cells is shown in Figure 3. At the lowest concentration, 4%-6% GD, there is a high TTC viability-value at 0° but relatively little cryoprotection at low freezing temperatures. With a higher cryoprotectant concentration, 8%-12% GD, even though it is more toxic, there is cryoprotection to a lower temperature, almost as a plateau to -23°, and reasonably high values beyond (A_{530} more than one-tenth the value--4.2 to 5.0--of untreated control cells), to -40°. The 8%-12% GD cryoprotectant concentration has, then extended the cryoprotective range. A further increase in concentration to 12%-18% GD gives a similar shape of freezing curve, but in this case the further-increased reagent toxicity caused a diminished viability-value throughout the temperature range, i.e., the 12%-18% GD, of 3 M total concentration, it beyond the optimum cryoprotective range.

In further experiments using a total molarity of 1.9 M as a base concentration, we tested various ratios of the reagents totaling 1.9 M. Figure 4 shows the effects of the compounds alone, and of mixtures of these in varying ratios, added to cane cells that were then taken to various low temperatures. We see that, for the various temperatures down to -40°, there is an increase in TTC-value when a mixture of protectants is used. The broad peak of cryoprotection lies at a GD ratio of about 8%-10 1/2% in the 1.9 M total mixture. This ratio of the protective compounds represents a mixture that is dominant in the molar concentration of D, the G:D ratio being about 1:3.

A singularly interesting aspect of the 0° curve in Figure 4 is the possibility that a mixture of cryoprotectants has the effect of alleviating toxicity, as compared to the toxicity of

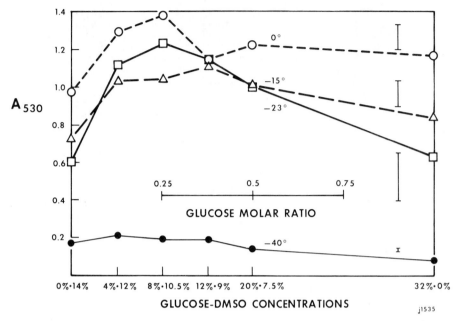

FIGURE 4. Effects on TTC-value brought about by varying G:D ratios. Total sample concentration of G + D = 1.9M. Percentage composition of the mixtures and molar ratios of the cryoprotectants are both indicated in the abscissa. Error bar indicates the maximum ± SE of the mean (from two to six sets of duplicate samples for each point) for each curve.

either reagent alone at 1.9 M. This reduced toxicity might constitute at least a portion of the improvement in viability that appears when a mixture of cryoprotective compounds is used. The data are not adequate to determine whether there is a *protective* effect from the combined mixture of cryoprotectants that goes beyond its effect in reducing reagent toxicity; but it would appear so, particularly at lower temperatures (e.g., in the -23° curve, where the viability peak values for combined protectant are twice the value for either single component). In any case, viability is improved by combining these cryoprotective agents, down to temperatures of -40° or beyond.

More significant than TTC-values is growth itself. Figure 5 shows an example of how the growth of cells is stopped by freezing, unless they are cryoprotected. This figure, and Figure 6, show the cryoprotective effects of two levels of the GD combination. In agreement with the results of Figure 3, at 0° (solid lines) 4%-6% GD is somewhat toxic to growth, and 8%-12% GD still more so. The freezing behavior is,

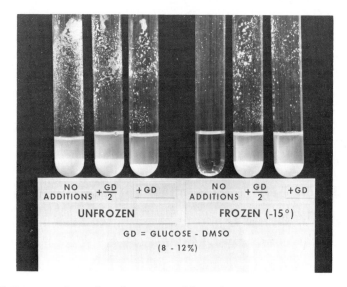

FIGURE 5.   Growth of cane cells after treatments.   Culture
tubes containing settled cells: nine days' growth.

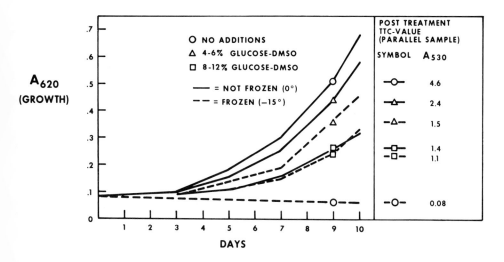

FIGURE 6.   Growth of cane cells after treatments.
Comparison with TTC-values.   Points on growth curves and TTC-
values obtained from duplicate samples, averaged.

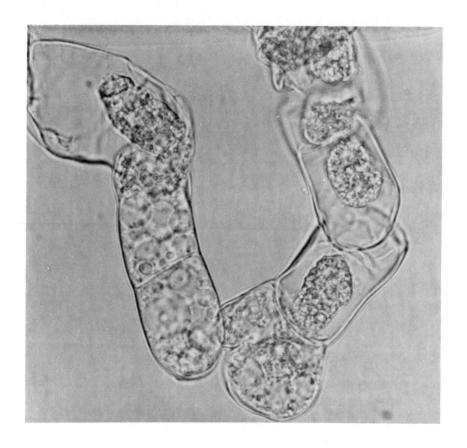

FIGURE 7. Appearance of unfrozen cells exposed to 8%-12% GD, then washed. (X675).

likewise, similar to that shown in Figure 3. In particular, the 4%-6% GD mixture allows a degree of loss of viability on freezing the cells to -15°; but when 8%-12% GD is used there is almost no further decrease in growth beyond the toxicity loss. That is, once the loss of growth from chemical toxicity is assumed, the 8%-12% GD is cryoprotective against any further loss from freezing to -15° (and also is protective to still lower temperatures). TTC-values of parallel samples run in the experiments are shown on the right of Figure 6, and indicate the same pattern of viability protection. It can be concluded, then, that the TTC assay procedure offers a rather reliable indication of the treated cells' capability for growth. Using the GD cryoprotectant combination, regrowth of these tropical cells was accomplished even when frozen to -40°.

It may be noted that under the microscope, a fraction of the cells show permanent plasmolysis (damage) after treatment with cryoprotective agents, e.g., 8%-12% GD, then washing (Figure 7). Freezing the cells to -15° or -23° appears not to further change the fraction of cells plasmolyzed. that is, those cells that were not damaged by chemical toxicity have been cryoprotected, as indicated by the data already discussed.

We also tested other cryoprotective compounds with the cane cells. Ethylene glycol was found to be a more toxic and less cryoprotective additive, on a molar basis, than glucose or DMSO, both alone and in combinations with them. Polyethylene glycol (Carbowax 6000), surprisingly, showed no toxicity to cells (Table II). It also gave no cryoprotection --a good reason to eliminate it from further trials! But, nevertheless, going ahead with other trials, when we combined a 10% solution of polyethylene glycol with 8%-10% GD, cryoprotection of the cells increased considerably. We are now incorporating this compound into our growth experiments.

TABLE II.   *Cryoprotective Effects of P and GD Combinations*

Treatment	$0°C$	$-34°C$[a]
	A_{530}	
Untreated cells	2.66	$(0.02-0.1)$[b]
+P (10%)	2.64	0.05
+GD (8%-10%)	1.55	0.14
+PGD (10%-8%-10%)	1.88	0.65

[a] Final temperature (stepwise).
[b] Range of values, from earlier experiments.

IV. CONCLUSIONS

We have seen that a combination of cryoprotectants can protect frozen cells from injury, even more than can any one of them alone at the same total molar concentration. We saw this in its most exaggerated form with polyethylene glycol alone (no effect), as against its effect when combined with glucose and DMSO (10%-8%-10% PGD), where the P further enhanced the effect of the GD combination itself. At least a part of the benefit from mixtures of cryoprotectants appears to arise from a decrease in reagent toxicity--an interactive effect whose basis one can only speculate about without present limited understanding of cryoprotective agents. However, we have cited some other studies that, too, are suggestive of such interactions (18, 20).

Having now found such synergistic effects with cane cells, we suspect that still other cryoprotectant treatments will supply additional benefits when applied in appropriate combinations. We presume, too, that aspects of what we are learning about enhanced cryoprotection can be extended to other, more hardy, species than the semi-tropical sugarcane, the goal being long-term frozen storage of the cultivars of many species. We are continuing investigations along these lines.

It seems likely that the pattern of effects being pursued may yield information as to which cryoprotective compounds act similarly (sparing action or additivity, up to a point), which act differently (including synergistic actions of protection or toxicity--an interesting nomenclature of Ruwart et al. (20) may be helpful here), and how these compounds function. Integration of these concepts with the large store of knowledge about natural freeze-hardening should stand as a grand objective.

ACKNOWLEDGEMENTS

We wish to express appreciation to P. H. Moore, of the United States Department of Agriculture, Honolulu, Hawaii, for supplying sugarcane cultures, H. Ginoza, at the same location, for suggestions about the usage of polyethylene glycol, B. E. Mackey for statistical analysis of the data, F. T. Jones for cooperation in preparing the photomicrographs, and L. Walker for diligent technical assistance.

Reference to a company and/or product named by the Department is only for purposes of information and does not imply approval or recommendation of the product to the exclusion of others which may also be suitable.

REFERENCES

1. Diamond, L. S., *Cryobiol. 1*, 95-102 (1964).
2. Djerassi, K., and Roy, A., *Blood 22*, 703-717 (1963).
3. Finkle, B. J., Sugawara, Y., and Sakai, A., *Plant Physiol. (Suppl.) 56*, Abst. 433 (1975).
4. Harlan, J. R., *Science 188*, 618-622 (1975).
5. Heber, U., *Cryobiol. 5*, 188-201 (1968).
6. Heinz, D. J., and Mee, G. W. P., *Crop Sci. 9*, 346-348 (1969).
7. Latta, R., *Can. J. Bot. 49*, 1253-1254 (1971).
8. Maximov, N. A., *Ber. Deutschen Botan. Gesel. 30*, 52-65, 293-305, 504-516 (1912).
9. Mazur, P., *Science 168*, 939-949 (1970).
10. McGann, L. E., and Kruuv, J., *Cryobiol. 14*, 503-505 (1977).
11. Meryman, H. T., *Cryobiol. 8*, 173-183 (1971).
12. Meryman, H. T., *Nature 218*, 333-336 (1968).
13. Meryman, H. T., in "Cryobiology," p. 63. Academic Press, New York (1966).
14. Morichi, T., and Irie, R., *Cryobiol. 10*, 393-399 (1973).
15. Nag, K. K., and Street, H. E., *Physiol. Plant. 34*, 254-260 (1975).
16. Olien, C. R., *Ann. Rev. Plant Physiol. 18*, 387-408 (1967).
17. Pribor, D. B., *Cryobiol. 12*, 309-320 (1975).
18. Pribor, D. B., *Cryobiol. 11*, 60-72 (1974).
19. Pribor, D. B., and Pribor, H. C., *Cryobiol. 10*, 93-103 (1973).
20. Ruwart, M. J., Holland, J. F., and Haug, A., *Cryobiol. 12*, 26-33 (1975).
21. Sakai, A., and Yoshida, S., *Cryobiol. 5*, 160-174 (1968).
22. Singer, S. J., and Nicolson, G. L., *Science 175*, 720-731 (1972).
23. Steponkus, P. L., and Lanphear, F. O., *Plant Physiol. 42*, 1423-1426 (1967).
24. Sugawara, Y., and Sakai, A., *Plant Physiol. (Suppl.) 56*, Abst. 433 (1975).
25. Taylor, R., Adams, G. D. J., Boardman, C. F. B., and Wallis, R. G., *Cryobiol. 11*, 430-438 (1974).

26.   Tumanov, I. I., Krasavtsev, A., and Khvalin, N. N.,
      Doklady Akac. Nauk SSSR, Bot. Sci. Sect. 127, 235-237
      (1959).
27.   Wilkes, G., *Bull. Atomic Scientists 33*, 8-16 (1977).

Part VII
Special Topics

PLANT COLD HARDINESS SEMINAR
SUMMARY AND GENERAL REMARKS

C. J. Weiser

Horticulture Department
Oregon State University
Corvallis, Oregon

I. SEMINAR SUMMARY

Seminar topics covered a wide spectrum of plant freezing stress responses, and the range of expertise of participants from nine nations was equally as diverse. The constructive, informal and integrative nature of the seminar deliberations gave rise to a particularly fine meeting which will have a significant impact on future freezing stress research.

Several cooperative research programs evolved among participants during the course of the meetings. These included collaborative cryopreservation research on potato callus; evaluation of bacterial ice nucleator activity in the freezing of tree fruit blossoms; and establishment of a genetics/biophysics working group to explore the freezing behavior of interspecific *Prunus* hybrids. The latter will be directed at genetically eliminating the deep supercooling phenomenon in hybrids of *Prunus* with commercial potential.

Major topics of consideration were the identification, definition and partitioning of freezing stresses; membrane properties and composition; biochemical, metabolic and microscopic changes associated with freezing, freezing resistance, and injury; the status of cellular water; environmental influences on cold acclimation; the role of bacteria in ice nucleation of plants in nature; deep supercooling freezing avoidance mechanisms and their ecological implications in some seeds and woody plant tissues; cryopreservation; and freezing stress selection criteria and breeding potentials.

Discussions were specifically directed at cereal grains, citrus, potatoes, deciduous tree fruits, Jerusalem artichoke, *Chlorella*, and Japanese and North American tree species. The following list itemizes some interesting topics and new findings:

Evidence of phospholipid degradation during freezing in black locust.

Implications that calcium may play a role in freezing tolerance and injury.

Evidence that highly structured water fractions observed in animal tissues do not play a role in plant freezing resistance.

Evidence that membrane damage in wheat occurs during freezing, not during thawing.

The need to separate dormancy from cold acclimation phenomenon in evaluating roles of proteins and nucleic acids in these processes.

The potentials for cryopreservation of plant cells and organs, and the need for expanded research.

The role of protein synthesis in *Chlorella* freezing resistance.

Red light promotion, and far red light inhibition, of growth cessation and cold acclimation.

Progress in breeding frost resistant potato cultivars.

The role of deep supercooling in tree distribution, and the distinct limits this phenomenon imposes on winter hardiness.

Affirmation of North American tree distribution, hardiness, and freezing pattern relationships in studies of freezing processes in Japanese species.

Dynamics of water redistribution in some dormant flower buds, and its influence on deep supercooling and lethal temperatures.

The potential for extending the range of fruit production in *Prunus* by genetically circumventing deep supercooling via interspecific hybridization.

The transitory nature of deep supercooling in some dormant flower buds and seeds.

Elimination of deep supercooling in lettuce seeds by puncturing the seed coat.

Extensive ion/solute leakage from frozen cells which can repair and remain viable.

Freezing damage to active membrane transport systems, and cellular ion/solute leakage from freeze stressed cells without loss of membrane semipermeability to water.

A decrease in the particle density of membranes as cells cold acclimate.

Inhibition of ice growth by cell wall mucilages which form a film around crystals.

Microscopically observable differences in ice crystal form and propagation in hardy and tender cells.

Frost hardy potato genotypes have more lipid bodies, fewer starch granules, and a higher incidence of double palisale layers than tender types.

Day-to-day fluctuations in hardiness of dormant flower buds during the winter.

Partitioning of nucleation, propagation, adhesion, and desiccation components of freezing stress energy.

Limited ice nucleation activity in axenically grown tender plants which supercool to -10° C. The role of certain bacteria in preventing such supercooling.

Bacterioccidal chemical treatments to tender plants which increase supercooling and lower lethal temperatures.

The close coincidence of water crystallization and freezing injury in citrus.

The close correlation of injury to isolated protoplasts caused by contraction and expansion induced either by freezing and thawing or by osmotic treatments.

The correlation of protoplast lysis with an absolute increase in protoplast surface area during deplasmolysis; implying that dissociation of a specific membrane component is involved.

II.   ENVIRONMENTAL STRESS STATEMENT--RESOLUTION

Drafted and adopted by seminar participants:

Environmental stresses, particularly drought, low temperatures, and freezing are dominant factors controlling the distribution and productivity of plants on the earth. Freezing and drought stresses largely determine where crops are grown, what crops are grown, how many crops can be produced per year, and the productivity of native forests and grasslands. These stresses strongly influence the levels of world food reserves and the quality, availability, and price of food commodities. Secondary impacts are manifest in world trade and social, economic and human health conditions.

For reasons that are not clear, the impact and importance of environmental stresses are not generally reflected in national and international research funding priorities and policies. This serious deficiency warrants prompt and critical review. Immediate and real benefits could be realized from environmental stress research directed at increasing and stabilizing crop productivity, attenuating losses, and optimizing utilization of petrochemical and water resources.

There seems to be a historical willingness to accept, rather than address, the devastating effects of climatic extremes. The projection that a 1° C decrease in the world mean temperature would decrease world rice production by 40% illustrates the negative impact of small climatic changes. The positive impacts of subtle microclimatic modification, or changes in genetic or physiological stress resistances are equally dramatic. A 2° increase in frost hardiness of citrus trees, winter cereals, potatoes, deciduous tree fruit blossoms and tender vegetables would greatly influence production of these crops. In wheat, for example, a 2° increase in hardiness could extend production to vast areas which now produce spring wheat with 25 to 40% lower yields.

Such small but highly significant improvements can be achieved through research on genetic stress resistance and technological manipulation of crop physiology and microclimate. These potentials are apparently poorly recognized by present policy makers.

III. RESOLUTION

Broad interdisciplinary research and international collaboration is needed to accelerate environmental stress research. Substantial and sustained support of such programs would yield predictable and achievable gains far outweighing the costs. Agriculturists, biologists, chemists and biophysicists from nine nations attending the Plant Cold Hardiness Seminar in St. Paul, Minnesota on November 2-4, 1977 strongly resolved that--*Clear focus and increased support be provided to environmental stress research needs and potentials by world governments, research agencies and institutions.*

SUMMARY OF THE PAPERS PRESENTED
AT THE FIRST EUROPEAN
INTERNATIONAL SYMPOSIUM
ON "WINTER HARDINESS
IN WOODY PERENNIALS,"
SEPTEMBER, 1977

T. Holubowicz

Academy of Agriculture
Poznan, Poland

At the International Horticultural Congress in Warsaw,
Poland in 1974 a Working Group on Frost Hardiness in Woody
Perennials was organized. The 1st Symposium on the Frost
Hardiness of Woody Plants took place in Poznan, September
12 to 19, 1977. During the first three days the participants
delivered and listened to the scientific papers, and then
there was a four day excursion to different orchards and pla-
ces of interest in Poland. Forty six participants from ten
countries took part in the Symposium: Poland-32, Soviet
Union-4, German Democratic Republic-2, Federal Germany-1,
Switzerland-2, Canada-1, Egypt-1, Romania-1, Sweden-1, and
USA-1.

The program included four invited papers and twenty contributed papers. The invited papers pertained to the review
of the latest literature concerning frost hardiness, and the
speakers discussed the eventual lines of future research.
Contributed papers gave an opportunity to get a first hand
knowledge of the research program going on in different laboratories, and for personal exchange of views. The Polish
participants who were able to speak both Russian and English
were very helpful in bridging the language barriers.

Dr. A. Kacperska-Palacz from the Institute of Botany,
Warsaw, Poland, gave in her invited paper on "Physiological
mechanisms of frost tolerance in plants", the most detailed
review of literature of this field. The speaker concentrated
on the following problems:

395

Kinds of freezing and freezing processes.
Freezing avoidance and supercooling.
Freezing tolerance and freezing injuries.
Acclimation events leading to the increased frost
    tolerance.
Protection against dessication injuries.
Acclimation as step by step process.

Since Dr. Kacperska-Palacz is present in St. Paul, Minnesota, I'm not going to summarize her paper.
Dr. Kurt A. Santarius from the Institute of Botany, University of Dusseldorf, West Germany, talked about "Biochemical
basis of frost resistance in higher plants". In the author's
opinion there is no universal mechanism of frost resistance,
and the different species have different types of protective
adaptations towards freezing stress. Cellular membranes are
particularly frost-sensitive and are the primary site of
freeze-thaw injury in plants. Since one of the sessions in
St. Paul was totally devoted to the problems of biomembranes,
it is not necessary to go into the details. I should only
mention that the main topics of Dr. Santarius' paper were as
follows:

Causes of frost injury.
Causes of frost resistance.
Cryoprotective compounds.
Changes in the biomembranes.
Supercooling of plant cells.

The role of cryoprotective-compounds and the mechanism of
solute protection was discussed by Dr. Santarius in great
detail. The experiments of the author and his co-workers with
isolated spinach thylakoids and mitochondria have shown that
various sugars and sugar alcohols, salts of organic acids,
amino acids and soluble proteins protect the membranes during
freezing. For complete protection, very different osmotic
concentrations of various cryoprotectants may be necessary. On
the other hand, membrane-toxic compounds interfere with the
protection afforded by cryoprotectants. According to the
author, the ratio between the cryoprotectant and the membrane-
toxic compound is decisive rather than the absolute concentration of these compounds in the membrane suspension. The most
important then is to find a "balance" between the different
solutions which will give membrane protection only at a given
concentration ratio between both compounds.
Our next speaker Dr. M. M. Tyurina from the Non-chernozem
Zone Pomology Institute, Moscow, Soviet Union, talked about
"Interaction between development of frost resistance and dormancy in plants". The author discussed the relationship of

those two phenomena in connection with photoperiod, initial stages of acclimation, frost hardiness development, attaining its maximum, dehardening and the resumption of growth. The sequence of necessary metabolic changes leading from one stage to another was underlined. Thus, in woody plants in which frost resistance is connected with rest, every stage of annual cycle appears to be at the same time the period of preparation for the next one. In the author's opinion the maximum of frost hardiness may be obtained even at the time when the rest period is coming to an end.

Dr. M. Solovieva from the Ukrainian Institute of Pomology, Kiev, Soviet Union presented the results of the investigations from her and other laboratories on the changes in the content of phosphoric compounds in the different parts of the fruit trees related to the frost resistance. The effect of temperature fluctuations, the soil and air moisture in relation to frost hardiness was also discussed. The author has found that the changes in phosphorus compounds affect the content of phenolic compounds in the tree tissues. The anthocyanins serve as natural filters for sun rays preventing the overheating of the cambium and phloem. These tissues are very often injured when the temperatures are very low. The injuries are especially evident early in the fall and at the end of winter when the fluctuations in temperature are very great. It is known that at this time one and two year old shoots are less damaged than the older branches and the trunk. As a consequence of cambium and phloem damage the transport of water and nutrients is greatly curtailed, and sun and frost scalds are evident. It is supposed that sun and frost scalds are not due to the ice formation but the destruction of the chlorophyll. Dr. Solovieva cited the results of her colleagues who have found the differences in the number of palisade layers, and the content of chlorophyll in leaves of different species more and less sensitive to frost. It seems that the species with thicker leaves, containing more layers of palisade cells were more resistant to low temperature. Also, the bark of the fruit trees and especially of the young shoots contains more chloroplasts than the trunk and older branches. During the acclimation, the synthesis of anthocyanins is more intensive in one and two year old shoots than in other parts of the tree. In the older branches the ability to synthesize anthocyanins decreases, and the symptoms of frost and sun scald are very frequent. It was also found that high rates of nitrogen fertilization prevented the synthesis of anthocyanins.

Dr. Solovieva talked also about the differences in the sensitivity to frost of different plant tissues, the changes in sensitivity during the course of the year, and the regeneration

after the injury.   The ability to heal the wounds, and the
beneficial role of auxins and gibberellins in the regeneration
process was especially stressed.

It was proposed that phosphorus and potassium fertilizers
increased the resistance of trunks to low temperature.   The
high soil moisture and high rates of nitrogen lowered considerably the frost tolerance of shoots, and the high soil
moisture was detrimental to the regeneration process.

As was mentioned before, twenty contributed papers were
presented at the Symposium.   They could be divided into four
thematic groups:

1. Development of frost hardiness and changes of some
    compounds connected with the acclimation and hardi-
    ness of plants--6 papers.
2. Factors affecting plants' resistance and tolerance--
    2 papers.
3. Modification of plant hardiness by using the growth
    regulators--8 papers.
4. Types of injuries and methods of their estimation--
    4 papers.

In the first group of contributed papers, Hellergren, Lund
Sweden, showed that the most effective climate for frost hardiness development in spruce and pine was short days and low
temperatures (2° C) followed by short days and room temperature (20° C).   By using different day length and temperature
the author came to conclude the importance of the photoperiod
prior to low temperatures for frost hardiness development.

Sobczyk and Kacperska-Palacz from Warsaw, Poland demonstrated marked increase of ATP content and of energy charge
during hardening of winter rape plants.   They studied activities of glyceraldehydro 3-phosphate dehydrogenase, pyruvate
kinase/glycolityc enzymes, NADP-isocitrate dehydrogenase/enzyme
of TCA cycle, glycose 6-phosphate dehydrogenase/regulatory
enzyme of pentase phosphate pathway, and APP-ase.   Results indicated that the increases of energy charge in the cold treated leaves was rather due to the limited utilization of ATP in
growth processes than to the cold induced synthesis of ATP.
It was also found in the preliminary study that cold pretreat-
ment increased stability of the glyceraldehyde 3-phosphate
dehydrogenase and pyruvate kinase in frost injured leaves of
rape plants.

Smolenska from the University of Warsaw, Poland and de la
Roche from the Ottawa Research Station, Canada presented some
data on the changes in lipid and fatty acid composition under
influence of low temperature treatment.

Smolenska concluded that adaptation of the rape leaves to
low temperature functioning and to freezing condition was correlated with an increase in content of phosphatidyl choline
and ethanolamine, predominantly esterified with linolenic acid.
De la Roche subjected seedlings of wheat to acclimation at
2^0 C in the presence of two inhibitors of linolenic acid synthesis and he had found that the stimulated synthesis of linolenic acid which occurs in cereals during low temperature acclimation is unrelated to the development of freezing
tolerance.

Sikorska and Kacperska-Palacz from the University of Warsaw, Poland presented the results of investigations involved
with phospholipid changes during hardening of winter rape
plants. The presented results indicated that there is some
"threshold" value for the phospholipid content which enables
cells to tolerate the frost evoked strain. It was found that
freezing brought about the phospholipid degradation which was
reversible only in the slightly injured leaves with the relatively high phospholipid content.

Maleszewski and Sosinska using isotope technique with
$^{14}CO_2$ have found that cold treatment resulted in an increase
of the incorporation of assimilated carbon into C_4-dicarboxylic
acid, alanine and proline. A lack of sensitivity of the gly-
colic acid pathway to temporary increases in the environmental
temperature was evident after prolonged cold treatment. Increased incorporation of ^{14}C into C_4-dicarboxylic acids in the
cold pretreated leaves was found to be correlated with the
increased ratio of PEP/RuDP carboxylase activities.

In the second group of contributed papers, Z. Gruca and
E. Pacholak, both from Poznan, Academy of Agriculture, Poland,
demonstrated that in raspberry canes and shoots of some clonal
apple rootstocks the most effective factors changing the frost
resistance level were diurnal fluctuations in temperature of
the air. They found very close relationships between water
content and frost tolerance of plants. During the winter the
lowest content of water in the plants was always correlated
with the highest level of their frost resistance. There was no
significant correlation between the water soluble proteins and
the content of mineral nutrient compounds in the shoots and
canes and their cold hardiness. Some results suggested that
high doses of nitrogen fertilizer may have some negative effect on the development of cold tolerance of raspberry canes.

Eight papers were presented in the group of modification
of plant hardiness by use of growth regulators. Among the
researchers using the same growth regulators the obtained
results were often contradictory and depended on the species or
cultivar, time of application, doses of used chemicals and so
on.

Ilyina and Burdasov from the Research Institute of Horticulture in Barnaul Siberia, USSR, showed that spraying cherry trees with Gibberellin at 50 and 100 ppm in June increased frost resistance of the flower buds but when they were sprayed with the same chemicals In August, it had reduced it. The treatment with Ethrel at 100 and 500 ppm in August increased frost resistance of flower buds.

M. Khamis and T. Holubowicz from Poznan, Academy of Agriculture presented the results indicating the Ethrel applied at the concentration of 1000 and 1500 ppm in July and August was harmful to peach trees. CCC at 1000 ppm increased slightly the cold tolerance of peach shoots. Alar and ABA gave ambiguous effects and the results varied depending on the time of application. It was found that CCC 1000 ppm sprayed July 13 induced earlier leaf fall, inhibited the elongation of the shoots, whereas ABA applied July 20 delayed the leaf fall and accelerated the shoots' growth. However, the content of lignin increased in the shoots sprayed with ABA, Alar and CCC.

Pieniazek, Holubowicz and co-workers from the Pomology Institute in Skierniewice and the Poznzn Academy of Agriculture, Poland using the *in vitro* technique for growth of the apple stem callus have found that by replacing sucrose with sorbitol in the basic Murashige-Skoog medium it was possible to change its killing point. Adding to the medium 20 ppm of CCC, GA, Alar, ABA and Ethrel changed the rate of growth and resistance of callus tissue.

Preliminary results with freezing the callus had indicated that ABA, Ethrel and Alar increased slightly the frost tolerance.

In the study of the influence of growth regulators on the frost hardiness of raspberry canes, B. Zraly from the Pomology Research Institute in Skierniewice, Poland had established that ABA and BA/Benzyl adenine increased frost hardiness of raspberry canes by prolonging hardiness period or the regeneration ability of canes after frost injury when they were applied in late winter (February 2) but they were not able to do the same when they were applied in the spring (March 23).

Rate of growth, flowering and bearing had strong effect on development of frost resistance of apple trees, according to the paper presented by Lobanov, Gogoleva and Tyurina from Mowcow and Barnaul Research Institutes, USSR. Trees with moderate flowering and annual fruit bearing that finished their terminal and radial growth in proper time were able to obtain the optimum level of winter hardiness. They increased greatly their frost resistance in the autumn, hardened well in winter, and they did not deharden rapidly in early spring.

Intensively flowering and heavily cropping trees were less resistant than those moderately and poorly flowering and cropping.

Authors had drawn the conclusion that time of growth cessation of the shoots was the main factor delaying or accelerating the beginning of the processes preparatory for hardening.

The control of growth and inducement of the development of hardiness were the aim of the experiments done by D. Ketchie from the Tree Fruit Research Center in Wenatchee, Washington, USA. Spraying with Gibberellin and defoliation during the summer had no positive effect on the development of frost resistance of apple trees, whereas cryoprotectants gave positive effect, and these types of experiments according to the author are very promising.

Dr. V. M. Burdasov from the Lisavenko Research Institute of Horticulture in Siberia, described the asphyxiation of bark and cambium at the base of the tree under snow as a little known and dangerous type of winter injury. This kind of injury was recorded in 133 species of cultivated and ornamental plants, including frost resistant species and cultivars of pear, plum, cherry, apricot, sea buckthorn and raspberry. The author described the different physiological anomalies resulting from the incomplete freezing of the soil after early snow fall in the autumn such as: saturation with water of the plant trunks, impeded gas exchange, browning and destruction of cells with the temperature rising above 0° C, the intrusion of the pathogenic microflora, and many others.

The weather conditions in summer play an important role in preparation for overwintering. Thus a cool and damp type of summer weather increases the incidence of asphyxiation, and a dry and hot one reduces it. It was found after a cool and damp summer that the buds of *Prunus ussuriensis* finished their rest already in October, whereas after a hot and dry summer they finished in December, and were better prepared for overwintering.

However, the main factor increasing the chances for asphyxiation is the temperature under the snow. The higher the temperature under the snow, the more intensive is the damage. Thus, when there is a mild winter and the temperature under the snow is well above 0° C, as was the case during the 1972–73 winter, the injuries were twice as great as during the 1974–75 winter, when the temperature was well below 0° C. The temperature under the snow cover depends on the time of the first snow cover formation, its height, cold winds so often experienced in that part of the Soviet Union, and the heat coming from the depths of the earth.

The author found 0.1% Ethrel solution application beneficial for improved maturation of Siberian apricot, *Armeniaca sibirica Lam.*, when it was sprayed on the trees at the end of the 1973 summer. The mortality of plants was reduced from 75 to 53%.

A soil management system in which Siberian apricot seedlings were grown on the earth banks with snow firmly packed in interrows proved to be the most practical method. The mortality of trees was reduced from 67 to 11%.

E. Zurawicz from the Pomology Institute in Skierniewice, Poland, described different types of frost injury symptoms on strawberry plants in the field. The obtained results showed that one of the most sensitive parts of the plant is the tip of the crown where flower buds are located, and the differences in the sensitivity between the flower buds--the more advanced in the development being the more sensitive.

This was usually due to the higher position of such buds above the ground level. This type of injury may cause a slight crop reduction and delays production of the first fruits. Heavier crop losses may be caused by severe winter injury to the main crown.

G. A. Gogoleva from the Moscow Institute of Pomology, USSR gave a summary of her laboratory work concerning the various components of winter hardiness. She said that winter hardiness is a multicomponent property of plants. The authors have found that in their climatic zone the essential components of winter hardiness were:

Time and rate of the development of ability to harden.
Potential level of frost resistance.
Loss of frost resistance after warm spells.
Ability to reharden after warm spells.
Regenerative ability.

Thirty-two apple cultivars were tested during three years in artificial conditions and then the results compared with the field survival in fifteen years. The cultivars which were shown to be hardy in laboratory freezing tests, based on recovery method of injury estimation, were also hardy under natural conditions, in about 80%. However, those cultivars which were more resistant in laboratory freezing tests than in the field were less resistant especially in more northern conditions. Gogoleva concluded that selection for winter hardiness of the new progenies by artificial cold stresses has to be carried out in several stages of breeding program and it may considerably accelerate the evaluation of more hardy cultivars before they reach bearing age.

Survival test as a method of frost injury estimation was the subject of Holubowicz', the Poznan Academy of Agriculture, Poland, paper. I compared the results of artificial laboratory freezing tests and the obtained frost injuries by three methods: survival test, diffusion of electrolites and measurement of electrical admittance. It was concluded that the method of electrical admittance of the shoots was less reliable than the other two methods for the evaluation of the injuries suffered by apple shoots during artificial freezing. Survival test and the diffusion of electrolites from the shoots gave more reliable results.

After sessions the participants took part in an excursion to the Pomology Experiment Station of Przybroda near Poznan, to the Pomology Institute at Skierniewice and to the Pomology Experiment Station Brzezna in the southern part of Poland. The excursion program gave the opportunity ot see Polish orchards and a large part of Poland.

At the end of the Symposium there was a round table discussion which gave an ample opportunity for the exchange of ideas about future work. It was suggested that the next symposium might take place in the Soviet Union after four to five years.

REFERENCES

A. *Invited Papers*

1. Kacperska-Palacz, A., *Physiological mechanisms of frost tolerance in plants*, Institute of Botany, University of Warsaw, Warsaw, Poland.
2. Santarius, Kurt A., *Biochemical basis of frost resistance in higher plants*, Institute of Botany, University of Dusseldorf, D-4000 Dusseldorf, West Germany.
3. Solovieva, M., *Winter hardiness and regeneration of frost injured fruit trees*, Ukrainian Institute of Horticulture, Kijev, USSR.
4. Tyurina, M. M., Gogoleva, G. A., Jegurasdova, A. S., and Bulatova, T. C., *Interaction between development of frost resistance and dormancy in plants*, The Non-Chernozem Zone Horticultural Research Institute, Moscow, USSR.

B. *Contributed Papers*

1. Group I

1. de la Roche, A. I., *Development of freezing tolerance in wheat without changes in lipid unsaturation*, Research Station, Research Branch, Agriculture Canada, Ottawa, Ontario, Canada.
2. Hellergren, J., *Frost hardiness development in spruce and pine*, Department of Plant Physiology, University of Lund, Lund, Sweden.

3. Maleszewski, S., and Sosinska, A., *Photosynthesis products in leaves treated with low temperatures*, Institute of Botany, University of Warsaw, Warsaw, Poland.
4. Sikorska, E., and Kacperska-Palacz, A., *Phospholipid changes during hardening of winter rape plants*, Institute of Botany, University of Warsaw, Warsaw, Poland.
5. Smolenska, G., and Kuiper, P. J. C., *Are the fatty acids changes involved in frost tolerance of winter rape plants?*, Institute of Botany, University of Warsaw, Warsaw, Poland and Biological Centre, University of Groningen, Haren/Gr, the Netherlands, respectively.
6. Sobczyk, E., and Kacperska-Palacz, A., *Effect of cold treatments on the energy metabolism in leaves of winter rape plants*, Institute of Botany, Univesity of Warsaw, Warsaw, Poland.

   2. Group II
1. Gruca, Z., *Factors affecting changes of frost resistance of some apple clonal rootstocks*, Institute of Horticultural Production, Poznan, Poland.
2. Pacholak, E., *Factors affecting frost tolerance of raspberry canes*, Institute of Horticultural Production, Poznan, Poland.

   3. Group III
1. Ilyina, Ye. I., and Burdasov, V. M., *An experience in using growth regulators with the aim of increasing frost resistance of cherry flower buds*, Lisavenko Research Institute of Horticulture in Siberia, Barnaul, USSR.
2. Ketchie, D., *The effect of gibberellin application and defoliation during the summer on cold resistance of apple trees the following winter*, Tree Fruit Research Center, Wenatchee, Washington, USA.
3. Ketchie, D., *The effect of dodecyl ether of polyethylene glycol/DEPG on cold resistance of apple and pear trees*, Tree Fruit Research Center, Wenatchee, Washington, USA.
4. Khamis, M., and Holubowicz, T., *Effect of the foliar application of Alar, Ethrel, ABA and CCC on the frost resistance of peach trees*, Institute of Horticultural Production, Poznzn, Poland.
5. Lobanov, E. M., Gogoleva, G. A., and Tyurina, M. M., *Frost resistance of alternately bearing apple trees in connection with endogenous growth rhythms and stored carbohydrate content in tissues*, Lisavenko Research Institute of Horticulture in Siberia, Barnaul, USSR (Labanov) and Non-Chernozem Zone Horticultural Research Institute, Moscow, USSR (Gogoleva and Tyurina).

6. Pieniazek, J., Holubowicz, T., Machnik, B., and Kasprzyk, M., *Apple stem callus frost tolerance and growth
   modification by adding sorbitol and some growth regulators
   to the medium*, Institute of Pomology, Skierniewice, Poland
   (Pieniazek and Machnik) and Institute of Horticultural
   Production, Poznan, Poland (Holubowicz and Kasprzyk).

7. Solovieva, M. M., *The effect of water content on the cold
   hardiness and recovery of fruit trees*, Ukrainian Institute
   of Horticulture, Kiev, USSR.

8. Zraly, B., *Winter hardiness of raspberry*, Research Institute of Pomology, Skierniewice, Poland.

4. Group IV

1. Burdasov, V. M., *Asphyxiation and ways of increasing as-
   phyxiation resistance in cultivated plants*, Lisavenko
   Research Institute of Horticulture in Siberia, Barnaul,
   USSR.

2. Gogoleva, G. A., Smagina, V. P., Mikheyev, A. M., Yefi-
   mova, N. V., and Tyurina, M. M., *Investigations on the
   components of winter hardiness in fruit plants*, Non-
   Chernozem Belt Horticultural Research Institute, Moscow,
   USSR.

3. Holubowicz, T., *Survival test as a method of frost injury
   estimation*, Institute of Horticultural Production, Poznan,
   Poland.

4. Zurawicz, E., *Winter injury of strawberries grown in
   Poland*, Research Institute of Pomology, Skierniewice,
   Poland.

Index

A

Acclimation
  ATP changes, 145
  callus cultures, 197, 208
  cell function, 148
  cold, 270, 274
  dessication injury, 147
  double low temperature exotherms, 235, 237, 238
  driving forces in process, 144, 145
  events leading to, 146
  flower buds, 221
  herbaceous plants, 139
  hormonal balance, 144, 145
  intracellular water, 167, 171
  mechanisms of, 139, 146, 147
  membrane changes, 209
  role of auxins, 208
  soluble protein, correlation with, 155
  unfrozen water, 170
  water content reduction, 165, 166
  water relation, 149
  winter wheat, 166
Anatomical, 268
Anthesis, 270
Arabinoxylan, 42, 44
Asphyxiation, 401
ATP, changes during acclimation, 145
ATPases, in relation to
  cryostress, 131
  freezing injury, 101, 107
  thermostress, 131
  water stress, 131
Auxin, role in callus acclimation, 208

B

Bacterial ice nucleation
  competition of INA with other bacteria, 245–250

distribution of bacterial ice nuclei in
  species, 255, 256
effect of chemicals on INA bacteria, 257
Erwinia herbicela, 249, 250, 251
exogenous ice nuclei, 255
frost damage
  in beans, 251, 253
  in corn, 252, 253
ice nucleation
  activity, 254
  as a function of temperature, 255
ice nucleus concentration, 254
prediction of frost injury, 255
Pseudomonas syringae, 249, 250, 251
replica freezing technique, 256
Bark
  hardening, 284
  slippage, 304
Bath test, 336, 337
Beta inhibitor, 275,
  maleic hydrazide, 275
Black-heart, 314
Bound water
  during acclimation, 170
  NMR analysis, 167
  plant tissue, 167
Breeding
  cold hardiness, 328
  temperate fruits, 313
Breeding potatoes
  cross breeding, 333
  field testing, 334
  genetic breeding, 333
  mass selection, 333

C

Ca^{++} efflux, freezing injury, 111
Callus
  cells, 355

cultures
    before and during hardening, 201
    cold acclimation, 197, 208
    effect of
        light on development of hardiness, 203
        2,4-D on growth rate, 200, 201, 202,
            208
    fracture face of
        plasma membrane, 205
        plasma membrane of hardened callus,
            206, 207, 209
    freeze thawing survival, 198
    from cherry twigs, 197
    from chrysanthemum, 197
    from *Jerusalem artichoke,* 197, 198
    hardening and dehardening cycle, 204,
        207
    hardiness, increase of, 202, 203
    procedure for, 198
    role of auxins in acclimation, 208
Carbohydrate accumulation, 307
Cell
    contraction, in freezing dehydration, 34
    damage, frost stress, 66
    freezing
        dynamic process, 30
        onion epidermis, 30
    membrane
        active transport system, 99, 100
        alteration
            by freezing injury, 104
            in freezing, 126
        ATPases, 100, 101, 107
        damage in freezing, 117, 134
        efflux of ions, 98
        fluid mosaic model, 106
        freezing injury, 93, 94, 98, 100, 101, 102,
            103
        lipids, 106, 107, 110
        permeability to
            K^+, 104, 105, 107
            nonelectrolytes, 104, 105, 106
            water, 98
        properties in freezing, 126
        proteins, 106, 107, 110
        reversibility of damage, 98
        semipermeability, 94
        transport studies, 104, 107
    size
        changes in algae, 179
        frost resistance, 62, 63
    solute concentration, relation with freezing
        avoidance, 12

surface area, protoplast expansion, 83, 84
wall thickness
    E. M. picture, 64, 65
    frost resistance, 62
    Solanum acaule, 63
    Solanum tuberosum, 63
Channel forming molecule, 107
Chemical, affecting freeze injury, 377
Chilling
    injury, 302
    resistance, phase transition temperature, 10
Chlorella ellipsoidea, frost hardiness studies,
    175, 176, 177
Chloroplast
    before and after frost hardening, 64, 65
    EM picture, 64, 65
    freezing injury, 110, 111
Chloroplast membrane
    cold
        acclimation, 77, 78
        hardening, 155
    freezing injury, 77, 78
Cold
    cabinet, 337
    hardening
        chloroplast membrane system, 155
        enzyme involvement, 158
        glycoprotein, 155
        membrane protein changes, 154, 155
        nucleic acids, 156, 157
        protein synthesis involvement, 153, 154,
            157
        soluble protein changes, 154
    hardiness
        adaptation, 267
        development of, 398
        permeability of membrane, 169
        polyethylene glycol, 166
        role of water, 165, 166
        undercooling, in relation to, 215
        unfrozen water, 170, 171
        water distribution, 167
        winter cereals, 165, 166
    injury
        electrolyte leakage, 219
        Pyrus, 219
Conductivity, 318
    method evaluation, 97
Cryodynamics
    energy of transition, 38
    Gibbs free energy, 38
Cryogenic seed preservation, 368
Cryopreservation, 3

Cryoprotectant, 38
dimethyl sulfoxide, 373, 379
effect on freezing damage, 375
ethylene glycol, 373, 379
polyethylene glycol, 373, 379, 386
on sugarcane callus, 379
Cryoprotection, 3
Cryoprotective traits, 37
Cryostress
in relation to salt stress, 13
and water stress in relation to salt stress, 12, 13
Cryptomerias, 289
Crystallization
energy, 46, 47
growth rate, 41
supercool water, 250
Cultural measures
environmental, 272
physiological, 274
Cytological transition, 355
Cytoplasm, relation to freezing survival, 354

D

Deacclimation, cold
low temperature exotherms, 235
seasonal variations of exotherms, 233, 234
water content, 233, 234
Dehardening
enzyme activity, 158
nucleic acid, 158
protein, 158
Dehydration, induced by extracellular freezing, 4, 23
Differential thermal analysis of
endosperm, 254
excised embryo, 246, 254
intact seed, 254
lettuce seed coat removal, 254
lettuce seed, intact, 246, 254
Dormancy and rest, 269, 273
DTA
electrolyte leakage, relation to, 218
profiles
of leaf buds, 217
in xylem pieces, 231, 236
Prunus, effect of prefreezing, 222
Prunus, 220
Pyrus, 216
Rosa taxa, 223
xylem injury, 231

E

Efflux of ions due to
cryostress, 14
thermostress, 14
water stress, 14
Electron transport system, effect of
antimetabolites, 180
surfactants, 180
Energy of melting, 43
Environment, relation of aminoacyl-tRNA synthetase activity, 58
Environmental
factor, biochemical change, 57
stress, potatoes, 334
Enzyme
activity in
frost hardening, 143, 147
relation to hardiness, 60, 61, 62
involvement in
cold hardening, 158
dehardening, 158
Espeletia schultzii, 5
Ethanolamine accumulation, 399
Ethephon, 300
Exosmosis of water, 11
Extracellular freezing
collapse of cell walls, 93, 96
injury, 34
in nature, 93

F

Flashed cell, 30
Freeze
injured cells
Ca^{++}
efflux, 111
role, 110, 111
$CaCl_2$ treatment, 99
cap plasmolysis, 100, 102
cell organelle, 66
efflux of ions, 97, 107, 110
electron microscopic picture, 103
EM observation, 66
evaluation, conductivity method, 97
K^+ efflux, 97, 101, 107, 110
leakage of ions, 96
membrane
lipids, 106, 110
proteins, 107, 110
methyl urea permeability constant, 106
microscopic observation, 95, 100

onion epidermis, 95, 100
permeability of cell membranes to water,
    98
plasmamembrane, 66
plasmolysis of, 95, 96, 100
post-thaw behaviour, 98, 99
potato leaf parenchyma, 102, 103
protoplasmic streaming, 95
semipermeability of cell membranes, 96,
    109
swelling of protoplasm, 100, 101, 102
tonoplast, 66
transport studies, 104, 106, 107
treatments for repair, 99
urea permeability constants, 106
protection
    artificial freeze protection, 299
    ethenol, 300
    flourinated compounds, 300
    glycerol, 300
    malic hydrazide, 300
    phosphoric acid, 300
thawing
    phospholipase activity, 127, 128, 130
    phospholipid changes, 122
tolerance
    artificial improvement, 374
    extracellular, 7
Freezing
effect on chlorophyll, 397
extracellular, 6
    in apples, 33
    in cortical tissue, 26
    in herbaceous plants, 25
    water, 169
intracellular
    in apples, 33
    water, 169
injury
    active transport system, 99, 101, 107
    ATPases, 101, 107
    Ca^{++} efflux, 111
    Ca^{++}, role, 110, 111
    cell membrane
        alterations, 104, 107
        properties, 93
    chloroplast, 110, 111
    conductivity method, 97
    hypothesis for cell death, 109
    K_{oil} oil/water distribution coefficient, 106
    LD_{50}, 94
    LT_{50}, 94
    leakage of ions, 93, 94, 96
    loss of turgor, 93
    mechanism, 99, 101
    membrane
        lipids, 106, 107
        proteins, 106, 107, 110
    methyl urea permeability constants, 106
    mitochondria, 110, 111
    permeability
        constants, 106
        properties, 168
        studies, 105, 106, 107
    plasma membrane, 112
        damage, 168, 169
    plasmolysis, frost, 96
    primary site, 99, 107
    pseudoplasmolysis, 96
    semipermeability of cell membranes, 99,
        100, 101, 102
    tonoplast, 111, 112
    urea permeability constants, 106
resistance
    broad leaved tree species, 232
    classification of deciduous trees, 228
    in relation to average annual minimum
        temperature, 232
    low temperature exotherms, 232
stress, analysis of
    plant hardening, 37
    plant recovery, 37
survival, in sugarcane cultured cells, 373
Frost
damage, scale, 338
hardening
    amino acids, 142, 143
    enzyme activity, 143, 147
    growth inhibitor changes, 144, 145
    light requirements, 140, 141
    lipids, 143, 146
    membrane permeability, 148
    metabolic changes, 142, 143, 144, 145
    proteins, 142, 143
    role of photosynthesis, 141
    stages of, 140, 141, 142, 143, 148
    sugars, 142
    temperature requirements, 140
    water stress, 142
    winter rape, 140
hardiness, effects of
    glucides, 189
    glucose addition, 185, 188, 191
    light, 190, 191
    plant hormones, 181, 182
    sugars, 187

plasmolysis, 66
resistance, 338
tolerance
  biochemical basis, 396
  events leading to, 146, 149
  growth
    cessation, 141
    retardants, 141, 144, 145
  membrane properties, 146, 147, 148, 149
  metabolic changes, correlation with, 142

G

Gene pool, 369
Genetic
  constitution, 267, 272, 274
  stock, 334
Germplasm potential, 314, 324, 326
Gibberellin, its effect on frost resistance, 400, 401
Glutamate dehydrogenase, 60
  activity during growth, 61
  in leaf, 61
    roots, 61
    stem, 61
    tuber, 61
Glutarate dehydrogenase, 62
Glycerol, 41
Gromicidin A, 107
Growth
  inhibitors
    ABA, 398
    phenolic compounds, 398
  regulators
    ABA, 398
    Alar, 398
    effect on freezing and Lignin content, 398
  retardants, frost tolerance, 141, 144, 145

H

Hardened cells
  EM picture, 194, 195
  O_2
    evolution, 186
    uptake, 186
Hardening
  changes
    of glucides during, 189
    in O_2 evolution, 186
      uptake, 186
    in structure of *Chlorella,* 193
  temperature, 10

Hardening and dehardening, callus cultures, 204, 207
Hardening process
  ATP change, 398
  chloroplast involvement, 182, 184
  effect of
    antimetabolites, 180, 192, 193
    DCMU, 182
    oligomycin, 181
    surfactants, 180, 192, 193
  lipid changes, 181
  metabolism of glucose, 186, 187
  mitochondria involvement, 182, 183
  protein synthesis, 181
  RNA synthesis, 181
Hardiest clones, comparison of several
  species, 322
Hardiness
  anatomical characteristics, 62, 63
  deep supercooling, 271
  double low temperature exotherms, 235, 237, 238
  increase in callus cultures, 202, 203
  in response to
    light, 57
    temperature, 57
  low temperature and high temperature
    exotherms, 234
  palisade layer, 62, 63
  regions in North America, 214, 215
  water content, 234
Hardy and nonhardy tissues
  microsomal fraction, 127, 130
  phospholipase activity, 126
  phospholipid degradation, 126
Heat sink, 46
Heat transfer per state of freezing, 41
Hormonal balance
  auxins, 2,4,5-T, 276
  during acclimation, 144, 145
  gibberellic acid, 276
Hybridization
  interspecific, 325, 329
  intraspecific, 325, 327
Hydration
  effect, 304
  of frozen cells, 24
  stress, 307
Hyperosmolality, 305

I

Ice
  branch growth, 23

branching of the ice front, 30
dendritic branches, 23
extracellular
    crystal growth, 18, 19, 25
    ice migration, 25
formation in relation to
    deplasmolysis, 76
    plasmolysis, 76
liquid–polymer interface, 44
nucleation
    activity, 254
    concentration, 254, 260
    effect of various treatments, 260, 261
    effect on frost damage, 262
    epiphytic INA bacteria, 261
    function of temperature, 255
    loci, 9, 10
Incipient indices, 319
Inheritance of cold hardiness, additive variance recurrent selection, 321
Inhibition of freezing kinetics, 41, 44
Intracellular
    freezing, 26
    avoidance, 3, 6, 8, 11
    comparison with intercellular freezing, 4
    in nature, 3
    in parenchymatous cells, 27, 29, 33
    nature of injury, 4
    liquid water
        calorimetry, 170
        content, 46
    water
        calorimetry, 170
        freezing of, 169
        NMR, 171
        properties of, 170, 171
Ion
    leakage
        freeze injured cells, 96
        freezing injury, 93, 94, 96
    pump, inactivation in stress cells, 14

K

K_{oil}, 106
Karyotype, effect of freezing, 355

L

Langmuir absorption isotherm, 42
Latent heat, 43, 44, 167

Leaf
    anatomy
        cross-section, 63
        frost resistance, 62
        hardy and tender plants, 62, 63
        palisade layer, 62, 63
    wilting or death, 5
Light
    effect on hardiness in callus, 203
    in relation to hardiness, 57
Liquid
    foam, 5
    nitrogen
        effect on
            genetic stability, 368
            seed viability, 368
            seeding vigor, 368
        freezing survival, 356–357
        preservation of seeds, 365
        storage, 346
        storage costs, 367
Living acclimated tissue, rate of ice formation, 9
Low temperature avoidance, 5

M

Maternal influence, 321
Measurement of hardiness
    exotherm analysis, 319
    indirect measurement, 320
    in situ freezing, 316
    laboratory freezing, 317
    natural freezing, 316
Mechanical refrigeration, 366
Membrane
    damage in wheat, 392
    frost tolerance, 146–149
    leakage, 304
    mechanical destruction of system, 32
    lipids
        freezing injury, 106, 107
        in relation to
            cell permeability, 10
            phase transition, 11
            temperature, 12
    permeability
        cold hardiness, 169
        frost hardening, 148
    proteins
        cold hardening, 154, 155
        freezing injury, 106, 107, 110

resistance, in relation to
  ice formation, 8
  to water efflux, 8, 9
  structure model, 104, 106, 376
Meristamatic
  areas, 354
  cells, 41
Methyl urea permeability, freezing injury, 106
Michaelis Menton Theory, 42
Microsomal fraction
  freeze thawing, 127, 128, 130
  in hardy and nonhardy tissues, 127, 130
  preparation, 119
Mild freezing of tender plants, 3
Minimum temperature zone, 325
Mitochondria
  freezing injury, 110, 111
  in relation to freezing, 396
Mitochondrial membrane
  cold acclimation, 77
  freezing injury, 77
Monovalent ions
  osmotic lysis, 86
  protoplast survival, 86
Mucilaginous polymers, 4

N

NAA, delay in bloom, 275
Net free energy, 38
Nitella, 19
  extracellular freezing, 20
Nitrogen fertilizers, effect in hardening the
  tissue, 399
NMR
  bound water, 167, 170
  intracellular water, 171
Nuclear pattern, changes in algae, 179
Nucleic acids
  cold hardening, 156, 157
  dehardening, 158
Nucleus, denatured, 23, 29
Nyctinastic leaves, 5

O

Oil/water distribution coefficient, 106
Operator–repressor mechanism, 307
Osmiophilic globuli, frost hardening, 66
Osmotically induced lysis
  monovalent ion, 86
  nonionic vs. ionic, 87

pH, 87
plasma membrane, 85
Oxygen
  evolution
    algal cells, 178
    effect of
      DCMU, 183
      oligomycin, 183
    frozen–thawed cells, 186
    mode during hardening, 184, 185
  uptake
    algal cells, 178
    effect of
      DCMU, 183
      oligomycin, 183
    frozen–thawed cells, 186
    mode during hardening, 184, 185

P

Palisade layers, 397
Parenchymatous cells
  flash and nonflash freezing, 27
  tissue of watermelon, 22
Pathological changes, 23
Pentaploid potatoes, 339
Permeability
  in relation to avoidance of intracellular
    freezing, 9
  of membrane to
    methyl urea, 106
    urea, 106
    water, 98
Phase transition temperature, relation to
  membrane lipid, 12
Phosphatidylcholine
  accumulation, 399
  degradation during freezing, 121, 124, 131
  pH effects, 131
Phospholipase
  activity determination in microsomes, 119
  in hardy and nonhardy microsomes, 126
  in relation to freezing, 134
  in situ activity, 126, 128
  inhibition of, 125, 129
  membrane bound, 126, 133
  pH activity profile, 124
  protective effect, 130, 131
  regulatory mechanism, 133, 134
  velocity of reaction, 126
Phospholipid
  changes after freeze–thawing, 122

degradation control in situ, 123, 130
degradation during freezing, 117, 119, 120, 392
extraction and analysis of, 118
phospholipase, 119, 124, 125, 126
protective effect of
 DMSO, 130, 131
 Mg^{++}, 130, 131
 sucrose, 130, 131
temperature causes degradation, 123
Phosphorus compounds, in relation to freezing, 397
Plasma membrane
 alteration
  by cold acclimation, 75, 76, 77
  by freezing, 75, 76, 77, 79
  during contraction, 82
  during protoplast expansion, 83
 barrier to ice formation, 8
 damage, 75, 78, 79
 disruption, 84, 87
 fracture face of, 205, 206
 in relation to surface area, 84
 lysis, 83, 84, 87
 site of freezing injury, 75
Plasmolysis, 9
 freeze injured cells, 95, 96
 normal cells, 95, 96
 permanent, 385
Poikilothermic nature, 5
Polyploids, 301
Polyribosomes, in relation to frost hardiness, 58
Pore size, 40
Potato wild species, *Solanum,*
 ajanhuiri, 334
 curtilobum, 334
 juzepczukii, 334
 others, 335
Potential energy of adhesion, 42
Preservation
 genetic degradation, 364
 long term, 374
 of seed, 361
 of seed germplasm, 361
 seed viability,
 ultracold, 363
Probil-analysis, 319
Protein, in relation to frost hardiness, 58
Protoplasm
 clotting, 19
 dehydration, 34
 stands, 21
 streaming, 19

Protoplast
 contraction, salt solution, 82
 expansion
  K^+ uptake, 105
  methyl urea uptake, 105
  salt solution, 83
  urea uptake, 105
 survival, monovalent ions, 86
Prunus buds
 DTA profile, 220
 DTA studies, 220
 undercooling, 219
Pseudo-plasmolysis freezing injury, 96
Pyrus
 cold injury, 219
 deep undercooling, 219
 DTA of leaf buds, 217
 killing temperature, 216

R

Radiation damage to seed, 366
Rapid cooling, 32
Reciprocal differences, 321
Red alga, 347
Relative enzyme activities, effect of low temperature, 7
Resistance, 268
 lettuce seeds, 242, 244, 247
 low water content of seeds, 241
 mechanism of avoidance, 247
 methods of achieving resistance, 7
Rest and dormancy, 269
Ribosomal RNA, in relation to temperature, 58

S

Sampling error, 273
Satellite system, 300
Sea urchin eggs, 34, 35
Seed
 deterioration, 363
 germplasm loss, 362
 imbibition, 244
 viability tests, 244, 362
Selection, 326
 for hardiness, 402
Semipermeability, 24
 in relation to thawing, 32
Shading treatment, 294
Shelter belt, 285
SH groups, relation to cell sap, 12

Solar radiation effect, 281
Solute accumulation, in relation to freezing
    tolerance colligative effect, 7
Spearman Karber method, 319
Starch, hydrolysis of, 8
Stomatal density, frost resistance, 62
Storage of genetic resources, 352, 353
Stress energy
    free energy for crystal growth
        along the cell wall, 38
        into the protoplast, 38
    osmotic activity, 38
    potential energy of adhesion, 38
    water potential, 38
Succinate dehydrogenase, 62
Sugar accumulation during hardening, 7, 9
Sunscald of plants, 3, 34, 314
Supercooling
    avoidance of freezing
        in flower buds, 241
        in xylem, 241
    degree of, 38
    deep, relation to northern limits of fruit
        production, 328
    differential thermal analysis, 241, 245
    endosperm envelope injury, 244
    exotherms, 243, 244
    structure of endosperm, 241
    supercooling nucleation point, 242
    xylem injury, in relation to, 227, 228
Survival
    of cortical cells, 350
    curve for xylem, 320
    of germplasm, 345
    of hydrated lettuce seeds, 246
    of red algal cells, 349
    test, 402
Susceptibility of tissue, 269
Suspension culture
    cell, 345, 347
    of sugarcane cell, 373

T

Tea plant, 282
Temperate fruits, breeding, 313
Temperature
    critical, 273, 274
    effective for freezing injury, 38
    in relation to
        freezing resistance, 232
        hardiness, 57
        phospholipids, 123

response curve, 273
ribosomal RNA, 58
viability curve of cane cells, 380
Tetraploid potato, 339
    crosses with diploid potato, 340
'Threshold' value for phospholipid content,
    399
Tissue culture cells, 347
Tradescantia cells
    cell to cell freezing, 27
    freezing in staminal hairs, 28
Translocation of sugar fractions, 307
tRNA, in higher plants, 58
TTC
    amino acid release, 199, 202
    reduction state, 319, 352, 354
    regrowth rate, 199, 202
    viability value, 381, 382, 383
Turgor loss, freezing injury, 93

U

Undercooling
    flower buds, 222
    freezing of water, 213, 214
    geographical distribution of species, basis
        of, 214, 215
    in Prunus, 219
    in Pyrus, 215, 219
    in Rosa, 223, 224
    limits of cold hardiness, 215
Unfrozen
    cell, appearance, 384
    tonoplast, 33
Unhardened cells
    EM picture, 195
    O_2
        evolution, 186
        uptake, 186
Urea permeability, freezing injury, 106

V

Vapor
    desiccation limit, 46
    diffusion, 45
    pressure, 45
        ratio of ice to liquid (p/p_0), 38
Varietal comparison, 315
Viability
    algal cells, 178
    browning, 217, 268, 282
    conductivity method, 94

decline variations, 368
electrolyte leakage, 218, 219
growth curve, 178
infiltration of tissue, 94
properties of the membrane, 93
protoplasmic streaming, 95
TTC reduction
and amino acid release, 199, 202
and regrowth rate, 199, 202

W

Water
activity
index, 40
of phase transition, 41
bound, 167, 170
content
deacclimation, 233, 234
hardiness, 234
efflux, ice formation, 77
freezing in plants, 167, 170, 171

NMR analysis, 167
permeability
freeze injured cells, 98
ice formation, 77
turgor changes, 111
potential, in relation to freezing, 44, 45
role in cold hardiness, 165, 166
Wild potatoes, 333
Woody plants
hardy cells, 24
specific tissue, 6

X

Xylem injury temperature
deciduous, 229, 237
double low temperature exotherm, 235, 237
evergreen, 229
in relation to water content, 229
limits of distribution of plants, 229
low temperature exotherms, 230, 237